RECENT ADVANCES IN DEVELOPMENTAL BIOLOGY

Recent Advances in Developmental Biology

Vol. 5
Invertebrate Reproduction and Development

5 –

A.P. DIWAN
N.K. DHAKAD

ANMOL PUBLICATIONS PVT LTD
New Delhi-110 002

ANMOL PUBLICATIONS PVT LTD
4374/4B, Ansari Road
Daryaganj, New Delhi-110 002

Recent Advances in Developmental Biology

First Edition 1995
ISBN 81-7488-039-9

PRINTED IN INDIA

Published by J.L. Kumar for Anmol Publications Pvt. Ltd. and
Printed at Mehra Offset Press, Delhi.

Preface

The fundamentals of Developmental Biology are presented within the framework of scientific discovery. Researches in Embryology have made almost incredible strides in the past few decades. Consequently, the existing concepts of the developmental biology have expanded. There has been a revolution indeed in this direction.

The text integrates the descriptive, experimental and biochemical approches into a conceptual framework for the analysis of development. All important points are illustrated diagrammatically. The title is not intended to be comprehensive, nor could it be at this length, but concentrates as putting across the basic principles of the subject as briefly and lucidly as possible. It does this with the aid of careful selected examples, some recent and other classic of the field and with numerous illustrations. The aim is to enthuse the reader with this active and exciting area of research and to lay a solid foundation on which further study of its various facts may be based.

Authors

Contents

1

Fertilization and Sexuality

Fertilization is the union of two cells to form a *zygote*. Cells which have this ability are called *gametes*. One essential aspect of fertilization is *karyogamy* i.e., the fusion of the two gametic nuclei (pronuclei) within the zygote to form a *synkaryon*. The doubling of the chromosome number which is connected with this is compensated by the nuclear events at meiosis. In all multicellular animals the gametes, which unite in pairs, are clearly differentiated from one another. They are called *eggs and spermatozoa* (or sperm cells) respectively. Eggs are formed by females and spermatozoa by males. Only in hermaphrodites are eggs and spermatozoa formed by the same individual. The phenomenon of the diversity of the fusing gametes or the individuals producing the gametes is called *sexuality*.

Since the gametes of all multicellular animals are sexually differentiated, the connection between sexuality and fertilization seems so self-evident that the gametes are referred to as "sex cells." Among the Protozoa, however, there are cases where neither the gametes nor the individuals forming them are morphologically different from each other. It must therefore still be tested whether there are differences of any kind between them which might correspond to sexuality in multicellular animals.

If the gametes are not morphologically differentiated we speak of *isovamety*, and of *anisogamety* if they differ either in size, form or structure. Gametes which differ in size are called *macro-* and *microgametes*. In many Protozoa, the microgametes resemble the egg cells of Metazoa while the microgametes resemble the spermatozoa. In this case anisogamety is referred to as *oogamety*. In some Protozoa all cells can become gametes under certain conditions. However, as a rule a *gamont* is formed first which produces the gametes as a result of a special reproductive act *(gamogony)*. The latter may involve binary or multiple fission. If gametes capable of fusing with one another are formed within the same alone, or if they arise from a common gamont, the organism is said to be *monoecious*. Species in which sexually compatible gametes arise from different clones or gamonts are called *dioecious*.

Sex determination can take place in various ways. If the realization of sex depends upon genes, we speak of *genetic or genotypic* sex determination. If the direction of differentiation is not due to genetic factors, sex determination is *modificatory*, or *phenotypic*. The potentialities of sexual differentiation are based upon the genetic "reaction norm" in either case. One important variant of phenotypic sex determination is the production of two sexually different daughter cells or nuclei from a common progenitor by *differential cellular or nuclear division*.

Fertilization is of widespread occurrence among Protozoa. Even if we disregard doubtful cases where karyogamy and meiosis have not been proved to take place, there is no class of Protozoa without at least some orders where fertilization has been found. Among the flagellates, the Phytomonadina, the Polymastigina and the Opalinina are examples, among the rhizopods the Heliozoa and the Foraminifera. Fertilization is generally characteristic for the classes Sporozoa and Ciliata.

On the other hand, it should not be overlooked that fertilization has not been found in many flagellates and rhizopods in spite of thorough search. Some species which have been cultured for years in the laboratory have reproduced only asexually, i.e., without fertilization and meiosis. Despite widespread occurrence, there is thus no reason to look upon fertilization and sexuality as "basic phenomena of life." Although all fertilization processes do lead to karyogamy sooner or later, the phenomena connected with them are of a surprising *diversity* among Protozoa. This includes the relation between gametes and gamonts, their mode of formation and their sexual differentiation as well as the occurrence of monoecy and dioecy. On the basis of the relationship between gametes and gamonts, three types of fertilization processes can be distinguished in Protozoa, viz. *gametogamy, autogamy,* and *gamontogamy.*

I. GAMETOGAMY

In the simplest case, gametes are formed which fuse as free-swimming cells and not within the gamonts. This mode, found in free-living Protozoa (Phytomonadina, Foraminifera) as well as in parasitic Protozoa (Polymastigina, Sporozoa) is called gametogamy.

Phytomonadina

In the phytomonads, fertilization has been proved for all well-known species and has been studied in detail in a number of forms. In all cases gametes fuse with each other of which at least one is free-swimming. Among the non-colonial species, especially those of the genera *Chlamydomonas, Dunaliella* and *Polytoma,* the gametes are often not formed by a special mode of reproduction but arise directly from vegetative cells *(hologamy).*

The fact that cells become gametes, i.e., that they acquire the capacity to fuse with other gametes, is due to external factors which may be different in the various species and which may collaborate in different ways. In *Chlamydomonas,* the determination of gametes is induced primarily by lack of nitrogen. If distilled water is poured on an agar plate containing non-flagellated cells reproducing by fission, they form flagella and become gametes. Depending upon species and sex, illumination is also of importance in this connection. In *Dunaliella salina,* an inhabitant of salt pits, decrease of salinity induces the formation of gametes.

External factors of this kind cause a *synchronization* so that many, if not all, cells of a population or a clone become sexually active at the same time.

In most phytomonads, however, external factors lead at first only to the formation of gamonts which, in turn, form gametes by successive multiple division *(merogamy)*. The nature of these external factors is not well known as yet.

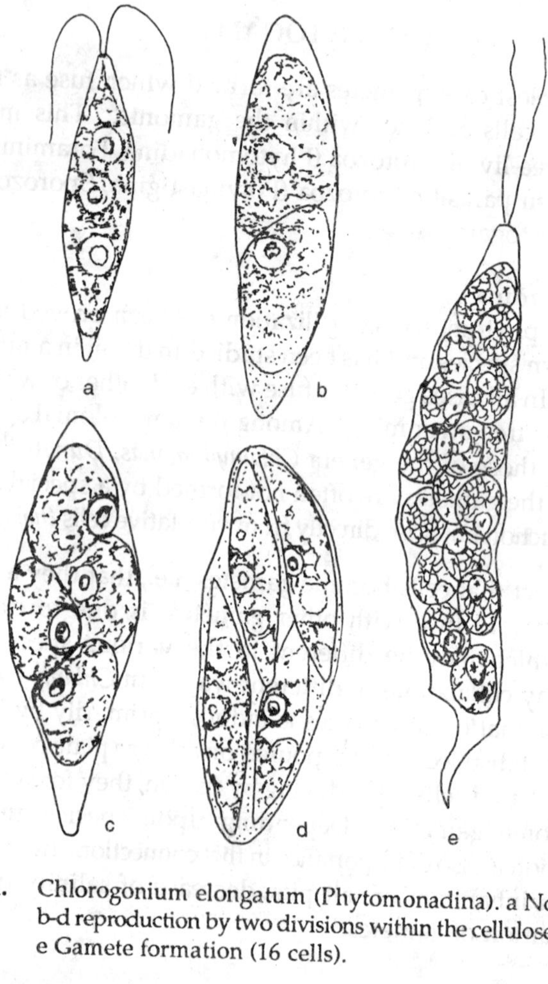

Fig. 1.1. Chlorogonium elongatum (Phytomonadina). a Normal cell. b-d reproduction by two divisions within the cellulose envelope. e Gamete formation (16 cells).

The gamonts of phytomonads cannot be distinguished from ordinary cells by morphological criteria. In *Chlorogonium elongatum,* which normally reproduces by successive multiple division, the gamont carries out 2-3 additional fissions so that the gametes are smaller than the daughter cells of asexual reproduction. Gametogenesis in *Chlamydomonas suboogama* is unusual enough to warrant special description. In this case, sexual reproduction beings with the determination of a gamont mother cell (progamont) which is enclosed in a gelatinous mass. It divides twice to form four daughter cells. Three of these always become macrogamonts while the fourth cell develops into a microgamont. The latter than forms four small microgametes. The macrogamonts do not divide but are directly transformed into macrogametes which are capable of fertilization. However, this change takes place only after the microgametes have left the gelatinous mass. The fusing macro- and microgametes are therefore normally descended from different gamont mother cells.

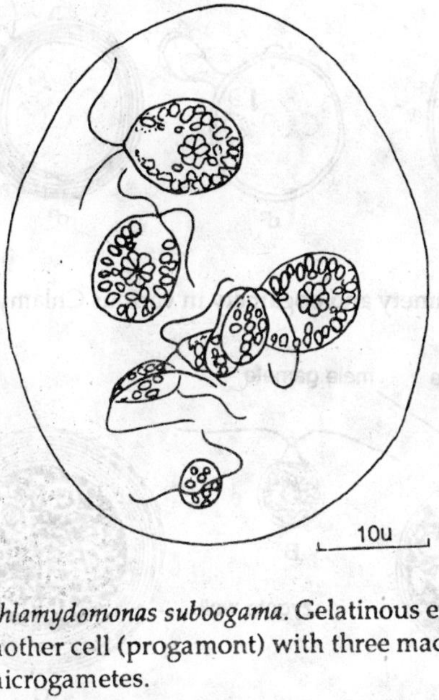

Fig. 1.2. *Chlamydomonas suboogama.* Gelatinous envelope of a gamont mother cell (progamont) with three macrogamonts and four microgametes.

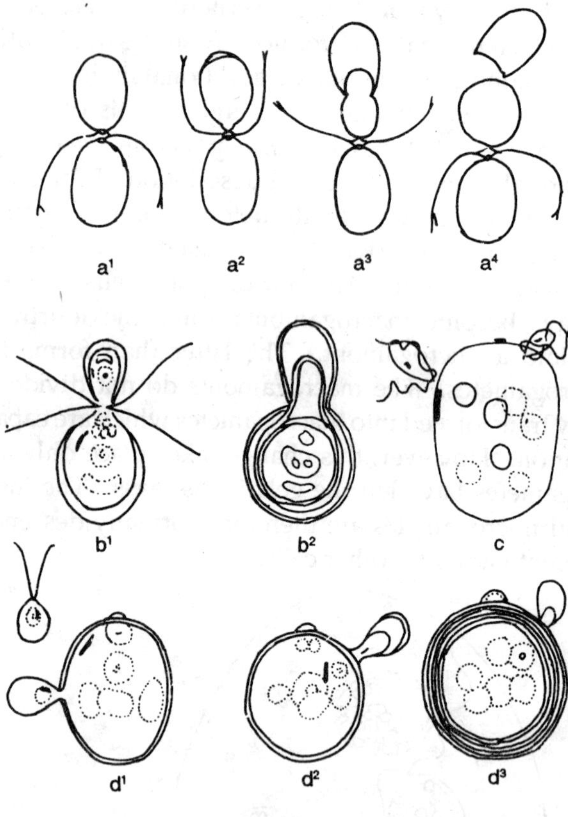

Fig. 1.3. Anisogamety and oogamety in various Chlamydomonadidae.

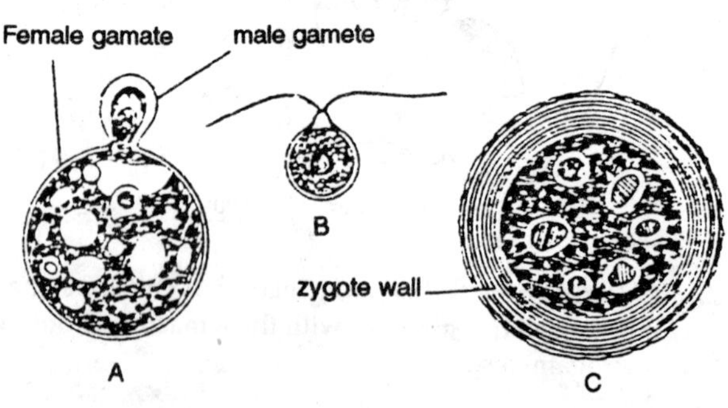

Fig. 1.4. *A-C. Chlamydomonas coccifera*, showing the process of oogamy.

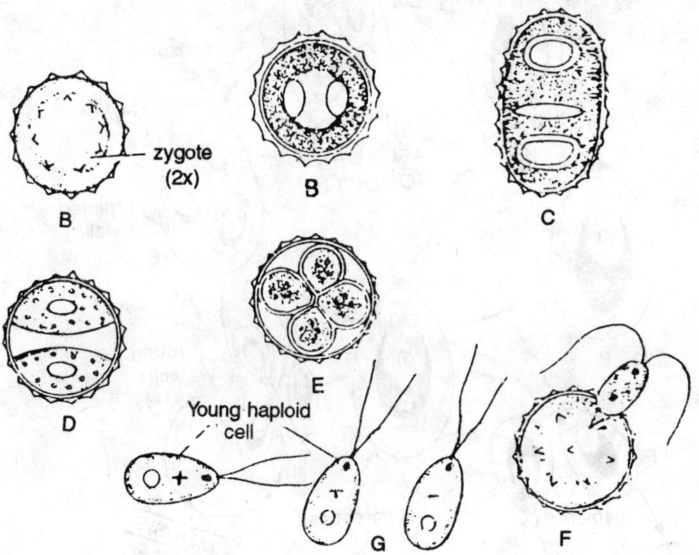

Fig. 1.5. A-G. Chlamydomonas ehrenbergii, showing tetrad formation and germination of zygote.

In some colonial phytomonads all the cells of a colony can become gametes, in other special generative cells are formed which are the only ones capable of reproduction. In some cases, *e.g., Pandorina morum,* it has been shown that gamete formation always sets in which colonies of different sex are brought together.

Sexual differentiation of gametes is of special interest in Phytomonadina because all transitions between isogamety and oogamety are found in this group. While some genera include only isogametic species *(Dunaliella, Pyramidomonas, Polytoma, Haematococcus, Brachiomonas)*, isogametic as well as anisogametic and oogametic species are known in the genera *Chlamydomonas* and *Chlorogonium.*

Most species of the large genus *Chlamydomonas* are isogametic. However, in some cases the fusing gametes exhibit

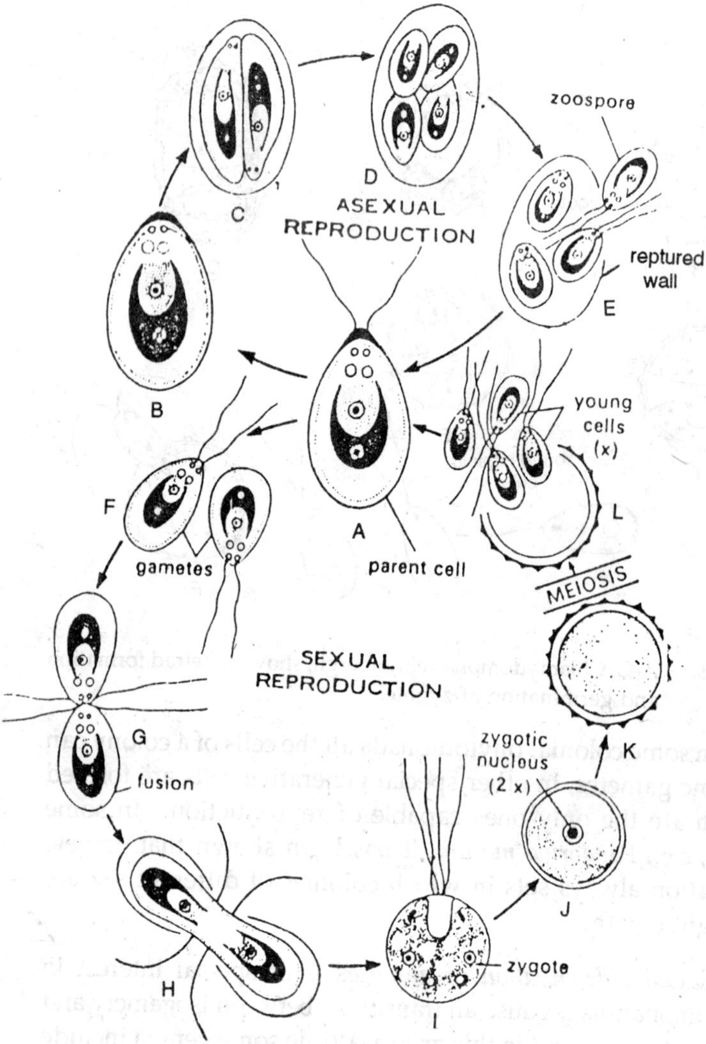

Fig. 1.6. Diagrammatic life-cycle of *Chlamydomenas*.

differences which become manifest only during fertilization. The gametes of *Chlamydomonas eugametos syn. moewusii* swim about four hours after fusion. They lie opposite each other

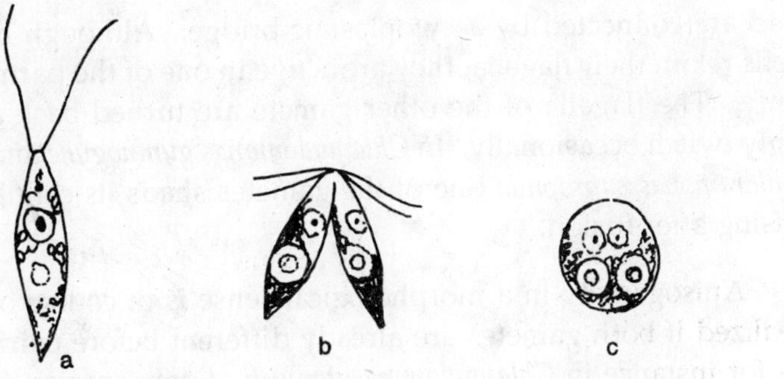

Fig. 1.7. *Chlorogonium elongatum*, a Free gamete. b fusion of isogametes. c. Zygote.

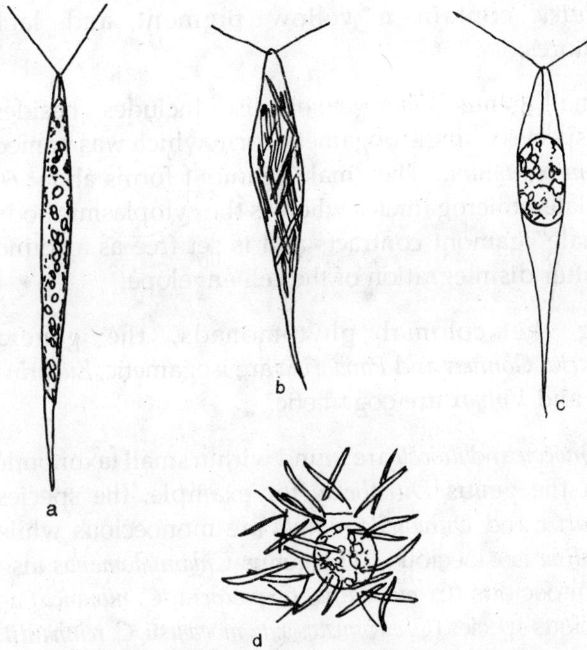

Fig. 1.8. Chlorogonium oogamum. a Normal cell. b gamont subdivided into numerous, slender microgametes. Microgametes within the cellulose envelope of the mother cell. d. Hatched macrogamete surrounded by microgametes.

and are connected by a cytoplasmic bridge. Although both cells retain their flagella, they are active in one of the partners only. The flagella of the other gamete are turned back and only twitch occasionally. In *Chlamydomonas gymnogyne* and in *Chloromonas saprophila* one of the gametes sheds its cellulose casing after fusion.

Anisogamety in a morphological sense is of course only realized if both gametes are already different before pairing, as for instance in *Chlamydomonas braunii*. Some species such as *Chlamydomonas suboogama C. pseudogigantea* and *C. coccifera* (d^1—d^3) can be considered oogametic since the macrogamete is a large, immotile cell, much bigger than the small motile microgametes. The microgametes of *Chlamydomonas pseudogigantea* contain a yellow pigment and lack chromatophores.

The small genus *Chlorogonium* also includes, besides isogametic species a single oogametic form which was named *Chlorogonium oogamum*. The "male" gamont forms about 64 small flagellated microgametes whereas the cytoplasmic body of the "female" gamont contracts and is set free as a round "egg cell" after disintegration of the cell envelope.

Among the colonial phytomonads, the genera *Stephanosphaera*. *Gonium* and *Pandorina* are isogametic, *Eudorina (Pleodorina)* and *Volvox* are oogametic.

Both *monoecy* and *dioecy* are found within small taxonomic groups. In the genus *Dunaliella*, for example, the species *Dunaliella parva* and *Dunaliella minuta* are monoecious while *Dunaliella salina* is dioecious. The genus *Chlamydomonas* also includes monoecious *(C. monadina, C. paupera, C. monoica)* as well as dioecious species *(C. eugametos syn. moewusii, C. reinhardi)*. *Haematococcus pluvialis* and *Stephanosphaera pulvialis* are monoecious. In the latter species, the eight cells of the mother colony carry out a variable number of fissions, giving rise to 4, 8, 16, or 32 gametes. These are spindle-shaped and are usually set free by rupture of the mother colony. Occasionally, especially if all cells of a colony have formed gametes, pairing occurs within the gelatine of the mother colony. Monoecy can thus lead to self-fertilization.

While *Gonium pectorale* is a dioecious species, *Gonium sociale* and *G. sacculiferum*, (both species with four cells in the colony), include dioecious as well as monoecious varieties or strains. Also in *Pandorina morum* a few monoecious clones could be isolated in addition to dioecious ones.

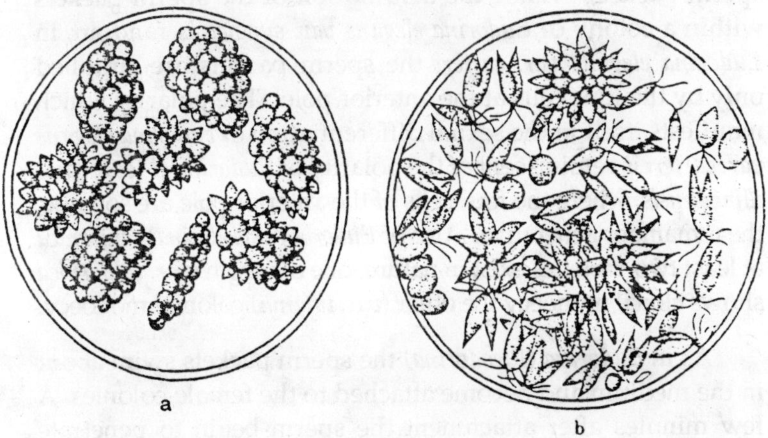

Fig. 1.9. *Stephanosphaera* **pulvialis. Formation and fusion of the isogametes within the gelatinous envelope of a colony.**

The genus *Eudorina*, which now also includes the species formerly grouped together in the genus *Pleodorina*, shows high variability. The species *Eudorina unicocca, Eudorina cylindrica, Eudorina (Pleodorina) illinoisensis* and the variety *elegans of Eudorina elegans* are dioecious. Some clones bring forth only female colonies and others only male colonies. While the cells of the female colonies are not distinguishable from normal cells, those of the male colonies carry out multiple fission. A packet of biflagellated microgametes or "spermatozoids" arises from each cell. These sperm packets at first swim about as a unit. Only in the vicinity of the female colonies do they come to rest

(b) and break up into the individual spermatozoids, which penetrate into the gelatinous mass and unit with the macrogametes or "eggs" (c).

In monoecious species two possibilities can be discerned. In *Eudorina conradii* only the clones can be designated as monoecious while the individual colonies are either purely "female" or purely "male". Whether dioecy or monoecy is present can therefore be determined only by clone culture. In the other case, monoecy may also extend to the individual colonies, i.e., some of the cells become eggs while others form sperm packets. While the distribution of the sperm packets within a colony of *Eudorina elegans var. synoica* is random, in *Eudorina elegans var. carteri* the sperm packets are supplied only by the four cells at the anterior pole. The polarity which manifests itself in the sexual differentiation of *eudorina elegans var. carteri* is reminiscent of the polarity of *Eudorina (Pleodorina) illinoisensis,* where the four cells of the anterior pole are somatic, the remainder generative. Also of *Eudorina (Pleodorina) californica* at least two varieties exist in nature, one of which *(var. californica* shows clonal monoecy the other *(var. tiffanyi)* colonial monoecy.

As in *Eudorina (Pleodorina)* the sperm packets swim about in the medium and become attached to the female colonies. A few minutes after attachment the sperm begin to penetrate and migrate within the gelatineous envelope.

Recently, fusion of sperm and egg has been observed in *V. carteri.* It is unknown, however, whether the sperm can fertilize the eggs of the same colony in case of colonial monoecy.

In all phytomonads with anisogamety or oogamety it is at once evident that sexual differentiation is confined to only two sexes. This *bipolarity,* familiar to us from Metazoa, holds for dioecious as well as monoecious species and varieties.

In the case of isogamety, however, only experimental methods can establish whether the copulating gametes are differentiated and if this differentiation is bipolar.

TABLE 1.1

Volvox. Distribution of monoecy clonal, colonial and dioecy in twelve species. After Starr (1968) [1655a]. Terminology changed as used in this context

	Monoecy		*Dioecy*
Volvox	*clonal*	*colonial*	
aureus	+		
barberi		—	
carteri			+
dissipatrix		—	
gigas			+
globator		—	
obversus			+
perglobator			+
powersii	+		
rousseletti			+
spermatosphaera	+		
tertius	+		

In *dioecious* species there is no doubt about sexual differentiation. That it is bipolar can be proved in many cases by combination experiments. If clones derived from the cells of a natural population or from the gones of a zygote are mixed together pairwise, fertilization will regularly take place in certain combinations and not in others. The pattern of combinations permits the assignment of all clones to one or another of two groups, designated as the "+" and "-" type.

Since it is highly improbable that this differentiation into two types is fundamentally different from sexual differentiation as expressed in cases of anisogamety and oogamety, the two types may also be looked upon as the "+" and "-" sexes.

The fact that a "+" gamete always fuses with a "-" gamete can be proved in some instances by *marking* one type of gamete. Marking may be phenotypic or genetic. If one of the two

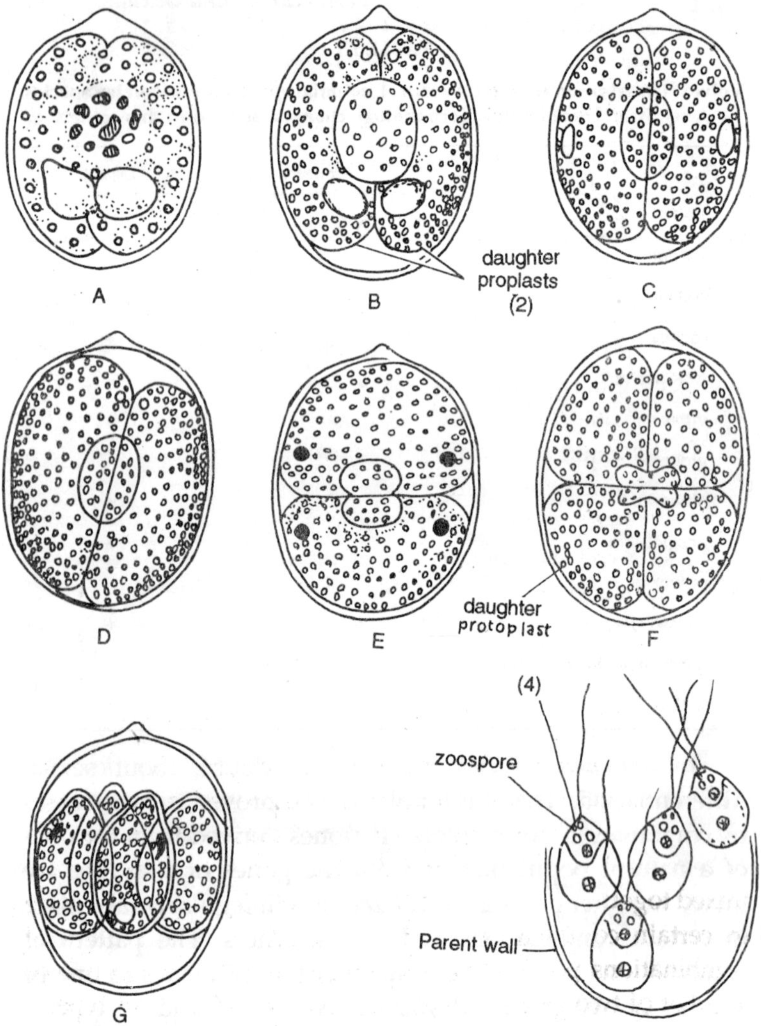

Fig. 1.10. *A-H. Chlamydomonas*, showing formation of zoospores.

Dunaliella salina clones to be used in a combination test is reared in a medium poor in nitrogen and phosphorus, the cells become red after sometime due to the deposition of . If such "marked" isogametes are brought together with those of the other sex, pairing will always be between a red and a green cell. It is, therefore, easily possible to mark one of the gametic types. If mutants are available, clones which are generally different can be combined and the bipolarity of the gametes follows from the hybridization analysis.

In all dioecious species examined so for sex determination is genetic. It is based upon the separation of a pair of alleles (segregation), one being responsible for the +, the other for the—sex. Since meiosis is zygotic half of the gones which arise during division of the zygotes belong to one sex, half of the other. If only one gone emerges from the zygotic sheath, while the other gonal nuclei *(Pandorina)* or cells degenerate, then it is with equal likelihood either from the + or from the - sex.

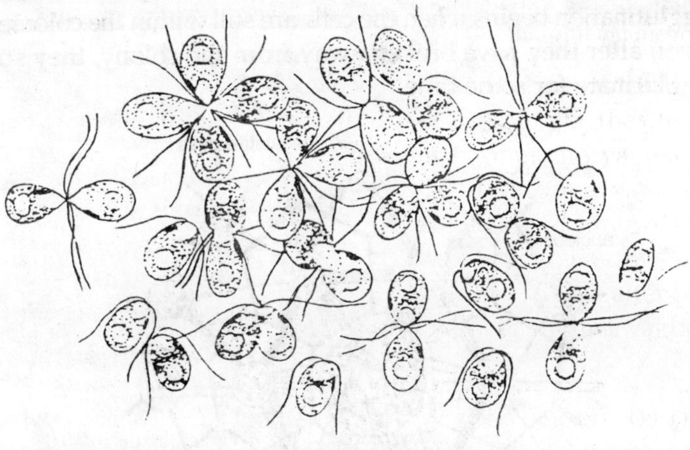

Fig. 1.11. *Dunaliella salina* (Phytomonadina). Fusion of isogametes which have been labeled due to different culture conditions. The darker ones shown are stained red, the light ones, green.

Sexual differentiation and bipolarity are much more difficult to prove in the *monoecious* phytomonads with isogamety. One possible approach is the Restgameten" (left-over gametes) method used successfully with *Acetabularia mediterranea.* Corresponding experiments on phytomonads have not, however, led to conclusive results as yet. Should it turn out that even gametes which are not sexually differentiated can mate with each other, them one could, of course, not speak of "monoecy."

An *agglutination reaction* takes place if sexually different isogametes of dioecious species are brought together. Groups of many gametes — often several hundred — are thereby formed whose adhesion is due to stickiness of the flagellar tips. These groups remain intact for approximately half an hour in *Chlamydomonas eugametos syn moewusii*, and for several hours in *Dunaliella salina*. Finally, they disperse into the individual pairs.

In *Pandorina morum* and *Astrephomene gubernaculifera* agglutination begins when the cells are still within the colonies. Even after they have broken away from the colony, they still agglutinate for some time.

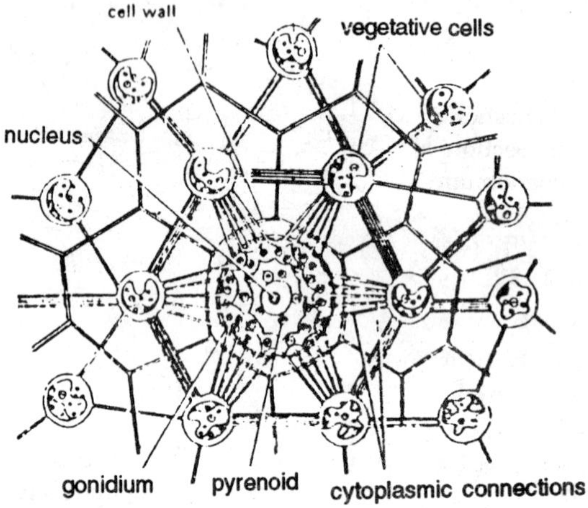

Fig. 1.12. Volvox aureus. A part of coecaobium showing a gonidium in surface view.

In oogametic phytomonads, whose macrogametes are immotile "egg cells", agglutination cannot play a role in the contact of the sex cells. The macrogametes probably exude substances which chemotactically attract the microgametes. However, detailed investigations are not available as yet.

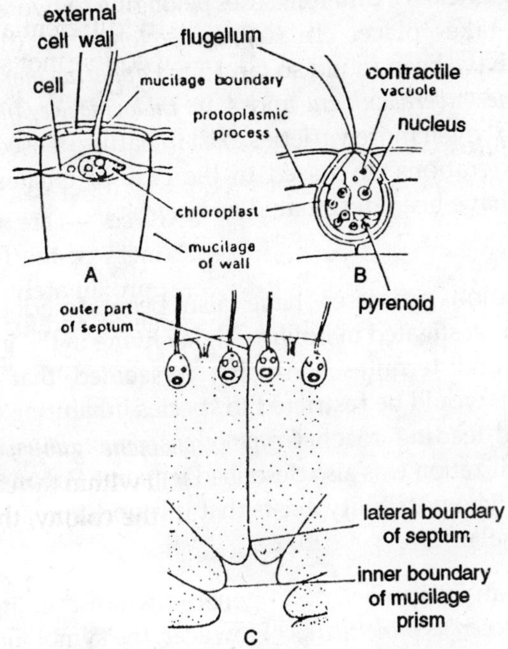

Fig. 1.13. A, Section of the periphery of coenobium of *Volvox globator;* B, Section with an enlarged cell; C, Section of a part of coenobium of *V. aureus.*

If the "egg cells" are not fertilized, they transform into "*parthenospores*" in some species (e.g. Volvox aureus). These are not distinguishable from zygotes : they have a double sheath and assume a reddish-brown colour. Under the same conditions which evoke the germination of zygotes, the "parthenospores" can also develop into colonies. Instead of "bisexual", a "monosexual" development or *parthenogenesis* thus takes place. In one strain of *Volvox aureus* parthenogenetic development is the rule.

Phytomonad species of which a fair number of clones have been isolated and cultured have been shown incapable of unlimited crossbreeding. Combination experiments indicated the existence of several "varieties" which are not distinct morphologically but are sexually isolated from each other : on combining gametes of different sexes belonging to two varieties, no mating takes place. In *Gonium pectorale.* 12 dioecious "varieties" have been found so far, 15 in *Pandorina morum*, 6 in *Astrephomene gubernaculifera* and 4 in *Eudorina elegans.* It is probable that even more varieties exist in nature. Based on the same considerations discussed in the case of ciliates, these "varieties" have been designated as *syngens.*

Polymastigina

Fertilization processes have also been described and thoroughly investigated in Polymastigina living in the intestine of roaches and termites. At first it seemed that sexual reproduction would be restricted to species inhabiting the gut of the wood-feeding roach *Cryptocercus punctulatus.* More recently, fertilization was also described for some Polymastigina of termites. The possibility exists that it is widespread in this order of flagellates.

More than 30 species, of 14 genera, live in the intestinal tract of *Cryptocercus punctulatus.* However, the symbiont fauna of this roach, which is discontinuously distributed over the Southeastern and Northwestern United States, is not the same everywhere, *Macrospironympha xylopletha*, for example, occurs only in Californian populations of the roach.

Sexual reproduction of the flagellates is induced by the *molting hormone (ecydson)* formed in the prothoracic glands of the host. As a rule, the roach molts only once a year. Fertilization of its flagellates is therefore limited to a comparatively short period of time. However, one must take into account that release of the molting hormone begins long before the actual molting act (ecdysis), i.e., before the old larval euticle is sloughed off.

As several examples show sexual reproduction starts at different times in the individual genera. In *Barbulangympha*, it begins approximately 50 days before the molting act, in

Trichonympha, on the other hand, only 5 days before. A certain concentration of the molting hormone, which differs in the individual genera and species, is apparently necessary to set the sexual processes into motion. Table 1.2 attempts to give a general survey of the diversity of the sexual processes which are even more complicated by the fact that some genera are haploid, others diploid.

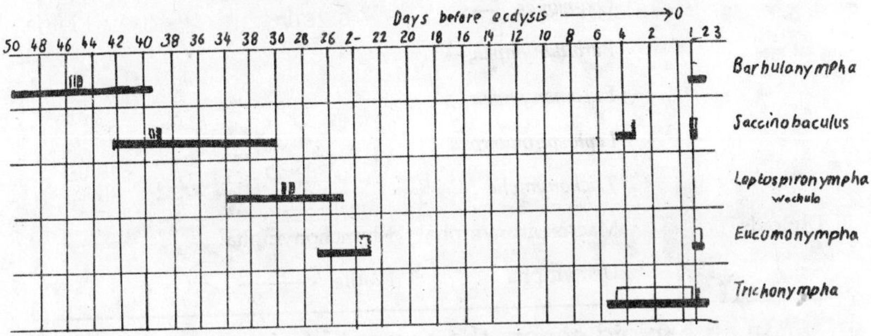

Fig. 1.14. Relationship between the sexual processes of some haploid Polymastigina of the roach *Cryptocercus punctulatus* and the molting period of the host. 5—0 days before, 0—3 days after molting (ecdysis).

In all cases, the effect of the molting hormone ideas first to the determination of a gamont of progamont. In most genera, no special morphological changes are associated with this transformation. Only in *Leptospironympha* do the gamonts differ conspicuously from ordinary cells. The flagellar bands which coil around the cell are broken down and the remaining flagellar tufts of the rostrum become transformed. In species of the genus *Trichonympha* the gamonts round off and surround themselves with a cyst envelope. During encystment, a complete resorption of all extranuclear organelles (flagella, parabasal bodies) takes place; only the two centrioles, the nuclear esp and parts of the rostrum remain. In *Saccinobaculus* and *Oxymonas* the axostyle is also resorbed.

TABLE 1.2

Sexual processes of Polymastigina living in the roach *Cryptocercus punctulatus*. Horizontal arrows indicate in which genera autogamy as well as gametogamy or gamontogamy has been observed

	Gametogamy	*Autogamy*	*Gamontogamy*
Haploids	*Saccinbaculus* →		
	Oxymonas →		
	Barbulanympha →		
	Eucomonympha		
	Leptospironympha		
	Trichonympha		
Diploids	Macrospironympha	Rhynchonympha	
	Urinympha	Notila ←——	

In all haploid genera the gamont carries out a *differential division* which leads to the formation of a male and a female gamete. The peculiar chromosomal events which are connected with this division in *Barbulanympha*. *Leptospironympha* and *Trichonympha* have been pointed out above. The various phases of gamete formation within the gamont eyst of *Trichonympha* take about 6 days and can be observed in life. After the centrioles, between which an extranuclear spindle has been formed, have moved to opposite poles of the nucleus new nagella are formed at the rostra (a). As soon as nuclear division is completed, the spindle is dissolved in the cytoplasm. Both daughter nuclei are now connected to their rostra by their own nuclear caps which are descended from that of the mother cell. The rostra can thus move independently within the cyst (b) together with their cell nuclei. With the elongation of the flagella their movement becomes more lively. At the same time, the formation of the new parabasal bodies takes place (c). Eventually, the cytoplasmic bodies of both gametes are completely separated (d). Soon thereafter the gametes leave the gamont cyst whose wall is locally dissolved.

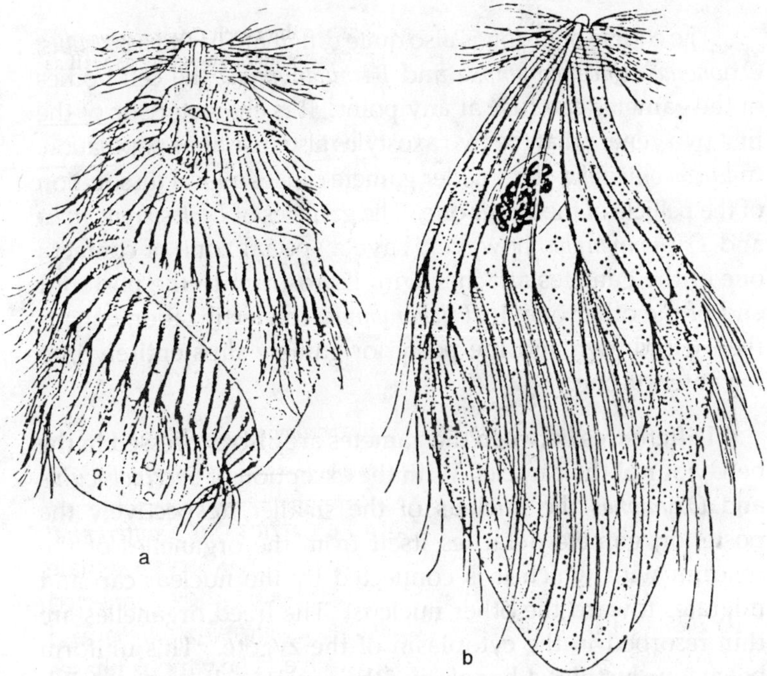

Fig. 1.15. *Leptospironympha uachula* (Polymastigina). a Normal cell. b Gamont.

With respect to gamete differentiation, there are all intermediates ranging from isogamety to anisogamety. Oogamety, however, does not occur. While there is no difference between the gametes in *Saccinobaculus, Oxymonas* and *Leptospironympha* the pairing gametes of *Eucomonympha, Barbulanympha* and *Trichonympha* are usually of different size. There are also structural differences in *Trichonympha* which normally becomes apparent only after emergence from the cyst. The larger gamete forms numerous small granules which become arranged in a ring at its posterior end. Within this ring is the "fertilization cone", a hyaline region which can be withdrawn and extended. The smaller gamete has granules scattered in its cytoplasm, but no special differentiations at its posterior end.

The course of fusion is also quite different. In *Saccinobaculus, Oxymonas, Eucomonympha* and *Barbulanympha* the cell bodies of the gametes can fuse at any point. It is characteristic of the first two genera that the two axostyles also fuse after population. In *Eucomonympha*, the smaller gamete causes a local dissolution of the pellicle of the larger one. The gametes of *Leptospironympha* and *Trichomympha*, however, have a special zone of contact : one of the gametes attaches with its rostrum to the posterior end of the other one. In *Trichonympha* the fertilization cone is then withdrawn and the posterior gamete slips entirely into the anterior one.

Differences between the gametes are also expressed in the behaviour of their nuclei. With the exception of *Saccinobaculus* and *Oxymonas* the nucleus of the smaller, respectively the posterior, gamete detaches itself from the organelles of the rostrum with which it is connected by the nuclear cap and migrates toward the other nucleus. The freed organelles are thin resorbed in the cytoplasm of the zygote. This uniform behaviour justifies a honology : the gamete whose organelles remain intact can be looked upon as "female" and the other one, whose organelles become resorbed, as "male". Once the pronuceis have touched karyogamy follows. This takes a variable amount of time. In *Swinobaculus*, the pronuclei lie next to each other for several days before they finally fuse. In others, fusion takes place within a few hours.

While in haploids meiosis follows fertilization, it takes place in diploids prior to the formation of gametes. *Macrospironympha xylopletha* is, indeed, the only diploid polymastigote of *Crystocercus punctulatus* which forms free swimming gametes. In this case, sexual reproduction is initiated by the determination of a "progamont" which divides into two gamonts. This cell division is connected with the first division of meiosis. Then each gamont encysts and it is within the cyst that the gametes are formed by a differential division which corresponds to the second meiotic division. The two gametes of *Macrospironympha* differ only slightly in size, but

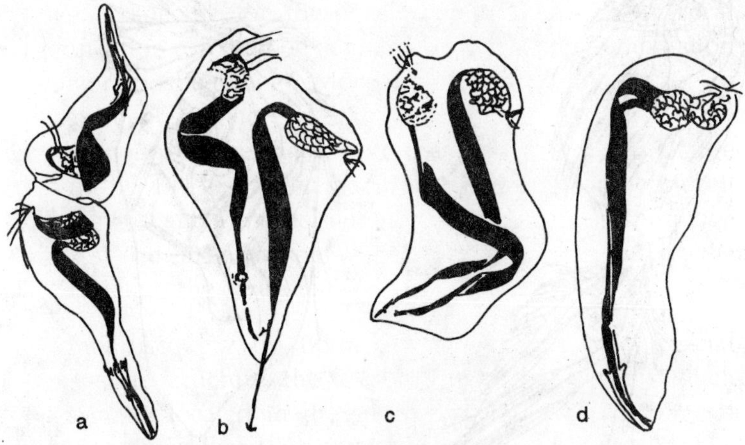

Fig. 1.16. *Saccinobaculus ambloaxostylus.* Different stages of fertilization. a The gametes fuse with the anterior ends. b laterla union. c. Fusion of the gametes completed. d Fusion of the axostyles, gamete nuclei united.

Fig. 1.17. *Eucomonympha imla.* Different stages of fertilization.

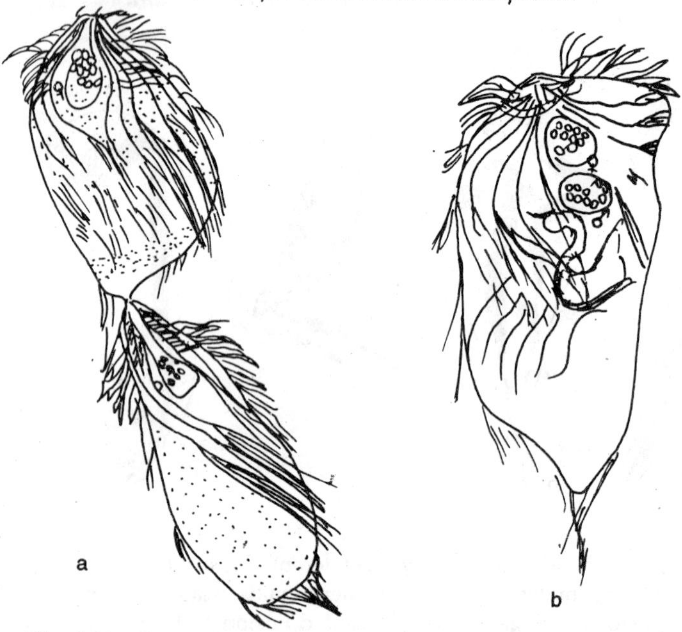

Fig. 1.18. *Leptospironympha* vachula (Polymastigina). Fertilization (a) and zygote (b).

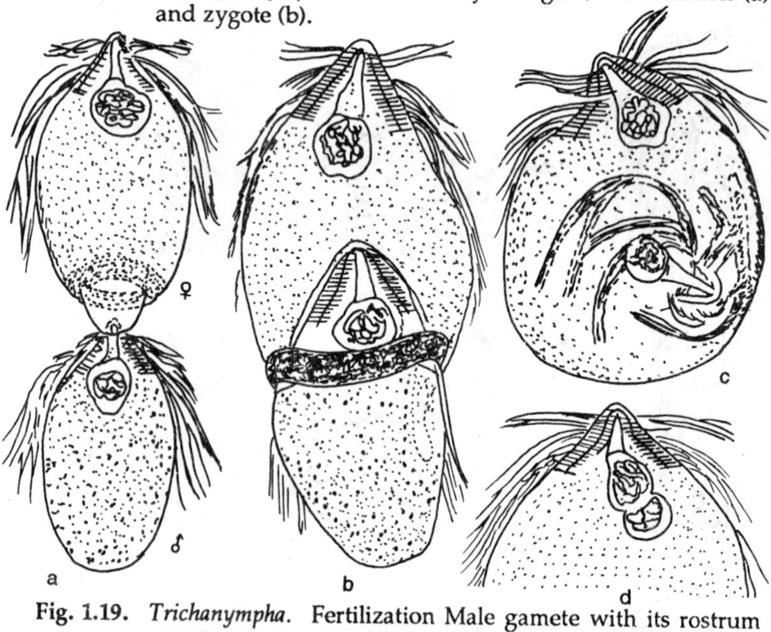

Fig. 1.19. *Trichanympha*. Fertilization Male gamete with its rostrum attached to the protruded "fertilization coue.

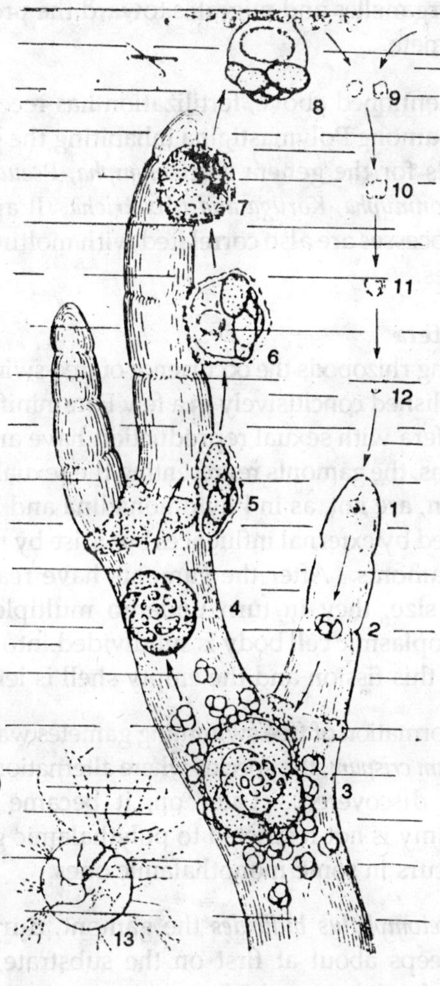

Fig. 1.20. *Tretomphalus bulloides*. Developmental cycle. 1 Amoeboid zygote. 2 Young (four-chambered) agamont. 3 Adult agamont forming agametes (gamonts). 4 Adult gamont. 5 The gamont surrounds fused with an envelope of detritus particles. 6 Formation of the "floating chamber." 7 Empty detritus envelope. 8 Gamont which has ascended to the surface of the sea, with swarming gametes. 9 Gametes. 10-12 Stages of fusion. 13 Gamont shown from below.

even in this case the pronucleus of the smaller gamete separates from its organelles and migrates toward the pronucleus of the larger gamete.

As mentioned above, fertilization has recently also been observed among Polymastigina inhabiting the gut of termites. This holds for the genera *Trichonympha, Pseudotrichonympha, Deltotrichonmypha, Koruga and Mixotricha.* It appears that the sexual processes are also correlated with molting of the host in these cases.

Foraminifera

Among rhizopods the occurrence of free-swimming gametes was established conclusively in a few Foraminifera only. Since Foraminifera with sexual reproduction have an alternation of generations, the gamonts representing the sexually reproducing generation, are not, as in Phytomonadina and Polymastigina, determined by external influences but arise by multiple fission of the agamonts. After the gamonts have reached a certain maximal size, they in turn undergo multiple fission. The entire cytoplasmic cell body is subdivided into gametes in the course of this fission and the empty shell is left behind.

The formation of free swimming gametes was first observed in *Elphidium crispum*, the species where alternation of generations was first discovered. Later on, it became apparent that gametogamy is not restricted to polythalamic genera but that it also occurs in some monothalamic ones.

In *Tretomphalus bulloides* the gamont, during its growth phase, creeps about at first on the substrate, e.g., an algal thallus. After it has reached a certain size, it surrounded itself with a sheath of detritus particles and forms a gas-filled floating chamber. Its specific weight thus decreased, it rises to the water surface where it releases the gametes. The empty shells of the gamonts are occasionally seen drifting in large numbers in the water near the seashore and may form a regular seam where they are washed ashore.

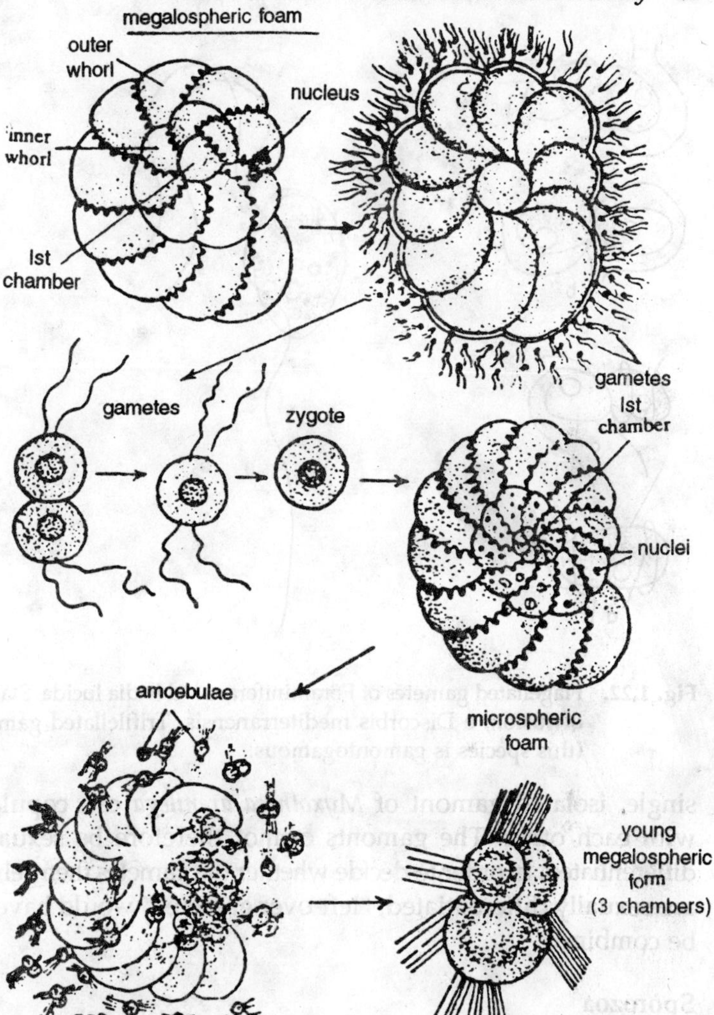

Fig. 1.21. Life history of Elphidium

The gametes formed by the gamonts of gametogamous species are very small and probably always biflagellated. Nothing is known concerning their sexual differentiation. It is likely that they are always isogametes. Their copulation has been described for *Iridia lucida*. The gametes formed by a

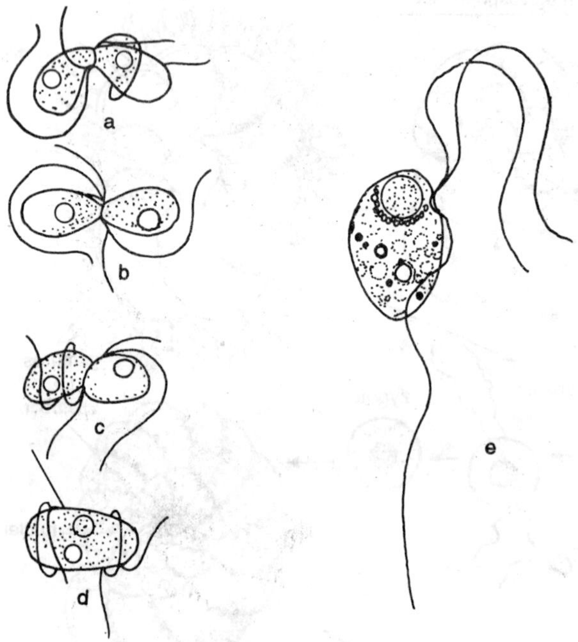

Fig. 1.22. Flagellated gametes of Foraminifera. a—d Iridia lucida. Stages of fusion, e Discorbis mediterranensis. Trifilellated gamete (this species is gamontogamous.

single, isolated gamont of *Myxotheca arenilega* can copulate with each other. The gamonts cannot therefore be sexually differentiated. In order to decide whether the gametes themselves are sexually differentiated, "left-over gametes" would have to be combined.

Sporozoa

With the exception of the Adeleidae, the formation of free-swimming gametes in Sporozoa is restricted to the Coccidia. The gamonts arise either directly from the sporozoites (Eucoccidia) or from the merozoites of a previous schizogony (Schizococcidia). While the "macrogamonts" are directly transformed into large, immotile macrogametes, the "microgamonts" undergo simultaneous multiple fission resulting in numerous small microgametes. *Oogamety* is thus characteristic for Coccidia. Sexual differentiation is often already recognizable it gamonts which may differ in size, form or structure *(anisogamonty)*. In *Eucocciducum dinophili*, the macrogamonts become large and cigar-shaped in the course of growth but assume an oval shape when they transform into fertilizable macrogametes. The microgamonts transform soon into roundedm cysts within which the microgametes originate.

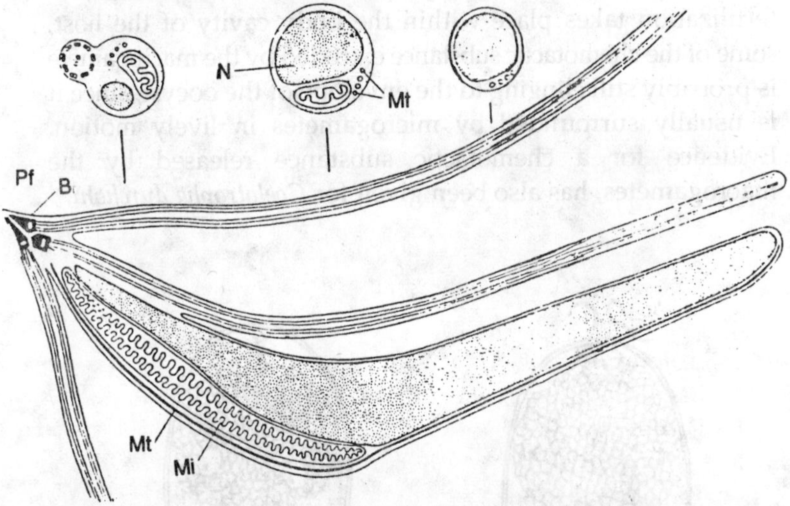

Fig. 1.23. *Eimeria maxima*. Diagram of microgamete in a longitudinal section and three cross sections. *N* Nucleus. *Mi* Mitochondrion. *Mt* Subpellicular microtubules. *Pf* Perforatorium. *B* Basal body.

While the microgametes of *Eucoccidium dinophili* and *Coelotropha durchoni* are cup-shaped and biflagellated, they have an elongated shape in most Schizococcidia. Species of *Eimeria* have microgametes with three flagella. As an example shows one flagellum may be more or less attached to the cell body, whose space is occupied mostly by the nucleus *(N)* and a large mitochondrion *(Mi)*. Microtubules *(Mt)* extend posteriorly beyond the mitochondrion. In *Aggregata eberthi* one of the three flagella is transformed into an undulating membrane. The thread-like microgametes of the Haemosporidae, on the other hand, seem to consist essentially only of a flagellum with the nucleus lying between the fibrillar bundle and the flagellar sheath.

In order to reach the macrogametes, the microgametes have to traverse a certain distance. It is very probable that the macrogametes release a substance which attracts the microgametes chemotactically. In *Eucoccidium dinophili,* where fertilization takes place within the body cavity of the host, some of the chemotactic substance extruded by the macrogamete is probably still clinging to the envelope of the oocyst, since it is usually surrounded by microgametes in lively motion. Evidence for a chemotactic substance released by the macrogametes, has also been given for *Coelotropha durchoni.*

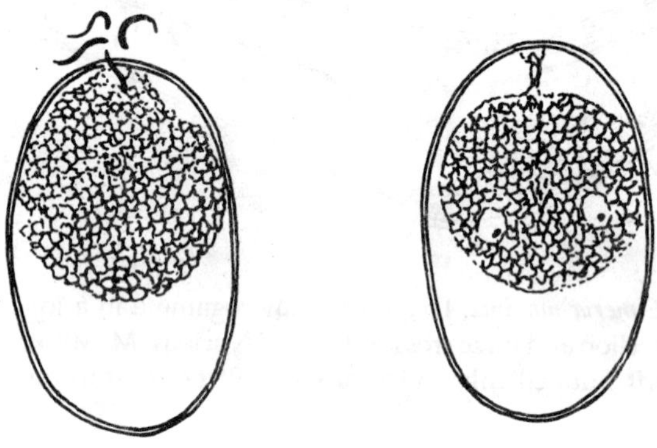

Fig. 1.24. *Eimeria propria.* Fertilization.

Prior to fertilization the nucleus of the macrogamete moves to the periphery of the cell. At this point, where the cell envelope is temporarily dissolved, the whole microgamete penetrates into the macrogametes. After karyogamy, the synkaryon becomes elongated. This stage, formerly referred to as "fertilization spindle", initiates the prophase of meiosis. Meiosis in turn leads to sporogony.

II. AUTOGAMY

Autogamy is a process of fertilization in which only the gametes or gamete nuclei derived from the same gamont participate. Autogamy is thus a matter of obligatory monoecy. Whether there is any sexual difference between the gametes or gamete nuclei has to be tested in each individual cause.

The possibility that a gamont gives rise to i.e., whole cells which fuse with each other, is realized some Heliozoa or Foraminifera.

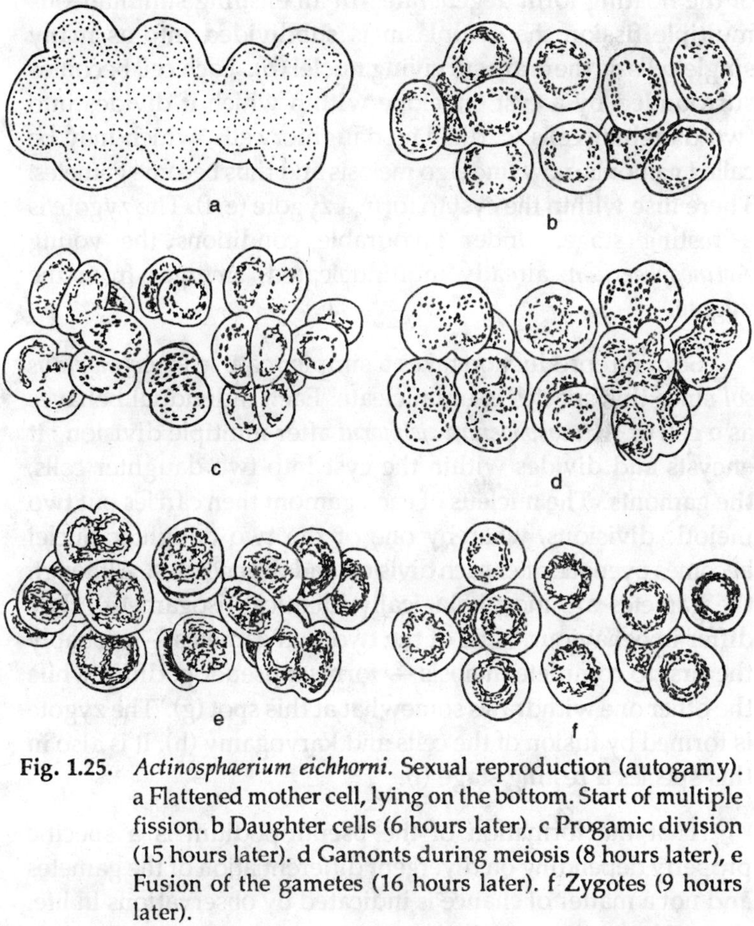

Fig. 1.25. *Actinosphaerium eichhorni*. Sexual reproduction (autogamy). a Flattened mother cell, lying on the bottom. Start of multiple fission. b Daughter cells (6 hours later). c Progamic division (15 hours later). d Gamonts during meiosis (8 hours later), e Fusion of the gametes (16 hours later). f Zygotes (9 hours later).

The only *Heliozoa* in which fertilization has been described in detail thus far are the two fresh-water species *Actinosphaerium eichhorni* and *Actinophrys sol.* Sexual reproduction takes place when food suddenly becomes scarce after a period of multiplication.

Actinosphaerium eichhorni, a large, multinuclear species, then sinks to the bottom, withdraws its axopodia and flattens to an irregularly shaped mass which is surrounded by a gelatinous covering. Most of the numerous (up to 500) nuclei of the floating form degenerate. In an ensuing simultaneous multiple fission the cytoplasm is subdivided into as many single cells as there are surviving nuclei (b). Each cell becomes surrounded by a cyst envelope within which it divides into two daughter cells (c, d). The daughter cells, which may be called gamonts, now undergo meiosis and thus become gametes. There fuse within the cyst to form a zygote (e, f). The zygote is a resting stage. Under favourable conditions the young *Actinosphaerium,* already multinucleated, emerges from the zygote.

Sexual reproduction takes a simpler course in *Actinophrys sol* since this species is uninucleate. Each individual behaves as a cell of *Actinosphaerian cichhorni* after multiple division : it encysts and divides within the cyst into two daughter cells, the gamonts. The nucleus of each gamont then carries out two meiotic divisions, whereby one of the two daughter nuclei becomes pyenotic after each division and degenerates. Although the gametes are morphologically identical (isogamety), they differ in behaviour. One of the two two gametes — probably the first to complete meiosis — forms a pseudopodium while the other one withdraws somewhat at this spot (g). The zygote is formed by fusion of the cells and karyogamy (h). It is also in this species a resting stage (i).

That the formation of the pseudopodium is a specific property depending on divergent differentiation of the gametes and not a matter of chance is indicated by observations in life.

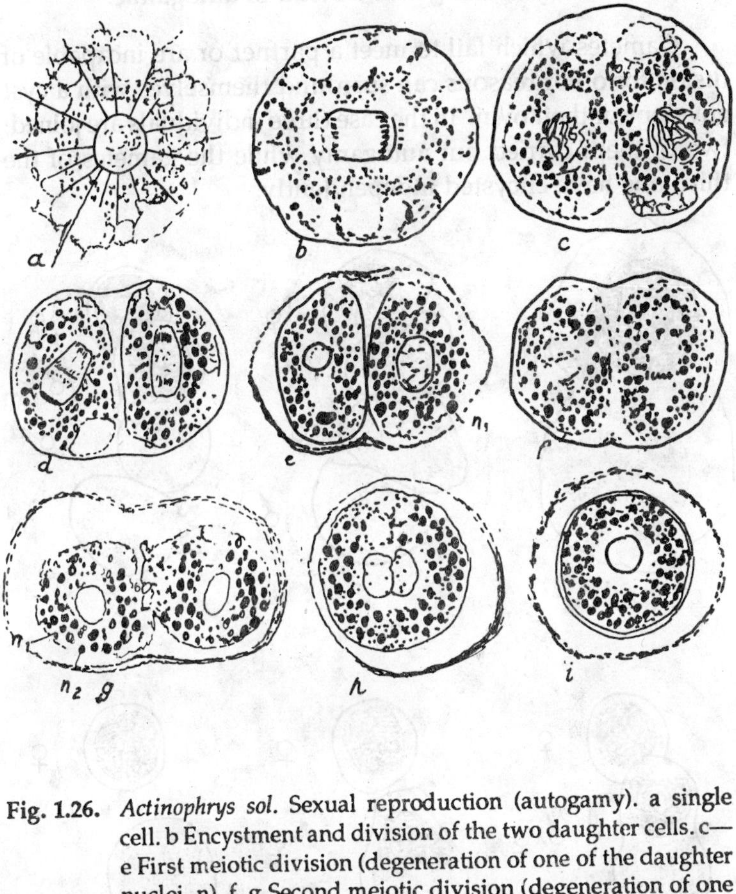

Fig. 1.26. *Actinophrys sol.* Sexual reproduction (autogamy). a single cell. b Encystment and division of the two daughter cells. c—e First meiotic division (degeneration of one of the daughter nuclei, n). f, g Second meiotic division (degeneration of one of the daughter nuclei, n_2) and sexual differentiation of the gametes (pseudopodia formation in one of the gametes). h Karyogamy.

If the pseudopodium is formed by way of exception at the side away from the partner, it is resorbed and reformed at the "correct" place. Occasionally, two individuals will join at the beginning of sexual reproduction and proceed to form a common cyst. If the gametes formed by them happen to be arranged in series with active and passive gametes alternating, the resulting fertilization may be allogamic instead of autogamic.

Gametes which fail to meet a partner or are incapable of fusion for other reasons can surround themselves with a cyst envelope of their own. In the case three individuals are joined; two of them carried out autogamy while the gametes of the third one have encysted independently.

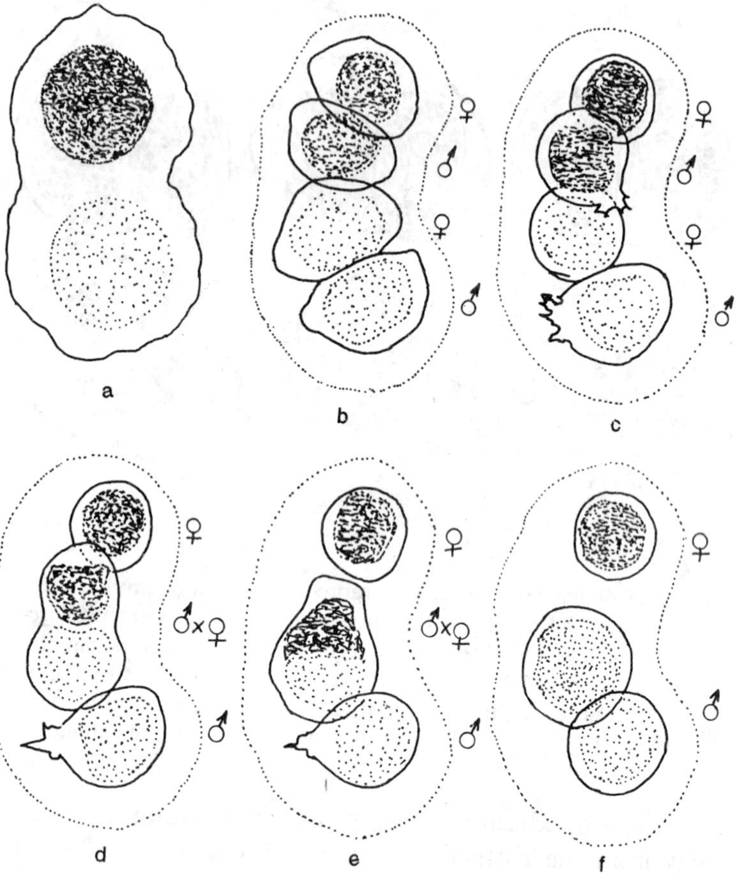

Fig. 1.27. *Actinophrys sol.* Fusion of two gametes, descended from different previously fused mother cells (diagrammatically indicated by dense and light stippling).

The encysted gametes can develop further by *parthenogenesis*. Although only a fraction germinate and of these many die off, indefinitely multiplying clone cultures can be obtained from the survivors. It is likely that the chromosome number in the survivors is regulated to diploidy.

If sexual differentiation of the gametes is accepted as a fact, then sex determination must be connected with the progamic division which follows encystment. Thus, this could be looked upon as a differential division.

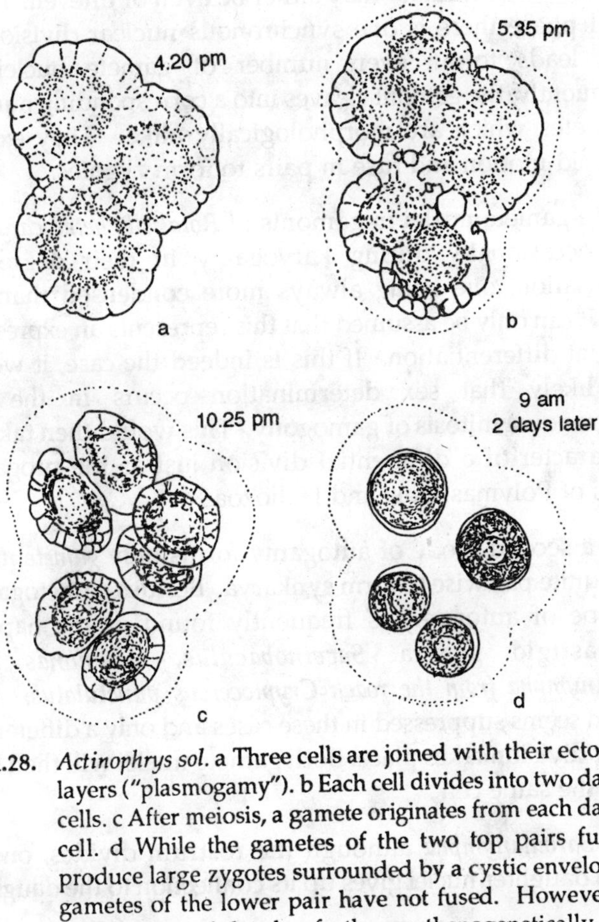

Fig. 1.28. *Actinophrys sol*. a Three cells are joined with their ectoplamic layers ("plasmogamy"). b Each cell divides into two daughter cells. c After meiosis, a gamete originates from each daughter cell. d While the gametes of the two top pairs fuse and produce large zygotes surrounded by a cystic envelope, the gametes of the lower pair have not fused. However, they encyst also and develop further parthenogenetically.

Among the *Foraminifera* one monothalamic species, *Allogromia laticollaris,* has been shown to be autogamic. Details of gamete formation and sexual fushion are not known.

More detailed studies are available concerning autogamy in the small polythalamic species of *Rotaliella.* As in all Foraminifera, the gamont is uninucleate. Gamogony consists of two phases and proceeds after the gamont has reached a certain critical size. In the first phase the nuclear divisions are strikingly asynchronous. Eventually, all nuclei reach a resting stage (d). Their number may either be even or uneven. In the second phase there is one synchronous nuclear division (e) which leads to an even number of gamete nuclei (f). Subsequently, the gamont cleaves into a corresponding number of gametes which are morphologically alike. They exhibit amoeboid motion and fuse in pairs to form zygotes.

The gamete nuclei in gamonts of *Rotaliella heterocaryotica* differ occasionally during karyogamy in their degree of condensation, one being always more condensed than the other. It can only be assumed that this represents an expression of sexual differentiation. If this is indeed the case, it would seem likely that sex determination occurs in the last (synchronous) mitosis of gamogony. This would then take on the character of a differential division just as the progamic mitosis of Polymastigina and Heliozoa.

In a second mode of autogamy forms only *gamete nuclei* which unite pairwise to form synkarya. Besides gametogamy, this type of autogamy is frequently found in the haploid polymastigid genera. *Saccinobaculus, Oxymonas* and *Barbulanympha from the roach Cryptocercus punctulatus.* Cell division seems suppressed in these cases and only a differential nuclear division takes place. Both gamete nuclei will then fuse within the same cell.

In *Barbulanympha* although the rostrum divides, one of the two daughter nuclei gives up its connection to the daughter rostrum with which it would otherwise remain in contact and migrates toward the other daughter nucleus. The two pronuclei thus exhibit during autogamy the very same difference in behaviour as the gamete nuclei during copulation.

A further simplification (which can scarcely still be considered "normal") consists in the total omission of nuclear division. What remains is the reduplication of the chromosomes without spindle formation, i.e., an *endomitosis*. This is evidently always the case if one of the entrioles degenerates prematurely. The "synkaryon" formed in this manner will later on carry out meiosis just as after gametogamy or autogamy.

In the diploid genera *Rhynchonympha* and *Urinympha* autogamy is the only mode of sexual reproduction. In *Rhynchonympha*, meiosis consists of two divisions, the first one being connected with cell division while the second (differential) division is only nuclear. The two nuclei then fuse in autogamy. In *Urinympha*, meiosis involves only one division. The two gamete nuclei fuse right away to form a synkaryon.

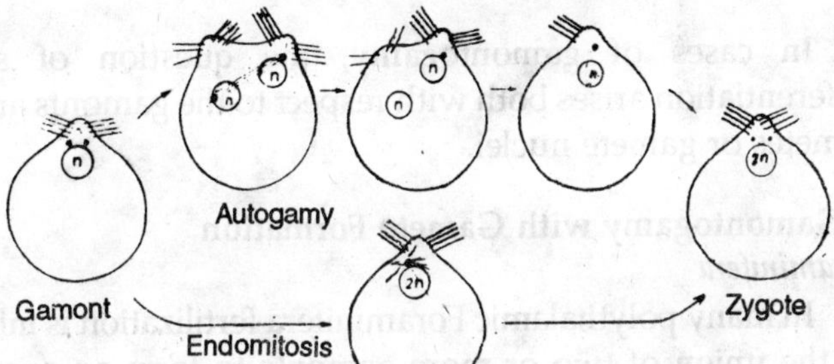

Fig. 1.29. *Barbulanympha*. Modifications of sexual reproduction due to omission of cell division (→ autogamy) and of nuclear division (→ endomitosis) of the gamonts.

In *ciliates*, where conjugation is the predominant mode of

fertilization, autogamy is also occasionally encountered. The micronucleus undergoes the same divisions as in conjugation; however, the migratory nucleus is not transferred into another cell but fuses with its stationary sister nucleus. In most cases (e.g. *Paramecium aurelia, Euplotes minuta)* autogamy occurs along with conjugation. Occasionally two cells attach as in conjugation but fail to exchange migratory nuclei, each partner carrying out autogamy. This modification of autogamy has been called "cytogamy". Both autogamy and cytogamy have been observed in *Paramecium polycaryum,* but conjugation has not. Some ciliates, e.g. *Tetrahymena rostrata,* carry out autogamy within a cyst.

Further studies will probably show that autogamy occurs in many ciliate species. Proof of autogamy requires close cytological control in clone cultures.

III. GAMONTOGAMY

If sexual reproduction is initiated by the union of the gamonts, we can speak of gamontogamy. Either the gamonts form gametes by multiple fission or the gamonts fuse directly and form only gamete nuclei.

In cases of gamontogamy, the question of sexual differentiation arises both with respect to the gamonts and the gametes or gamete nuclei.

1. Gamontogamy with Gamete Formation

Foraminifera

In many polythalamic Foraminifera fertilization is initiated by the union of two or more gamonts to form an *aggregate.* After gamogony, the gametes fuse in the common space of the aggregated gamont shells.

It is highly probable that gamontogamy in Foraminifera has developed from gametogamy. In the genus *Discorbis* there are gametogamous *(D. vilardeboanus)* as well as gamontogamous species *(D. opercularis, D. patelliformis, D. mediterramensis).* The latter form several hundred flagellated gametes which swim about within the shells of the gamonts. Indeed, the gametes of all gamontagamous species have three flagella. This also holds for *Glabratella sulcata* which is related to the genus *Discorbis.*

The remaining gamontogamous Foraminifera studied thus far form, on the other hand, but a few amoeboid gametes and are in this respect like the autogamous species. They include *Patellina corrugata, Spirillina vivipara, Metarotaliella parva, Metarotaliella simplex* and *Rubratella intermedia.*

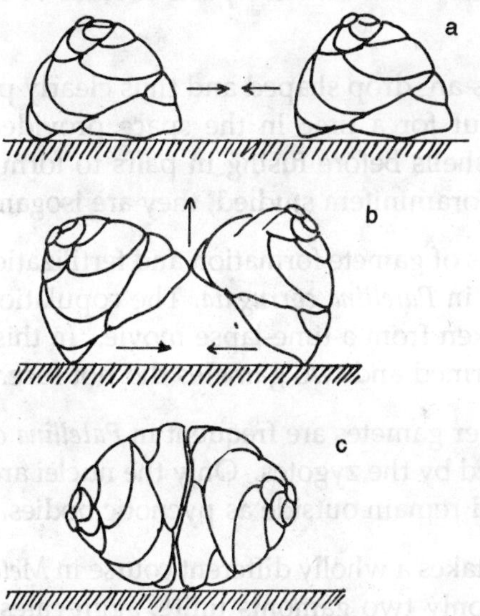

Fig. 1.30. *Discurbis mediterransis.* Aggregation of two gamonts.

In the simplest case only two gamonts mate. They creep toward each other, touch with their ventral surfaces and extrude a substance which firmly cements them together. In the course of gamogony the inner walls separating the chambers of the gamonts dissolve, providing a common space within which the gametes can pair. In many species aggregates of *several gamonts* are regularly found. In *Patellina corrugata,* where the number of aggregating gamonts ranges from 2-14, the gamonts cover each other by the rims of their tests and form a thin organic membrane which fastens them to each other and to the substrate. All further events take place in this space roofed over by the gamonts.

As soon as the gamonts have united some asynchronous nuclear divisions take place. In this way each gamont acquires between two and five nuclei depending upon its size. At this stage the protoplasts emerge from their shells and round off. Each protoplast is then subdivided into as many parts as the number of nuclei. Each of these parts divides to form two gametes.

The gametes are drop shaped and thus clearly polarized. They move about for a time in the space provided by the empty gamont shells before fusing in pairs to form zygotes. As in all other Foraminifera studied, they are isogametes.

Many details of gamete formation and fertilization can be observed in life in *Patellina corrugata.* The copulation of two gamete pairs taken from a time-lapse movie. In this case six zygotes were formed and two gametes did not fuse.

Such left-over gametes are frequent in *Patellina corrugata.* They are ingested by the zygotes. Only the nuclei are usually not ingested and remain outside as pycnotic bodies.

Gamogony takes a wholly different course in *Metarotaliella parva.* Usually, only two gamonts unite, aggregates of three being rare. The first phase of gamogony takes place in the initial chamber (a, b). even distribution of all nuclei throughout the cytoplasm follows (c). The last mitosis of gamogony, though synchronous in each gamont, begins a little earlier in one of the gamonts (d). This is also the gamont which begins with the amoeboid form changes which take place within the common space of the shells. Each gamont spreads into the shell of its partner, only a part of the plasmatic body remaining in its own shell. In this manner the gametes of both gamonts become thoroughly mixed (e-h). They then fuse in pairs to form zygotes (i).

The diagrammatic representation gamonts of equal size, each forming the same number of gametes. However, in *Metarotuliella parva* the gamonts are often of different size and form different numbers of gametes. Some gametes may thus be left without a mate, as described for *Patellina corrugata.*

Since neither the aggregating gamonts nor the copulating gametes are morphologically different in Foraminifera, the problem of their sexual differentiation can only be solved by indirect methods.

Combination tests with the gamonts of *Discorbis mediterranensis* and *Patellina corrugata* led to the result that the gamonts are differentiated into two types designated as the "+" and "-" sex respectively. When two gamonts are put into a culture dish they either move toward each other and attempt to form an aggregate or they behave indifferently. After separation of the aggregating gamonts, their behaviour toward other gamonts can be tested. By combination experiments of this type it can be decided whether all the gamonts derived from the same agamont are of one or the other sex.

In aggregates of *Patellina corrugata,* which frequently comprise several gamonts, both sexes are always present, though in varying proportions. Aggregates of three must of course include two gamonts of the same sex.

The assumption that the gametes of *Patellina corrugata* are also unable to combine randomly is supported by the observation that in many of the aggregates left-over gametes are found which do not fuse with each other. Through analysis of time-lapse motion pictures the developmental history of zygotes and residual gametes could be traced back to the gamonts. In the triple aggregate six zygotes were formed and two gametes were left over. Analysis showed that gamont 1 had formed six gametes whereas gamonts 2 and 3 had given rise to four gametes each. Every gamete of gamont 1 took part in the formation of the six zygotes, besides all four gametes of gamont

2 and two of the gametes of gamont 3. The two left-overs were from gamont 3. Thus, if gamont 1 is of the "—" sex and gamonts 2 and 3 of the "—" sex only gametes from gamonts of different sexes have united, the two left-overs belonging to the same sex.

In an aggregate of two, if one separates both gamonts by glass fibers, the gametes being produced by them cannot copulate—neither those of the different gamonts nor those of the same gamont. After some creeping around all gametes of each gamont finally unite to form a multinucleated plamodium which later dies.

Hence, *Patellina corrugate* is a dioecious species, probably with genetic sex determination.

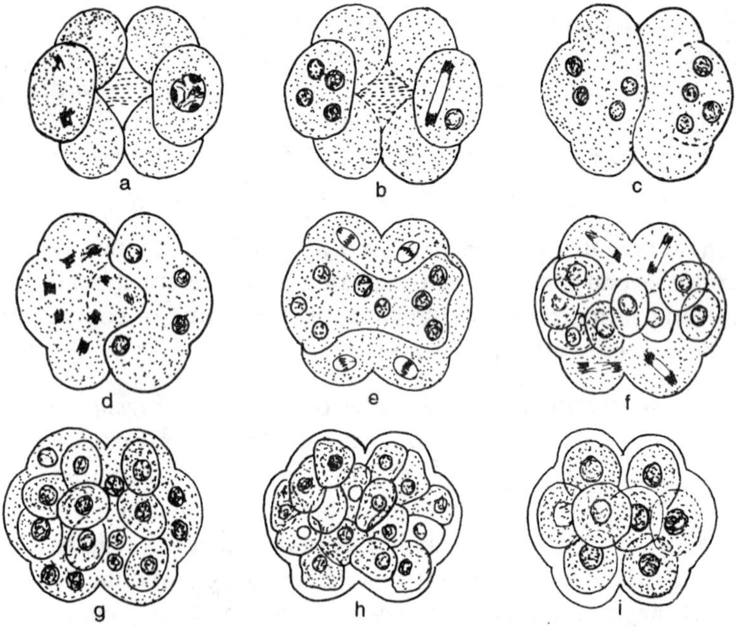

Fig. 1.31. *Metarotaliella parva*, Diagram of gamogony and fertilization.

There is also evidence in *Rubratella intermedia* that a gamete cannot mate with every other gamete. This foraminiferan forms only aggregates of two. The gamonts may be of equal or of unequal size. In the first case, the the gametes and their nuclei are also of equal size (a—d). In the second case, a mechanism of regulation becomes effective : Although both gamonts form the same number of gametes, those of the little gamont are smaller and their nuclei are more condensed. At karyogamy, a large and a small nucleus fuse (a'—d'). Thus, it is evident that mating takes place only between the gametes of different gamonts.

However, conclusions regarding the sexual differentiation of gametes cannot simply be extended to include the gamonts, as shown by experiments with *Metarotaliella parva*.

It could be established conclusively that gamonts of this foraminiferan are *not sexually differentiated* before mating. Any two gamonts which are put into the same culture dish can form an aggregate.

The possibility of combinations is thus not restricted by the differentiation of types.

The question arises whether each gamete can also fuse with every other gamete. It cannot be solved by cytological study since a regulation, as described above for *Rubratella intermedia*, is absent in *Metarotaliella parva*. If gamonts of different size fuse, they form different numbers of gametes, some of which are left without a partner at pairing. Statistical analysis of the numbers of zygotes and residual gametes in aggregates showed, however, that only the gametes of different partners can pair with each other. Two residual gametes from the same gamont can never fuse. Gamont-specific differentiation of the gametes is also indicated by the results of combination experiments. If gamonts of different size are brought together to mate, the number of zygotes (or agamonts) formed depends solely on the smaller partner. If large gamonts of different sizes are combined with smaller partners of constant size, the number of emerging agamonts also remains constant.

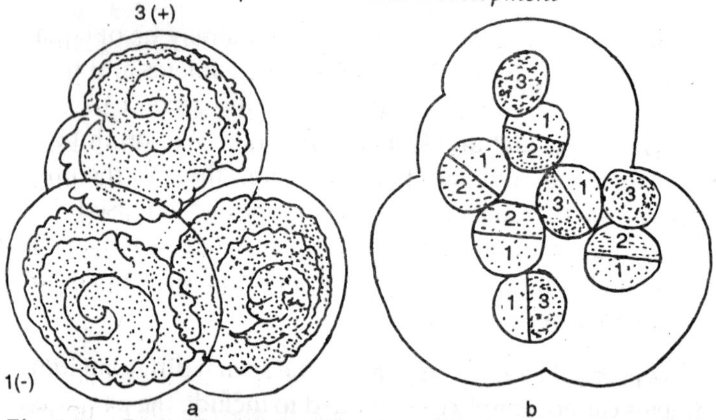

Fig. 1.32. *Discurbis mediterransis.* Aggregation of two gamonts.

Experiments with aggregates of three indicate that the second possibility is realized in *Metarotaliella parva.*

Combination experiments with gamonts of *Metarotaliella parva* show that they pass through three different stages in their individual development in which they differ in their mating behaviour. These may be designated as the juvenile (I), the mature (II) and the old (III) stage.

These differences in graphs. When *gamonts of the same stage* are combined every stage can be characterized by a definite type of mating curve (a).

It is characteristic of the *juvenile stage* (type 1) that in a I x I combination a latency period (to) elapses before mating sets in. The latter commences only after one of the partners has reached an age of at least 2 days.

In the *mature stage* (type 2), which lasts from about the second to the seventh day, pair formation commences as soon as the partners are brought together (II x II).

Old gamonts (type 3), differ from those in the mature stage in a progressive decrease of the speed of mating in III x III combinations. At the same time, the proportion of non-mating gamonts increases steadily.

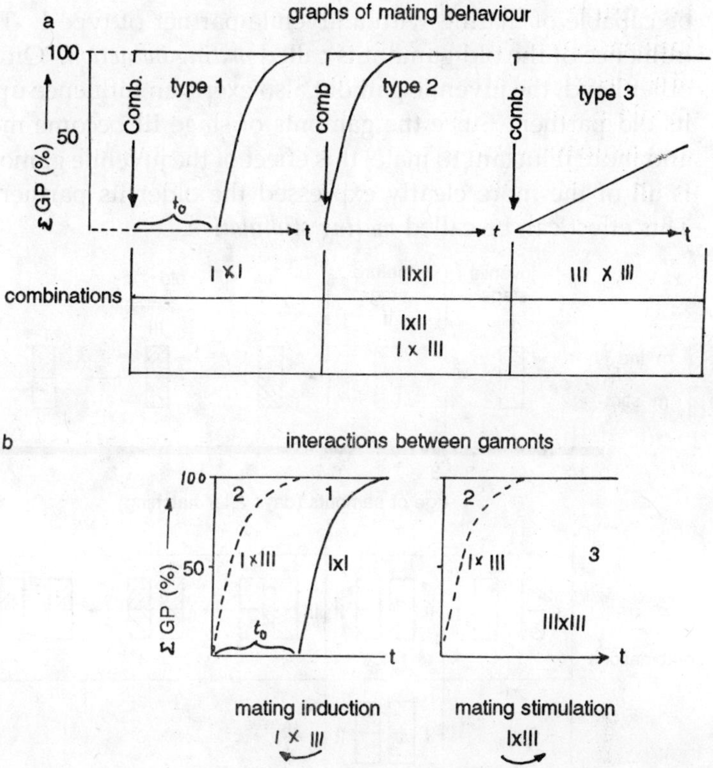

Fig. 1.33. *Metarotaliella parva*. Diagram of the mating behaviour of gamonts of different stages of age (a) and the reciprocal interaction in a combination of young and older partners (b). Explanation in the text.

When gamonts of *different stages* are combined (I x II or I x III), the mating behaviour corresponds to that of mature gamonts (II x II).

This result indicates a *reciprocal interaction* between gamonts of different stages. In a combination of young and old gamonts (I x III) both partners must influence each other (b). The old gamont effects the disappearance of the latency period (to) in its juvenile partner. Mating-begins immediately upon combination at a time when the young gamont would not yet be capable of mating with a juvenile partner of type I. This influence of the old gamont is called *mating induction.* On the other hand, the juvenile gamont also exerts an influence upon its old partner. Since the gamonts of stage III become more and more reluctant to mate, this effect of the juvenile gamonts is all of the more clearly expressed the older its partner is. This effect can be called *mating stimulation.*

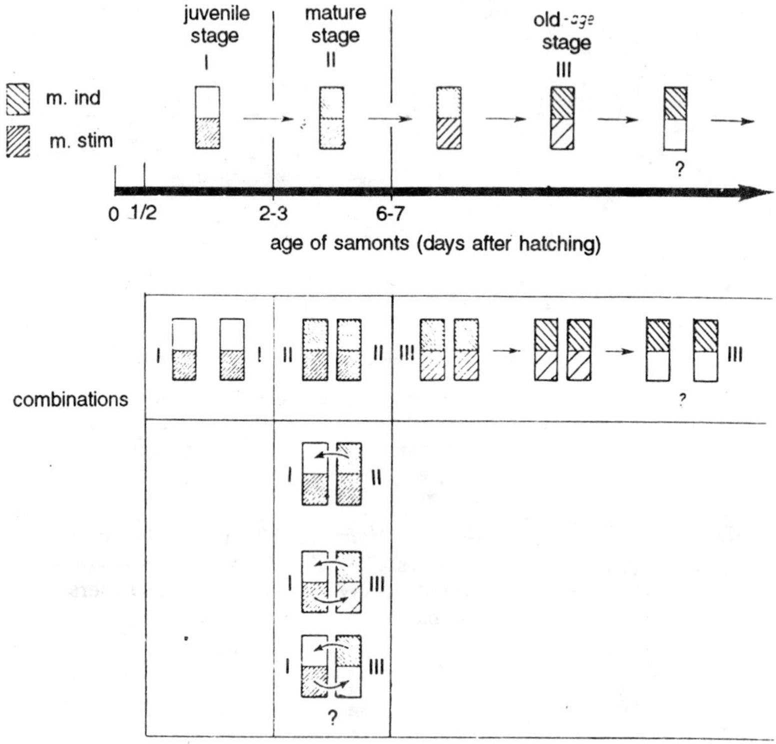

Fig. 1.34. *Metarotaliella parva.* Change of aggregation properties during the development of the gamonts. ***m. ind.*** = ability of mating induction. ***m. stin.*** = ability to stimulate mating. Curved arrows = proved partner effects. For the combination in which mating is possible, the partial functions of the two partners are shown close together; for the combinations in which mating is not possible, they are shown far apart.

Since the combination of young and old gamonts leads to the same course of mating as the combination of two mature gamonts, it must be assumed that we are dealing with a *phenomenon of complementation* due to the effects of each partner upon the other.

Based on mating behaviour we can thus recognize two properties of the gamonts which appear in succession in the course of development.

In juvenile gamonts only the ability to stimulate mating is present. Therefore, mating between two gamonts is not yet possible in this phase. With transition into maturity the gamonts acquire the ability of mating induction, i.e., they can now mate with each other and with older as well as younger gamonts. In the old stage the ability to stimulate decreases steadily; the gamonts thus become more and more reluctant to mate. The ability to induce mating, however, remains unaffected.

In order to obtain mating in an aggregate, both properties must be present. However, it is not necessary that each partner has both properties as is the case in a mating of mature gamonts (II x II). Even if one partner is deficient in one characteristic (I x II), or if the properties of both supplement each other (I x III), mating can take place as between gamonts of the mature phase.

Sporozoa

Among the *Sporozoa* both the gregarines and the adeleids show gamontogamy. While the gamonts of eugregarines arise directly, by growth, from sporozoites, those of schizogregarines and of adeleids develop from the merozoites of a previous schizogony.

Mating of gamonts can take place in *eugregarines* either before or after termination of the growth phase. These differences seem to depend upon the species. It must still be clarified to what extent readiness for mating depends upon internal factors such as the growth stage reached and which external conditions might be involved. It has been shown in some cases that the hormonal situation of the host plays a role in the mating of gamonts.

The gamonts may attach side by side, as in many Monocystidae, or lengthwise. In some species, e.g., in *Stylocephalus longicollis* the gamonts attach by their protomerites. In most cases, however, one gamont fastens by its protomerite to the deutomerite of the partner. Thus, an anterior and a posterior gamont (designated as primite and satellite, respectively) can be distinguished. Chains of more than two gamonts or several satellites attached to one primite can also occur. The further development of such aggregates is not known.

In *schizogregarines* the mating of gamonts is delayed until a certain degree of infection of the host is attained by way of schizogony. Sexual reproduction is therefore probably triggered by conditions which are brought about by infestation with the parasites. Since the duration of the gamontic growth phase is variable in schizogregarines, the number of gametes produced varies accordingly.

After uniting, the gamonts flatten against each other, thus assuming a hemispherical shape and surround themselves by a common cyst envelope. The cysts of intestinal parasites are soon passed with the feces of the host. Species which inhabit the body cavity or tissues continue development within the host up to sporogony.

Encystment is followed by gamogony, which is initiated by a phase of nuclear multiplication. Most nuclei finally move to the surface of the cytoplasm, which has frequently been subdivided before into single portions (cytomeres). During gamete formation, which takes place synchronously in both gamonts, the whole cytoplasm is not generally used up. The remaining masses fuse in the center of the gamont cyst to form the so-called residual body. It may still contain a few nuclei.

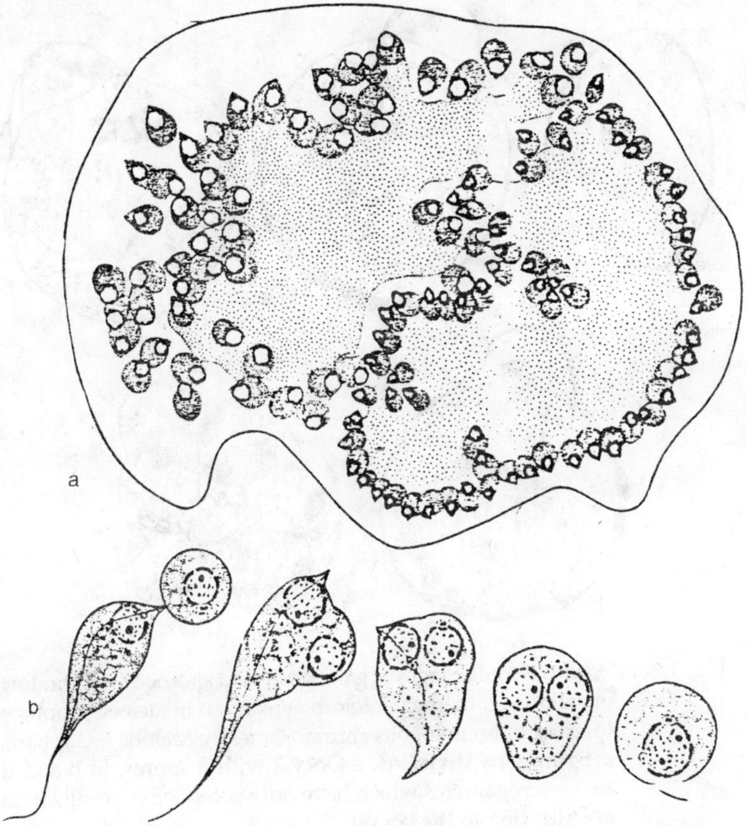

Fig. 1.35. Fertilization processes in eugregarines, a Section through the gamont cyst of a species of *Monocystis*. After Brasil, 1905. b Fusion of the gametes of *Stylocephalus longicollis*.

In many gregarines, especially the schizogregarines, neither the aggregating gamonts nor the copulating gametes are morphologically different from each other. In other cases, the gamonts differ clearly in form, cytoplasmic structure and staining with vital dyes. Anisogamety, if present, is easily recognized

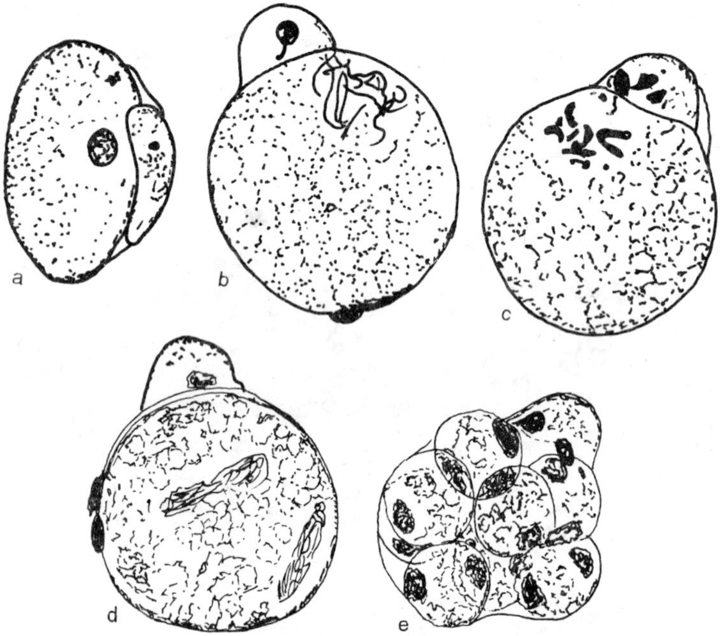

Fig. 1.36. *Adelina deronis.* Stages of gamogony and sporogony. a Omiting of macro and therogametoic. b Synkaryon in meiotic prophase (pairing of homologous chromosomes). c Meiotic metaphase. d Sporogony divisions. e Oocyst with 8 spores. In b and d two microgametes which have not succeeded in fertilization are attached to the oocyst.

in gregarines because the two gamonts form hundreds of gametes lying side by side within the same cyst. Occasionally, there are only minor differences in the form or size of the nuclei. However, there are also oogametic species where the gametes of one type are immotile while those of the other type can move about. A very peculiar differentiation is found in the gametes of *Stylocephalus longicollis*, the "egg cell" being smaller than the motile gamete.

Although there is no doubt that sexual differentiation in gregarines is bipolar, the mode of sex determination is by no means clarified. Since meiosis takes place with the spore, the method of single-spore infection does not yield clear results as in Coccidia.

Although the *Adeleidae* differ from other Coccidia with respect to gamontogamy, the direct transformation of the macrogamont into the macrogamete and the occurrence of multiple division only in the microgamont follows the typical coecidian pattern. The gamonts may either unite at the onset of the growth phase or after its termination. In either case the paired gamonts are of different size. The lens-shaped microgamont adheres to the macrogamont and gives rise to only two or four microgametes, one of which unites with the macrogamete. Meiosis of the cynkaryon and the onset of sporogony follows immediately.

The adeleids are therefore not only oogametic, which is characteristic for all Coccidia, but there is a difference in size of the gamonts as well (anisogamonty).

2. Gamontogamy without Gamete Formation

In a special type of gamontogamy the mating gamonts do not produce gamete cells but only gamete nuclei.

Among the *flapillates,* this has only been described for the diploid polymastigid. The gamonts within arise by differential cell division from a "progamont" are not different from ordinary cells (a), Sexual differentiation, however, is expressed in the behaviour of the nuclei. As soon as the "male" and "female" gamonts have fused with each other (b), the axostyles also fuse just as in gametes of *Oxymonas* and *Saccinobaculus.* At the same time, the nucleus of the male gamont separates from its associated organelles (axostyle, flagella). A new set of organelles is formed from the intranuclear centrioles of the nucleus of the female gamont (c). The old axostyle which has been formed as a fusion product is resorbed in the cytoplasm. Each gamont nucleus now divides and reduces the chromosome number. Two females gamete nuclei, each connected with axostyle and flagella, and two motile male nuclei are formed in this manner (d, e). The latter migrate toward the female gamete nuclei and fuse with them, forming two synkarya (f). The "double zygote" then undergoes cytokinesis, leading to the formation of two diploid cells.

3. Conjugation

In ciliates, the process of fertilization is called "conjugation". Actually, this term denotes two rather different phenomena : in one case (isogamonty) both gamonts are morphologically identical; in the other case, they are either different from the beginning or become different in the course of mating (anisogamonty). Both possibilities will be discussed separately.

(a) Isogamonty

In the majority of ciliates the conjugants are morphologically identical. The cells touch first by their anterior ends and then lie side by side along their whole length. The first change in the micronuclei is an enlargement (a). While the nuclear volume increases the chromosomes undergo the processes which are generally characteristic of meiotic prophase. Reduction of the chromosome number is achieved in the course of two micronuclear divisions which follow in rapid succession (b-d). In this manner four haploid daughter nuclei are formed, three of them becoming pycnotic to be resorbed in the cytoplasm (e). The remaining nucleus now divides once more (f). This nuclear division, which is designated as postmeiotic, gives rise to the two gamete nuclei (pronuclei). One of these, the *stationary nucleus,* remains within the same conjugant while the other one, the *migratory nucleus,* wanders into the mating partner. In each cell, the stationary nucleus fuses with the migratory nucleus of the other conjugant to form a synkaryon (g). Conjugation this involves *mutual fertilization,* the mating partners separating afterwards. The cells are now referred to as *exconjugants.* In the latter, the synkaryon divides into two daughter nuclei, one becoming the micronucleus while the second develops into a new macronucleus (h-k). The old macronucleus dissolves in the course of conjugation. Usually it disintegrates into numerous fragments which are sooner or later resorbed in the cytoplasm.

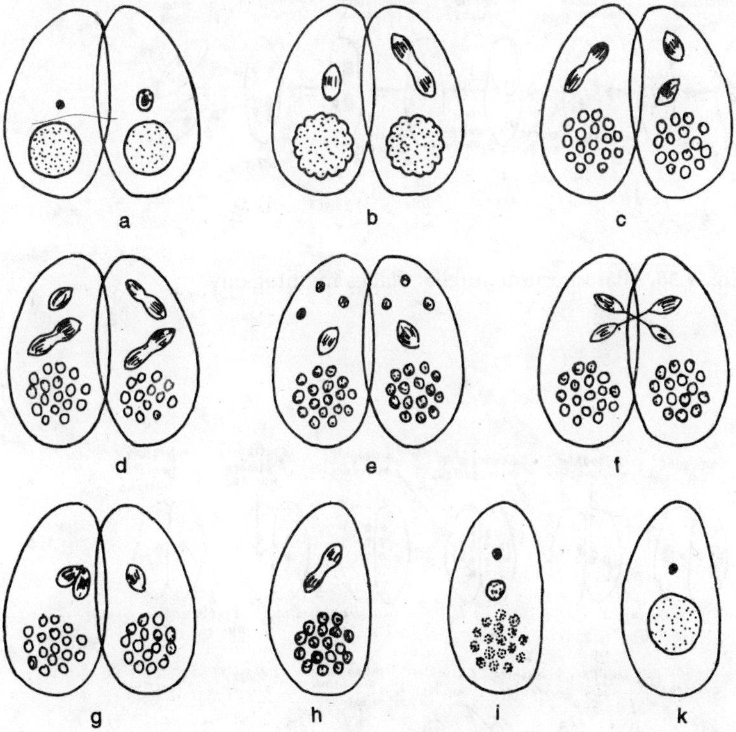

Fig. 1.37. Diagram of conjugation (isogamonty) in ciliates.

In contrast to *Notila proteus* whose two "male" nuclei of one gamont migrate toward the two "female" nuclei of the other gamont, the eiliate form two gamete nuclei of different behaviour in each gamont. Only one of these, which may be designated as "male", migrates into the partner and fuses with the "female" gamete nucleus to form the synkaryon. According to the behaviour of their gamete nuclei, the isogamontic ciliates can be looked upon as *monoecious*.

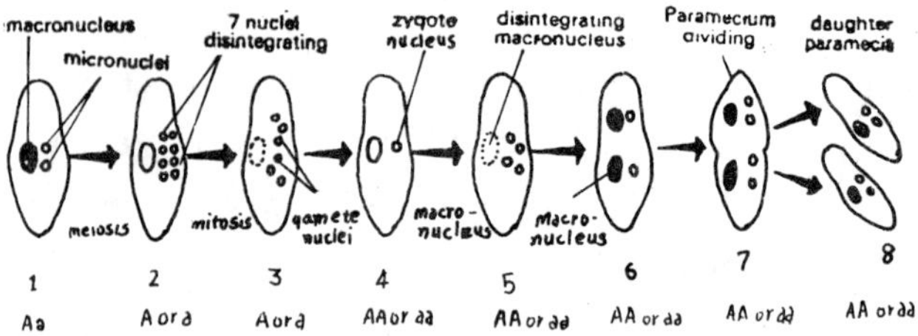

Fig. 1.38. Paramecium aurelia. Stages in autogamy.

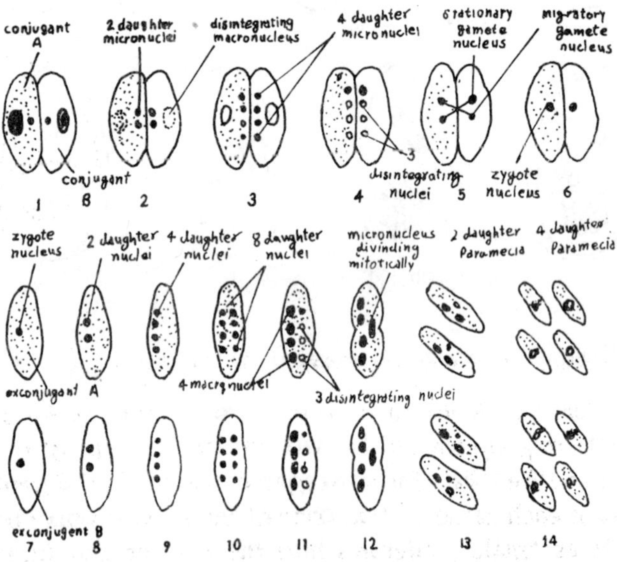

Fig. 1.39. Paramecium. Stage in conjugation.

A most remarkable variation of progamic events is found among species of *Tracheloraphis* and *Trachelocerca*. Recent studies have shown that several of the gone nuclei formed by the two meiotic divisions can carry out a postmeiotic division.

In *Tracheloraphis phoemicoprerus*, which has six micronuclei, a variable number of gamete nuclei originate in each conjugant. Some of these differentiate into migratory, others into stationary nuclei. After exchanging migratory nuclei, each conjugant forms several synkarya all out one, however, are destined to become pycnotic. In *Trachelocerca coluvex* the two micronuclei of each conjugant give rise to eight gone nuclei, four of which divide again. Four of the eight descendants of this division become pyenomic while the remaining four differentiate into two migratory and two stationary nuclei. Two synkarya are thus formed in each conjugant (b).

In *Euplotes vannus* (c) the micronucleus of each conjugant first undergoes a "premeiotic" division. After meiosis, there are then eight gone nuclei, two of which participate in a postmelot division leading to four gamete nuclei. Two of these degenerate, however, while the remaining two differentiate into a migratory and a stationary nucleus. Genetic studies showed that the two functional nuclei can either be daughter nuclei of the same division or descendants of different gone nuclei. At least in *Euplotex vannus* it is therefore improbable that the occurence of a postmeiotic division is a prerequisite for the differentiation of a stationary and a migratory nucleus.

That the migratory nuclei are not simply pushed into the partner by the elongating spindle indicated by cases where the postmeiotic division is parallel to the border between the conjugant. One example of this is provided by the suctorian *Heliophrya (Cyclophrya)*. As in all Suctoria conjugation starts with the formation of lump-like swellings which elongate and connect both conjugants with each other. The gene nucleus which survives and gives rise to the gamete nuclei always moves into the left corner close to the border between both conjugation. Its division leads to the formation of a stationary nucleus at left and a migratory nucleus at right (c, d). One gets the impression that influences of the membrane separating both cells are involved in the determination of both gamete nuclei. In addition to their different behaviour, the gamete nuclei also show occasional structural differences as in some other ciliates. After local dissolution of the membrane in front of each migratory nucleus, it fuses with the stationary nucleus of the same side while the membrane is reconstructed behind it (e, f).

In the free-swimming ciliates no uniform mode of attachment of the conjugants exists. In *Paramecium aurelia* the cells attach after the agglutination reaction described at first in a definite region of their anterior ends, the so-called holdfast region. At this stage of mating they can still be separated with ease. Final fusion follows in a region close to the cytostome where the so-called *paroral cone* is formed during meiosis. The paroral cones of both conjugants lie on top of each other and are the sites at which the pellicles first dissolve and through which the migratory nuclei pass.

Whereas in *Paramecium* a wide cytoplasmic connection is only occasionally formed between the two conjugants this can be the usual manner of fusion in other ciliates. Paroral cones are then not formed. Frequently, the entire region of the peristome partakes in fusion of the partners. In other cases, the peristomes remain free during conjugation. In *Cycloposthium bipalmatum*, a ciliate living in ruminants, the two conjugants fuse only at the rims of their peristomes. The migratory nucleus has a long cytoplasmic appendage and must traverse the closed space between the peristomes in order to reach the other partner. It is similar in behaviour and appearance to the spermatozoon of an animal.

The *metagamic* events are likewise variable. It is by no means the rule that the first division of the synkaryon gives rise to a micronucleus and a macronuclear anlage as in

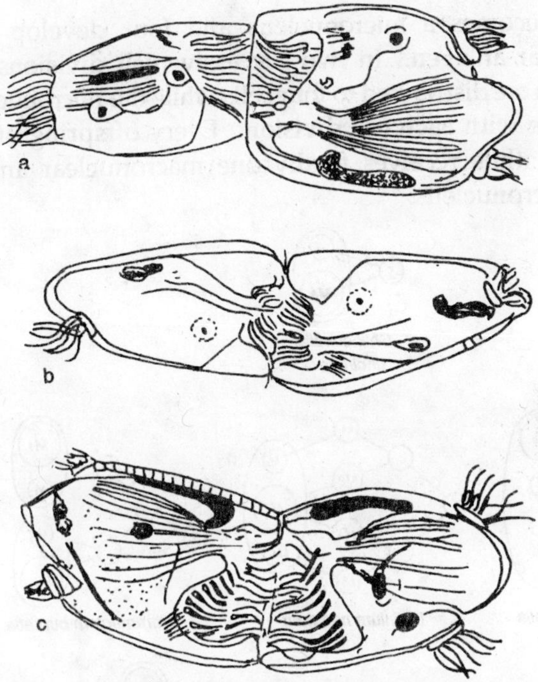

Fig. 1.40. *Cycloposthium bipalmatum.* Stages of conjugation. Both conjugants touch with the rims of their peristomes. a Third micronuclear division. b, c Transfer of the migratory nuclei with their cytoplasmic appendages through both peristomal spaces to the partner.

Chilodonella uncinata. Frequently, this differentiation takes place only after the second metagamic mitosis. If a cell division occurs simultaneously, as in *Paramecium aurelia,* each daughter cell receives a micronucleus and a macronuclear anlage. The micronucleus then divides again. In other cases, two of the four daughter nuclei degenerate *(Euplotes patella)*. It can also happen that two macronuclear anlagen fuse with each other *(Didinium nasutum)*. In *Paramecium bursaria,* three metagamic divisions take place; however, the first one is meaningless since one of the two nuclei degenerates. The three metagamic divisions of *Paramecium caudatum,* on the other hand, lead to the formation of eight daughter nuclei. Three of these die while on becomes a micronucleus and four develop into macronuclear anlagen. In two successive cell divisions the four anlagen are distributed to four cells while the micronucleus divides once with each cell division. Every offspring of the exconjugant thus receives finally one macronuclear anlage and one micronucleus.

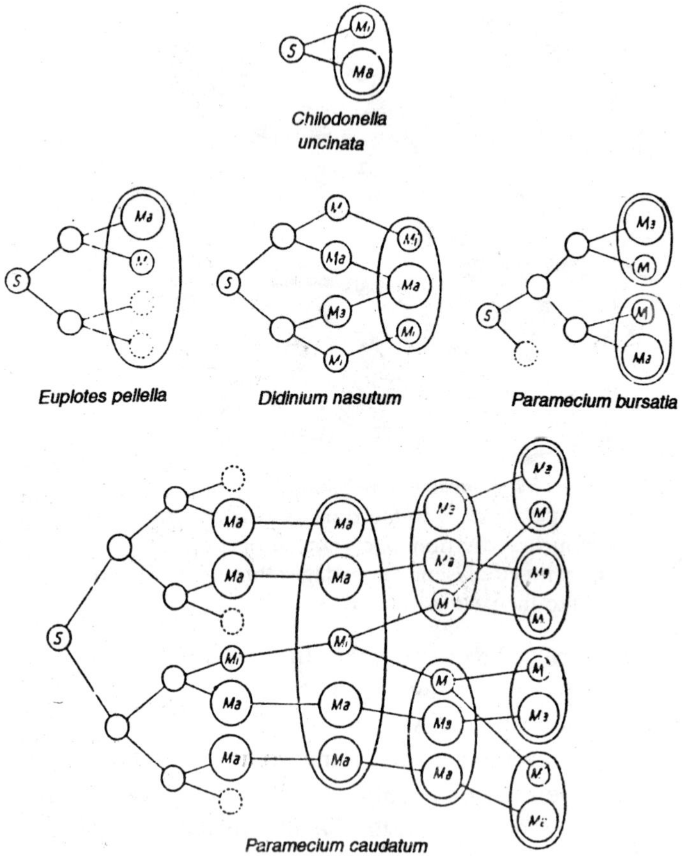

Fig. 1.41. Various possibilities of metagamic nuclear divisions with regard to the differentiation of micronuclei *(Mi)* and macronuclear anlagen *(Ma)*. S Synkaryon.

Even in one and the same species variations from the typical course of conjugation have frequently been observed.

Thus, the third progamic micronuclear division may not be carried out in *Paramecium trichium*, two daughter nuclei of the second division becoming the gamete nuclei. One race of this species is said to have only one progamic division. In *Paramecium caudatum* more than three metagamic divisions of the synkaryon may take place before determination of micronuclei and macronuclear anlagen. The number of the latter may also vary from two to ten. It is also claimed that one of the two daughter nuclei of the synkaryon degenerates while the other one carries out three metagamic divisions.

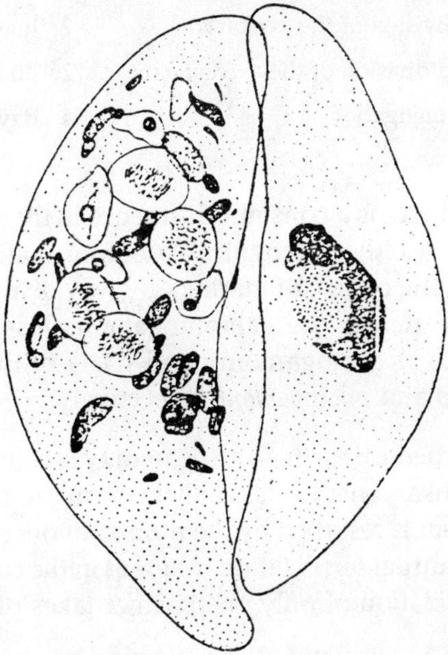

Fig. 1.42. *Paramecium caudatum*. Re-conjugation. Reconjugant (left) with four micronuclei which are in meta-or anaphase and four degenerating macronuclear anlagen. Normal partner (right) in prophase of the first progamic division.

In some ciliates fragments of the old, disintegrated macronucleus are exchanged at conjugation.

While the exconjugants are usually incapable of renewed conjugation, "re-conjugations" are occasionally found between individuals whose nuclear apparatus has not yet reformed and normal individuals.

The duration of conjugation varies considerably among ciliates. In most eases it takes several days. The time needed for the various phases is tabulated below for *Paramecium bursaria*.

1st micronuclear division	14	hours
2nd micronuclear division	1	hour
3rd micronuclear division	3	hours
Exchange of pronuclei and karyogamy	about 20	minutes
1st, 2nd, and 3rd division of the synkaryon	2	hours each
Reorganization and division of the exconjugant	72	hours
Total duration of conjugation	4	days

(b) Anisogamonty

In some ciliates there is a constant difference in the size of the partners which is either recognizable before mating or becomes manifest in the course of mating. By way of analogy with other Protozoa such as the gamontogamous Adeleidae we can designate this as anisogamonty and speak of the two partners as the *macro* and *microgamont* respectively.

In all closely studied cases, this anisogamonty is connected with the fact that a synkaryon is formed only in the macrogamont while the microgamont is resorbed by the macrogamont during mating. Instead of mutual fertilization inherent in the concept of conjugation, fertilization of only one partner takes place.

Anisogamonty is especially pronounced in the *peritrichs* where the microgamonts are the result of a special reproductive act. This can occur in two different ways.

In the *solitary* genera (e.g. *Vorticella, Opercularia, Urceolaria, Opisthonecta, Lagenophrys)* the gamonts are formed by a differential division which is unequal. The larger cell may become the macrogamont while the smaller one differentiates into the microgamont. The latter forms a special circlet of cilia

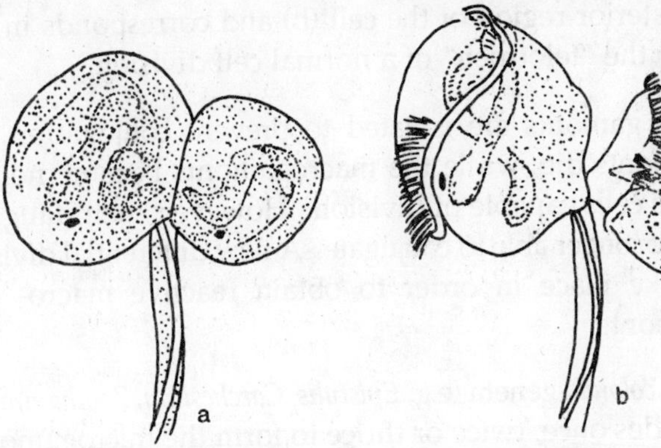

Fig. 1.43. *Vorticella campanula*. Unequal division, leading to the origin of a macroconjugant (left) and a microconjugant (right, with aboral circlet of cilia).

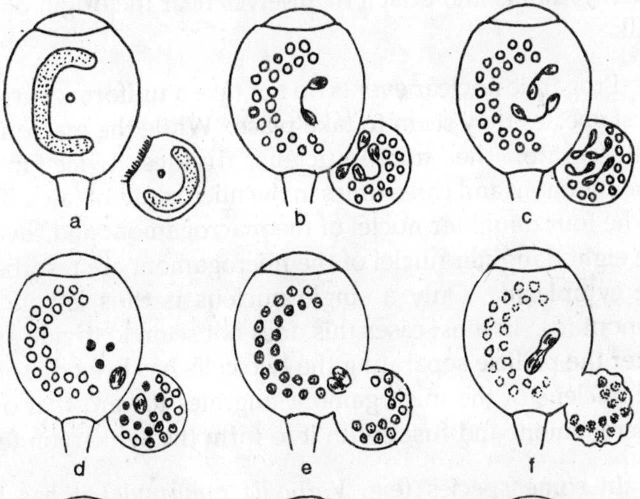

Fig. 1.44. Diagram of sexual processes (anisogamonty) in peritrichs.

in the posterior region of the cell (b) and corresponds in this respect to the "telotroch" of a normal cell division.

If the gamonts are isolated to prevent conjugation, the microgamonts die, while the macrogamonts become normal vegetative cells capable of division. However, their daughter cells are no longer able to conjugate. A new differential division has to take place in order to obtain reactive macro- and microgamonts.

In the *colonial* genera (e.g. *Epistylis, Carchesium, Zoothamnium)* a cell divides once, twice or thrice to form the microgamonts.

While the macrogamonts remain attached to their stalks, the microgamonts are free-swimming. They seek out the macrogamonts and attach themselves near the origin of their stalk.

Progamic nuclear events do not take a uniform course. In most species they seem to take place. While the macronuclei disintegrate, the micro-nucleus divides twice in the macrogamont and three times in the microgamont (a-c). Three of the four daughter nuclei of the macrogamont and seven of the eight daughter nuclei of the microgamont are resorbed in the cytoplasm. Only a single nucleus is thus left in each gamont (d). In most cases this does not seem to divide again. After the pellicle separating the two cells has been dissolved, the nucleus of the microgamont migrates toward that of the macrogamont and fuses with it to form the synkaryon (e).

In some species (e.g. *Vorticella campanula*) it has been observed that a postmeiotic division takes place in each gamont. However, one of the two nuclei becomes pycnotic, leaving a "stationary nucleus" in the macrogamont and a "migratory nucleus" in the microgamont. Of one species *(Vorticella monilata)* it is said that both sister nuclei remain and that a mutual nuclear exchange takes place. A synkaryon is formed in the macrogamont only; the two pronuclei of the microgamont will not fuse and degenerate thereafter.

These modifications of progamic events indicate that unilateral fertilization is derived from mutual fertilization.

As the microgamont is resorbed by the macrogamont, the synkaryon gives rise to eight daughter nuclei in the course of three metagamic divisions. One of these becomes the micronucleus and the remaining seven becomes macronuclear anlagen. These are then — similar to *Paramecium caudatum* — distributed to the descendants of the macrogamont in the course of subsequent fissions, each cell receiving finally one macronuclear anlage.

Although it may seem reasonable to consider the type of fertilization found in peritrichs as an adaptation to the sessile mode of life, it should not be overlooked that anisogamonty also occurs among free swimming ciliates. It has been observed in species of the genera *Metopus* and *Urostyla* that both partners are at first of equal size and indistinguishable by morphological criteria. However, after fusion one of the mating partners is gradually resorbed by the other one. In *Metopus sigmoides* a residue remarks which detaches and dies, but in *Urostyla weissei syn. hologama* resorption of one partner by the other is complete. In both species resorption is followed by encystment.

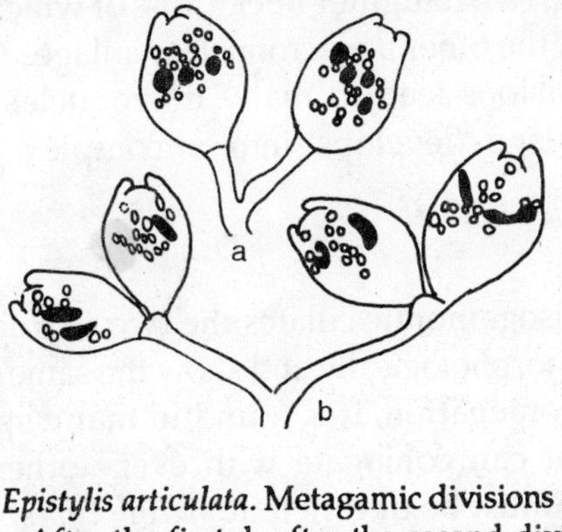

Fig. 1.45. *Epistylis articulata*. Metagamic divisions of the exconjugants. a After the first, b after the second division (macronuclear anlagen black).

In most *Suctoria* conjugation is not essentially different from that of other ciliates. After exchanging migratory nuclei the two partners separate again. Only in *Tokophrya cyclopum in Choanophrya infundibulifera* and in *Ephelota gemmipara* has conjugation been modified to unilateral fertilization, one partner being torn from its stalk and resorbed by the other one.

Both partners of *Ephelota gemmipara* may be of equal or of unequal size. The gamonts form at first lump-like swellings at their adjoining surfaces which carry out rhythmic pumping movements. As in other suctorians, they are recognizable before the partners touch each other. Evidently, their formation must be due to chemical interaction at a distance. Initial contact is established by the food capturing towards and it is by their and that the gamonts approach in a hand-over-hand fashion. Connection is finally made at the "mating lumps" (e, d). After some hours one of the partners, in cases of unequal size probably always the smaller one, detaches from its stalk and is resorbed by the other. In very case a size difference becomes manifest during mating. One partner can thus be designated as the macrogamont, the other as the microgamont. Each cell normally forms only one gamete nucleus. The two gamete nuclei fuse to a synkaryon which divides within the macrogamont into two daughter nuclei, one of which becomes the micronucleus, the other a macronuclear anlage. Continued micronuclear divisions lead to many micronuclei while the macronuclear anlage develops into a complex, branched macronucleus.

(c) Mating Types

Although in isogamontic ciliates the two partners cannot be distinguished morphologically and show the same behaviour in the course of conjugation, it is found in many species that not every gamont can conjugate with every other gamont. The two partners must be of different *mating types*.

The existence of different mating types is proved in the same manner as sexual differentiation in the case of isogamety.

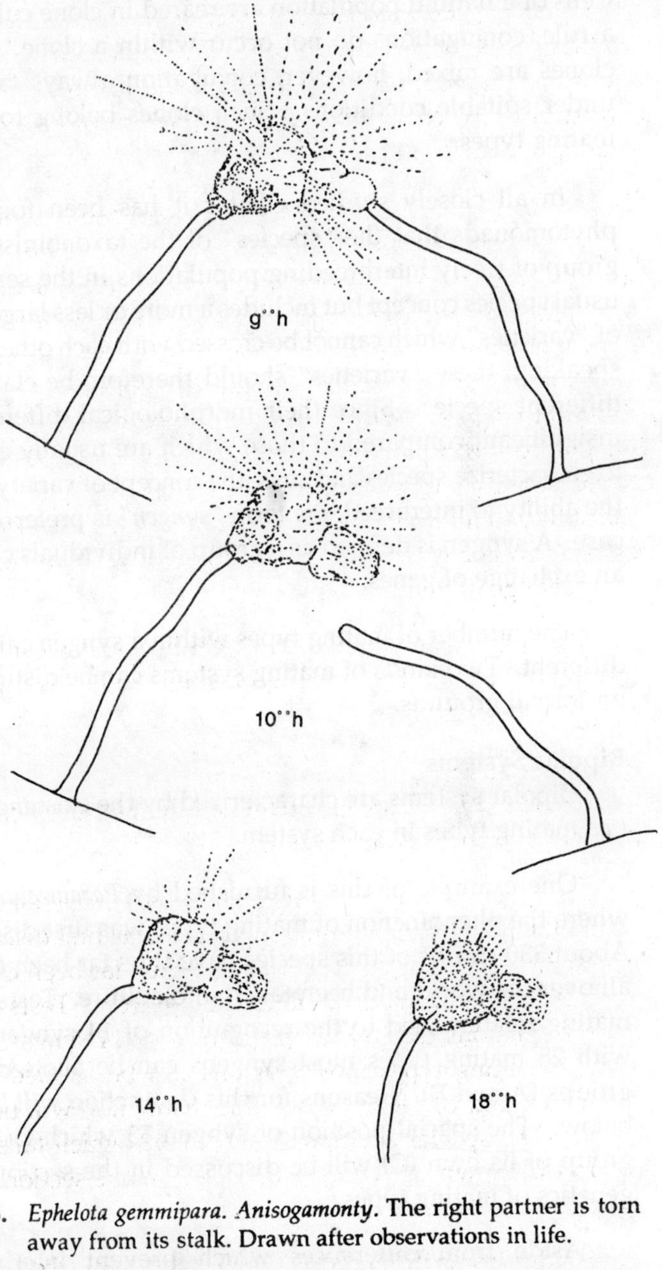

Fig. 1.46. *Ephelota gemmipara. Anisogamonty.* **The right partner is torn away from its stalk. Drawn after observations in life.**

Cells of a natural population are reared in clone cultures. As a rule, conjugations do not occur within a clone. If certain clones are mixed, however, conjugation always takes place under suitable conditions. Such clones belong to different mating types.

In all closely studied ciliates it has been found as in phytomonads that the "species" of the taxonomist is not a group of freely interbreeding populations in the sense of the usual species concept but includes a more or less large number of "varieties" which cannot be crossed with each other. Strictly speaking, these "varieties" should therefore be classified as different species. Since their morphological differences are insignificant compared to those which are usually employed to characterize species and since the concept of variety includes the ability to interbreed, the term "*syngen*" is preferred in this case. A syngen is defined as a group of individuals capable of an exchange of genes.

The number of mating types within a syngen can be very different. Two kinds of mating systems can be distinguished on formal grounds.

Bipolar Systems

Bipolar systems are characterized by the existence of *only two* mating types in each system.

One example of this is furnished by *Paramecium aurelia* where the phenomenon of mating types was first discovered. About 330 strains of this species have thus far been collected all over the world and been taken into culture. Tests of their mating relations led to the recognition of 14 syngens (1-14) with 28 mating types most syngens can be allotted to two groups (A and B). Reasons for this distinction will be given below. The special position of syngen 13 which falls into a group of its own (C) will be discussed in the section on the genetics of mating types.

Aside from differences which prevent interbreeding

(incompatibility), the syngens differ in the form and size of the individuals, the rate of fission, temperature dependence, the spectrum of realizable antigen properties, the behaviour at conjugation and other characteristics. Most differences can only be defined statistically or by means of special tests. Although the geographic distribution is only imperfectly known, it is already clear that certain syngens (e.g. 1 and 2) are cosmopolitan and others are restricted to certain areas.

The syngens are in general strictly isolated from each other. Normal conjugation which involves the major part (about 95%) of the individuals and which leads to viable clones of exconjugants occurs only if mating types of the same syngen are combined. Combinations of mating types from different syngens usually do not lead to any reaction. In some cases mating is incomplete and does not lead to an exchange of migratory nuclei (Inc.) or a certain number of pairs is formed (10% or even 95%). However, exconjugants or the clones derived from them die out sooner or later.

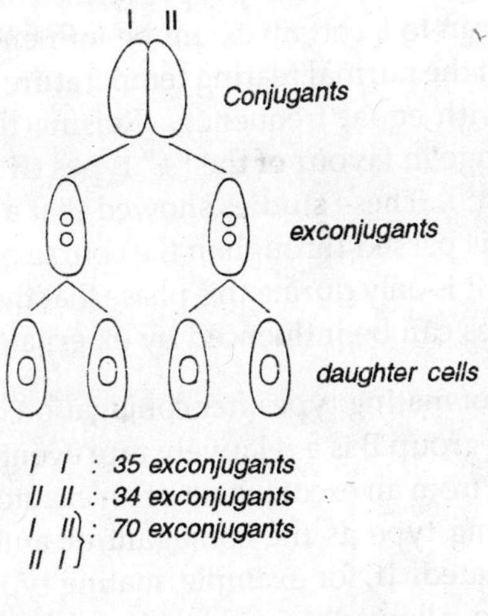

Fig. 1.47. *Paramccium aurelia* (group A, syngen 1). Change of mating type after conjugation. Below the two daughter cells of the left exconjugant the possible distribution of mating types and the frequency of this distribution in 139 examined exconjugants are presented.

As the two macronuclear of *Paramecium aurelia* are distributed to the two daughter cells in the course of the first metagamic division it seems reasonable to assume that *determination of the mating types* takes place during the development of the new macronuclear anlagen. This assumption has been verified in a number of ways.

In one of the strains of syngen 1 more than two macronuclear anlagen are frequently formed in each exconjugant. Final fixation of the mating type is accordingly connected with a later fission when all macronuclear anlagen have been distributed and when they grow into new macronuclei in the descendants. The connection between mating type differentiation and macronucleus is also indicated by the fact that a cell with a macronucleus regenerated from a fragment of the old macronucleus which disintegrated during conjugation or autogamy always has the same mating type as the conjugant or autogamont from which this fragment originated.

The ratio of the two mating types formed after conjugation or autogamy can to a certain extent be influenced by external conditions. At the normal rearing temperature of 19°C the two types occur with equal frequency. Raising the temperature leads to a change in favour of the "+" types (II in syngen, I, VI in syngen 3, etc.). These studies showed that a relatively brief *sensitive phase* is passed through in the course of macronuclear development. It is only during this phase that the determination of mating types can be influenced by external conditions.

A change of mating type after conjugation or autogamy in the syngens of group B is a relatively rare event. As a rule, the clone resulting from an exconjugant or an exautogamont shows the same mating type as the conjugant or autogamont from which it originated. If, for example, mating types VII and VIII (syngen 4) are allowed to conjugate with each other, an exconjugant from the VII conjugant will bring forth a clone belonging to mating type VII, and from the VIII conjugant an exconjugant clone of mating type VIII will again arise.

However, in a relatively small number of cases a sudden change of mating type after conjugation or autogamy can also take place in syngens of group B. Further investigation proved that this change is regularly connected with the development of the new macronuclear anlage and can be influenced by temperature.

The difference between the two groups can therefore not be a basic one. The higher instability of mating type differentiation within group A is evidently due to the fact that cytoplasm plays only a minor role as a determination factor, whereas in group B a definite cytoplasmic condition is evoked by the macronucleus which then, for its part, determines a certain mating type for the developing macronuclear anlage. The fact that such cytoplasmic condition can be maintained over many cell generations but can also on occasion suddenly change into another state will be treated further in the discussion of the so-called antigenic properties. At any rate, determination of mating types in both groups is by way of modification. As the exconjugants and their descendants have the same set of genes determination of the mating type cannot be due to genetic differences.

Four syngens could so far be isolated from *Paramecium multimicronucleatum*, which is closely related to *P. aurelia*. In addition to strains of syngen 2 in which the mating type of the daughter cell clones remains constant (III or IV), so called cycler strains occur whose clones show a *daily periodic change* of the mating type. Some realize mating type III at night and mating type IV during the day or vice versa.

The mating type of the "cyclers" at a given time can be tested with clones whose mating type remains constant ("non-cyclers"). The results of crossbreeding indicate that the difference is due to a single pair of alleles.

Diurnal periodicity is controlled by the alternation of light and darkness (timer) : however, it persists in continuous light or dark culture (endogenous rhythm).

Of *Paramecium caudatum* 16 syngens are known so far, which differ partly in the size of the cell and the nucleus. Autogamy is lacking but can be artifically induced. Instead, selfing (see below) takes place frequently in the clones, which impedes analysis of the mating system and genetic investigations. In some syngens (3, 12), determination of the mating types seems to be due to genetic factors.

Paramecium woodruffi and *P. colldinsi* also have bipolar mating systems. Determination of the mating types is by way of modification.

Multipolar Systems

Most of the ciliates which have been examined thus far show a multipolarity of their mating system : in all syngens more than two mating types occur, every combination leading to confugation.

In *Paramecium bursaria,* where this mode of differentiation was first discovered four mating types have been found in some syngens (1, 3), and eight in others (2, 4, 5, 6).

TABLE 1.3

Paramecium Bursaria. Reaction Scheme of Mating Types in Syngens 1 and 3 (a), Resp. 2, 4, 5 and 6 (b)

	I	II	III	IV
I	—	+	+	+
II		—	+	+
III			—	+
IV				—

a

	I	II	III	IV	V	VI	VII	VIII
I	—	+	+	+	+	+	+	--
II		—	+	+	+	+	+	--

III			—	+	+	+	+	–
IV				—	+	+	+	–
V					—	+	+	–
VI						—	+	–
VII							—	–
VIII								–

b

The mating system of *Tetrahymena pyriformis* has also been very thoroughly investigated. Until 1965, 12 syngens could be established, each comprising a definite number of mating types. Fifty mating types were found altogether. Syngen 1 with 7 mating types is cited as an example.

Multipolar systems are also found in hypotrich ciliates of the genera *Oxytricha, Stylonychia, Euplotes* and *Uronychia*. The largest number of mating types within a syngen, a total of 48, was found in *Stylonychia mytilus*.

Detailed studies on the determination of mating types have only been carried out with some syngens of the species mentioned above. In syngen 1 of *Tetrahymena pyriformis*, it takes place in a manner similar to that of *Paramecium aurelia* (group A and B), i.e., through determination of the developing macronuclear anlage by way of modification. Since autogamy does not take place in *Tetrahymena pyriformis*, a change of mating type can, as a rule, only take place after conjugation. As in the syngens of group A of *Paramecium aurelia*, the daughter cell clones in which different macronuclear anlagen develop can also realize different mating types. If a sufficiently large number of clones of such daughter cells are available, it can be shown that they are distributed among seven mating types. The individual types occur with fairly constant frequency under the same conditions. Thus, determination involves essentially the selection of one mating type from a spectrum of seven possible ones.

It has, however, been proved for certain syngens of *Paramecium bursaria* (1, 2) and *Tetrahymena pyriformis* (2, 8), a species of *Glaucoma*, all species of *Euplotes* which have been investigated so far, and *Uronychia transfuga*, that the mating types are determined directly by genetic factors and not by modification. The way this is achieved is different in the individual species and will be discussed in detail in the chapter on genetics.

Selfing

Although, as a rule, conjugation is only possible when clones belonging to different mating types are mixed together, in certain strains or under certain conditions intraclonal conjugation can also take place. This phenomenon, designated as *selfing*, still has to be examined thoroughly.

It may at first seem that the incompatibility barrier of mating type differentiation is abolished in the event of selfing, so that each cell can conjugate with any other one. However, a more detailed analysis of selfer clones showed that intraclonal conjugation is preceded by differentiation of a complementary mating type and that the partners belong to different mating types. The problem thus narrows down to how a complementary mating type can appear in a daughter cell clone which would normally differentiate uniformly.

In some strains of *Paramecium aurelia* intraclonal conjugations could be shown to be due to the fact that daughter cells of exconjugants or exautogamonts changed their mating type only after a few fissions and not right away. Since their descendants differed in this respect, the same clone contained individuals which belonged to the old mating type and others which had already expressed the new mating type.

However, selfing cannot, as a rule, be explained as a phenomic lag of this sort. Instead, it is due to the fact that a differentiation which had been constant over many cell generations has become unstable. If single cells are isolated from "selfer clones," selfing occurs again in the subclones.

It could be shown that selfer clones of syngen 7 of *Paramecium aurelia,* which formerly were of mating type XIII, suddenly changed to mating type XIV. Since the change never took place in the opposite direction, it is thought that the mating type substance corresponding to mating type XIII represents a precursor of the substance corresponding to mating type XIV.

That irreversibility of this kind is not the general rule is shown by the example of the "cyclers" of *Paramecium multimicronucleatum* whose mating type changes with a diurnal periodicity. Since during the time of change individuals of either mating type occur side by side, selfing is possible.

While with such rapid changes of mating type the assumption that the state of determination of the macronucleus changes wholly can scarcely be avoided, studies of selfer clones of syngen 1 of *Tetrahymena pyriformis* have led to a totally different interpretation. If single cells are isolated from selfer clones and allowed to multiply, selfing takes place again under good nuritional conditions. In starved cells selfing is suppressed and the subclones finally become "pure" with respect to a certain mating type. It was shown that only two mating types are formed in most selfer clones and that certain combinations occur preferentially. Since mating type differentiation is bound to the macronucleus, it seemed likely that the appearance of different mating types in selfer clones is due to *heterogeneity of the macronucleus.* The subunits of the macronucleus might become differentially determined during growth of the macronuclear anlagen. Stabilization of subclones would then be due to progressive *assortment of the subunits* and concomitantly increasing uniformity of the macronuclei. On the basis of this hypothesis and quantitative data obtained in the analysis of selfer clones it has been estimated that a macronucleus of *Tetrahymena pyriformis* should contain about 45 subunits after fission.

Although this hypothesis is attractive to explain the situation in *Tetrahymena pyriformis,* it certainly cannot be used to account for the phenomenon of selfing in other cases. In *Euplotes patella,* cells of certain mating types release specific substances which induce selfing among the cells of other mating types. If clones of different mating types are mixed, cells of different as well as of the same types will conjugate with each other as shown by labeling experiments. The concept of "mating type" thus is different in this case from its original definition. Curiously enough, the closely related *Euplotes eurystomus* shows precisely the opposite behaviour. Intraclonal conjugations are the rule in this species. However, if clones of different "mating types" are-mixed together, selfing between cells of the same type is suppressed and mating occurs predominantly between individuals of different types.

Since it is highly improbable that substances which either induce or suppress selfing play any role at all in the natural biotope of the species, the suspicion arises that we are dealing merely with "laboratory effects". Their analysis is, however, of no less importance in order to understand the physiology of conjugation.

In *Euplotes crassus* selfing can take place if clones heterozygous with respect to the mating type alleles have reached a certain age. This phenomenon will be discussed later.

While it is common to all cases of selfing discussed so far that differentiation as such remains constant despite on intraclonal change of mating type, the so-called *chemically induced conjugation* seems to be due to total abolishment of the incompatibility barrier. If paramecia are treated with certain substances such as K. Mg, heparin, EDTA cells of the same type, even those of different syngens or species of the aurelia group, can be brought to conjugation. The agglutination reaction does not take place in this case. Aside from the fact that all possible "irregularities" occur with respect to the relative positions of the conjugants, conjugation as such takes a perfectly normal course.

However, the exconjugants or the cells derived from the are not capable of living.

Abolishment of the incompatibility barrier is probably due to a change of the ciliary envelope, EDTA (ethylene diamine tetracetic acid) might, for example, bind calcium incorporated in the ciliary envelope.

Even though there is no doubt that mating type differentiation is of widespread occurrence among ciliates, the possibility cannot be excluded that species without this differentiation exist. In such species each cell should be able to mate with any other one provided that all external and internal conditions otherwise required for conjugation are realized.

(d) Physiology of Conjugation

Even if two clones which are combined belong to appropriate mating types, conjugation does not always follow. Besides "complementarity" as expressed in mating type differentiation, certain conditions must be fulfilled before mating is possible. Light and temperature are among the *external factors* which influence the readiness to conjugate. Even the time of day may play a role. A definite state of nutrition is indispensable since starved or overfed individuals generally will not conjugate.

In addition to these conditions whose importance may vary in the different syngens or strains, the readiness for conjugation depends on a more or less prolonged *immaturity period* in most ciliates. This means that the descendants of exconjugants are not immediately ready for another conjugation but must pass through a certain number of cell divisions before they enter the *period of maturity*.

The immaturity period of *Paramecium aurelia* is relatively short. There are strains whose clones are ready for conjugation again as soon as reorganization of the nuclear apparatus is completed. Even in the other strains the immaturity period lasts maximally 9 days (= 35 cell divisions). Once the state of maturity is reached, it may be interrupted briefly by autogamy but cannot be abolished by it.

2

Reproduction in Hydra and Other Coelenterates

The hydras are represented by several species. The common species are the green hydra, *Chlorohydra viridissima,* formerly called *Hydra vividis;* the brown hydra, *Hydra obgactis,* the white a Pinksih hydra, *Hydra gangetice;* and the commonly occuring Indian speices, *Hydra valgous* — the colourless species. Since the times of A. Trembley (first half of the 18th Century) freshwater hydra have been used in laboratories as a suitable species for studying the mechanisms of regeneration and asexual reproduction (for a review of earlier papers. Since the middle of the 20th Century there have been many detailed investigations on the mechanisms of cell differentiation and the control of morphogenesis. It is impossible to review so much work in his short chapter. Investigations make use of the adult, intact, regenerating, and budding hydra as well as mutant strains, their chimaeras, and cell cultures.

This contribution does not pretend to provide a comprehensive description of *Hydra* as an experimental animal used in developmental biology. Specifically, we do not present any information regarding the mutant strains of hydra. The following information should be regarded as introductory necessary for those investigators using *Hydra* for the first time. More detailed information can be found in the references as the end of the chapter or obtained from the scientific centers mentioned above.

Sexual reproduction and, in particular, embryogenesis of the hydra are less accessible to research. Nevertheless, *Hydra* is a suitable species for studying the early stages of gametogenesis.

Along with freshwater hydra, various marine coelenterates, particularly hydroid polyps, are being increasingly used in research. They are used in studies dealing with the mechanisms of cell differentiation, regeneration, and morphogenesis. Furthermore, they may be used as test systems for screening various physiologically active substances or as indicator organisms of sea water pollution.

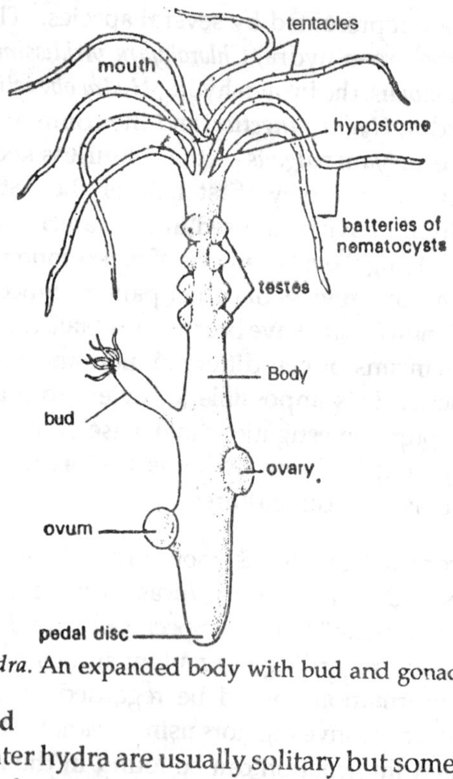

Fig. 2.1. *Hydra*. An expanded body with bud and gonads.

Species Used

Freshwater hydra are usually solitary but sometimes form temporary colonies. They belong to the genus *Hydra*. The following species of the genus *Hydra* have been recognized. *Chlorohydra viridisima, H. gangetica, H. oligactis, H. brauezi, H.*

baicalensis, H. vulgaris, H. attenuata, H. circumcincta, H. oxycnida, H. oxycnidoides and H. littoralis.

The most characteristic external feature of *H. oligactis* is a clear division of the body into trunk and stem parts, while in *H. attenuata* and *H. littoralis* there is no distinct separation of the body on the two regions.

Another feature useful for species diagnosis is the order of tentacle origin in the growing buds. In *H. hymanae* buds all the five tentacles arise quite simultaneously; in *H. oligactis* two lateral tentacles arise early and assume a slender shape before the other tentacles appear; in *H. attenuata* the tentacles arise in a slightly staggered sequence.

Of marine hydroid polyps the various *Obelia* species (= Gonothyrea, Laomedea) should be mentioned : *O. loveni, L. flexuosa, O. longissima, O. geniculata* (family Campanulariidae, order Leptolida, class Hydrozoa), and *Dynamena* (= Sertularia) *pumila* (family Sertulariidae, same order and class). All these species are colonial.

Distribution Collection and Preservation of Hydra

The hydras are solitary freshwater animals and are cosmopolitan in their distribution. They occur in lakes, ponds, streams and seasonal ditches, where weeds and other vegetation are commonly present. Though absent in water which is foul and too warm, hydras flourish well in cool, clear and relatively permanent and stagnant waters. They may be found attached to and hanging downwards from the underside of solid objects in water such as leaves, sticks, stones, weeds, etc. Their bodies extend at right angles to the substratum. When hungry, their body and tentacles are stretched to the maximum limit and swayed in water in order to capture any prey that comes in contact with them. When disturbed, the body at once contracts into a minute circular and jelly-like knob which escapes detection by inexperienced observers. Removed from water it collapses into a soft and shapeless mass.

Hydra may be collected easily during early winter from shallow water of ponds, lakes and streams, about 30 cm. in depth. As these animals live habitually clinging to aquatic vegetation, the quantities of vegetation may be picked and put in jar filled with pond water.

After a while the hydra move to the illuminated side of the vessel or attach themselves to various objects placed into the jar (plants, branches, etc.) and spread out, thus becoming clearly visible. Alternatively the hydra can be taken directly from the water. Individual plants may be picket out of the water from the shore, stones, bridge, raft, or boat, or examined from close range in the upper water layer. Parts of plants covered with hydra are taken and placed into small jars of water and transported to the laboratory. Minimal experience in recognizing hydra in their natural environment is usually sufficient to make locating them a simple matter.

It should be remembered that hydra do not, by any means, live in every body of water. Generally it is recommended that cultures grown in the laboratory for many years be used, especially as they are better adapted to artificial conditions.

Asexual budding of hydra in nature proceeds throughout the summer. Sexual maturation and the formation of gonads in nature takes place predominantly in autumn, although in *H. vulgaris* gonads may appear during the summer as well.

The marine hydroid polyps belonging to the species mentioned above occur widely in the littoral areas of most seas in cold and temperate climatic zones. In the Soviet Union the best area to collect them is in the littoral zone of the White Sea, where they live in brown algae, stones, and on piles *(O. longissima)*. Vegetative reproduction (budding) under natural conditions is at its most intense in June or early July. In late July the gametes mature, while vegetative reproduction is partially or completely arrested. It is returned in the autumn; under artificial conditions budding may occur in winter too.

We can preserve the specimens after fixing them is Bouins fluid for atleast 30 minutes. Also gradual washing in 30%, 50% alcohol we can preserve them in 70% alcohol for a long.

Culture of Hydra

If one does not require a *Hydra* culture with a controlled rate of increase, one may follow the recommendations described below.

The animals are kept in glass jars varying in size from half a liter to large aquariums, with moderate amounts of *Elodea* which *Hydra* prefers for attachment. Sand on the bottom is not necessary. Tap water stored in the laboratory for at least 24 h or even better, pond water is used. The water should be clean and fresh; if it does not become foul it can be kept for weeks without change. Using a pipette, the jars should be regularly cleansed of various dead residues. The animals should be kept away from direct sunlight and at room temperature. They should be fed on small crustaceans (1-2 Daphnia per Hydra per day) or mosquito larvae.

Under these conditions the animals usually appear healthy and budding occurs, although sometimes, for no apparent reason, they go into decline (they stop feeding and do not respond to touch; later their tentacles undergo resorption and the animals die). If this happens the water should be changed and the animals kept at a lower room temperature. They should be given food after they have resumed the capacity to feed.

Decline may be the result of infestation by parasites such as infusorians (e.g., *Tricodina, Cerona).* If this is the case, the culture should be cleansed of these parasites. The infusorians may be removed from the body of *Hydra* with a brush; alternatively this can be done by "rinsing" the animal in water. The parasites can also be removed using a pipette; alternatively the animals can be placed into a medium lethal to the parasites but which only temporarily and reversibly affect hydra, for example lukewarm water or a weak solution of Nile Blue sulfate.

After such manipulations the *Hydra* should be transferred to clean water and checked to see if the parasites have gone. The other enemies of *Hydra,* such as Turbeliaria, *Microstomum* or molluscs *(Helix* and *others),* as well as the larvae of Chaoborinae should not be present in aquariums containing *Hydra* cultures.

Conditions For Obtaining Cultures with a Controlled Reproduction Rate

Loomis and Lenhoff recommended the following conditions for growing *Hydra.* The culture should be set up on flat 15 mm culture dishes containing one the following culture media: CN, KNC, or M. CN is similar to that previously described as L-solution; it is useful in experiments such as those involving glutathione, in which K^+ affects the animals negatively. Solution KNC is suitable for all brown *Hydra,* whereas M-solution prepared with either NaCl or $NaHCO_3$ works for both the green and the brown *Hydra.*

Special precautions should be taken against heavy metals in the water for preparing the above solutions. It is recommended to pass the distilled (deionized) water produced by reverse osmosis through a tank of mixed bed resin. If one plans to use *Hydra* only occasionally and does not need a constant culture, one may detoxify the heavy metals present in tap water with Na_2 EDTA (final concentration about 20 mg per liter of culture medium).

TABLE 2.1

Composition of Stock and Regular Strength Culture Solutions

Culture solution	*pH*	*Components of stock solution*	*Dilution (to prepare culture solutions)*	*Final concentration*
CN	7.6	(a) 0.1 M $NaHCO_3$	1 in 1000	10^{-4} M
		(b) 1.0 M $CaCl_2$	1 in 1000	10^{-3} M

KNC	7.6	(a)	0.1 M $NaHCO_3$	1 in 1000	10^{-3} M
			0.1 M KCI		
		(b)	1.0 M $CaCl_2$	1 in 1000	10^{-3} M
M	7.8	(a)	1.166 M Tris-HC1 buffer	6 in 1000	10^{-3} M
			0.166 M $NaHCO_3$		10^{-3} M
			0.016 M KCI		10^{-4} M
			0.016 M $MgCl_2$		10^{-4} M
		(b)	1.0 M $CaCl_2$	1 in 1000	10^{-3} M

a) Three liters of stock M solution (a) may be made by mixing 1000 ml 0.5 M Tris-HCl buffer pH 7.6-7.8, 500 ml 1.0 M $NaHCO_3$, 50 ml 1.0 M KCI, 50 ml 1.0 M $MgCl_2$, and 1400 ml distilled water.

b) Prepare Tris-HCl buffer from a 1 M solution of Tris adjusted to pH 7.6-7.8 with a mixture of equal volumes of water and concentrated HCl. Dilute the solution to almost 2 liters, check the pH, and readjust it as required ; then bring it to a final volume of 2 liters. Refrigerate this stock buffer when it is not being used.

In and some other modifications of culture media are recommended (in no case above 10 mOsm).

The *Hydra* are fed using the larvae of the brine shrimp *Artemia*. To obtain a permanent culture of larvae, 1/4 teaspoon of dry *Artemia* eggs (about 0.7 g) should be placed onto the surface of 500 ml of Nacl solution (3.5 g/liter) in a flat dish. The incubation should be at room temperature. After 48 h the hatched larvae should be collected using a net with fine cells washed in the L-medium and these larvae added to the *Hydra* cultures. Onto and Campbell recommended that between 12 and 15 larvae be added daily to 60 ml plastic Petri dishes containing *Hydra cultures*.

Twenty four hours after feeding , the L-medium should be decanted (while the *Hydra* are still attached to the walls) and changed with a fresh portion.

Under these conditions the exponential increase in the number of *H. littoralis* and *H. vulgaris* has been observed over a number of days. If rapid growth is not necessary the *Hydra* may be left in a medium without feeding for several weeks at

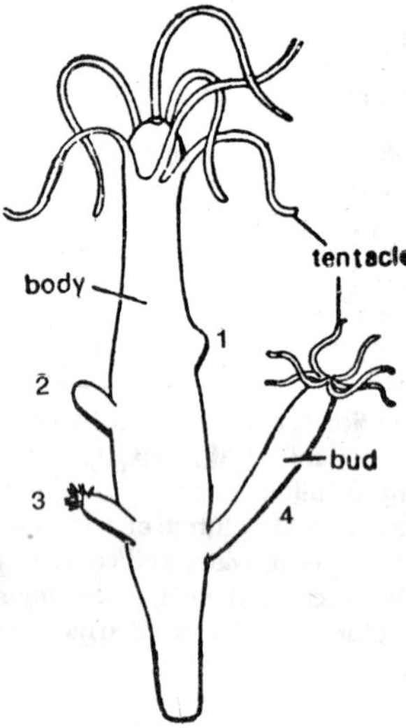

Fig. 2.2. *Hydra* showing stages of budding.

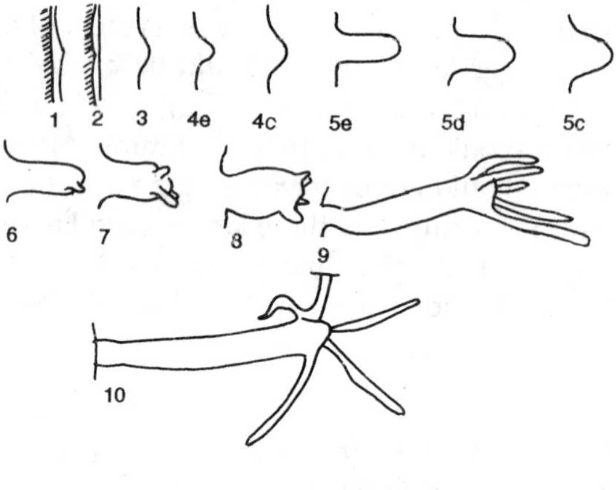

Fig. 2.3. Stages of bud development in *Hydra attenuata*. Buds are shown in profile with stage numbers below the sketches. Stages 1 and 2 were not followed. In some stages, buds are in various states of contraction: c) contracted; e) elongated; d) distended with fluid. Crosshatching in stages 1 and 2 represents endoderm.

room temperature or for several months in a refrigerator. Within 48 h of returning to room temperature and beginning to feed, the *Hydra* start budding again. Sexual differentiation under these conditions is described below.

For culturing large numbers of *Hydra*, Lenhoff recommends two methods: a simple tray method requiring no special equipment, and a special vertical-plates method.

Development Of Buds In Freshwater Hydra

The fed *Hydra* undergo intensive budding. Buds are laid down as tubercules in the lower part of the trunk. In *H. attenuata* the first buds are formed opposite each other, while subsequent buds are, as a rule, located spirally; the angle between the consecutive buds is about 120°.

The consecutive stages of bud development according to Otto and Campbell are described below.

Stage 1 : Condensation. The ectoderm thickens in the region of the future bud; this can be seen best when the animal contracts.

Stage 2 : Peak. A small mass of endoderm penetrates the thickened ectoderm.

Stage 3 : Mound. Both the ectoderm and endoderm of the bud protrude from the wall of the parent in a simple low mound. The endoderm intrusion has become blunter.

Stage 4 : Shield. The profile of the contracted bud, is shaped like a shield. A peak occurs at the distal end of the bud. The bud has elongated slightly since Stage 3. The length of the bud from top to base is less than the diameter of the parent. The extended bud does not have a shield shape.

Stage 5 : Bulb. The bud has a bulbous shape when it is distended with fluid, with the proximal part being narrower than the distal part, where the tissue appears to be stretched. When the bud contracts, it has a similar shape to that of Stage

4. However, it is longer than in Stage 4; its length is at least equal to the parent's diameter.

Stage 6 : Rudiment. Tentacle rudiments which are shorter in length than their breadth mark this stage. In *H. attenuata* there are generally two initial rudiments opposite one another on the lower surface of the bud.

At this and the preceding developmental stage a mouth appears in the bud for the first time and a feeding response to glutathione can be noted.

Stage 7 : Tentacle. The initial tentacle rudiments are longer than their breadth. Three additional tentacle rudiments are usually present, two on the upper side of the bud and one on the lower side between the two initial rudiments. The base of the bud is slightly constricted when the bud is extended.

Stage 8 : Constriction. The proximal region of the bud immediately adjacent to the parent has constricted and its diameter is less than two-thirds that of the bud gastric region.

Stage 9 : Basal disc. The ectodermal cells in the basal region of the bud are translucent as a result of basal disc differentiation. This basal region is greatly constricted, but the tissues are still connected to those of the parent.

Stage 10 : Independent. The bud remains on the parent, but no tissue connection between the bud and its parent can be seen. Basal differentiation appears complete.

The same authors give the following data on the timing of *H. attenuata* developmental stages at 20-23°C:

Stages	*Time, h*
1-2	Cannot be accurately determined
2-3	3
3-4	6
4-5	6

5-6	10
6-7	8
7-8	9
8-9	20
9-10	8
10 (prior to the separation of the bud from its mother)	Varying duration upto several days

Sexual Reproduction

No two authors express the same opinion regarding the conditions leading to sexual reproduction. Tardent, on the basis of a 2-year observation of *H. attenuata* cultures, came to the conclusion that there is a spontaneous cycling of appearance and disappearance of sexuality, which is unaffected by temperature, feeding conditions, or the presence of phosphates. The individual polyp does not change sex for a long period of time land as a rule its sexual progeny is of the same sex. Spontaneous and generally irreversible sex changes are possible.

Loomis and Lenhoff observed the formation of gonads in all dense cultures of *H. littoralis* when they were fed with *Artemia* larvae and cleaned daily. If feeding is carried out every 30 min at a temperature of 21°C, gonads appear 10 days after the beginning of culture. To stop gonads from appearing, the water must be changed frequently and the culture should be kept in sparse aeration, which also inhibits the transition to sexual reproduction.

Aisenstadt observed long-term sexual reproduction in *H. oligactis* proceeding for about 6 months, beginning from late September after the animals were changed over from their usual feed of *Daphnia* and *Cyclops* to pieces of sludge-worm (*Tubifex*). Pieces of the sludgeworm were given with forceps to each *Hydra* 2-3 times a week. The animals were kept at room temperature.

Davidson stimulated transition to sexual reproduction in

H. hymanae by transferring the animals from 24°C into a medium with a temperature of 15°C or 4°C for 5-25 days. The latter case brings particularly good results: 50% of the embryos are hatched by the 45th day and there is a high survival rate.

Hydra reproduces sexually by the fusion of gametes, generally in the autumn. The gonads are developed temporarily from the interstitial cells of the epidermis which accumulate to form protruding bulges in the body wall. In some species the male and female gonads occur in the same individual, which is then known as *hermaphrodite* or *monoecius.* Usually the testes develop towards the distal end and the ovaries towards the basal end of the body. Most species *hydra oligactis),* however, are unisexual or *dioecious,* i.e., the individuals are either male or female and can be distinguished. The male is smaller in size and bears one to eight testes, while the larger female has only one or two ovaries. Even in the hermaphrodite species, self-fertilization is avoided, because the spermatozoa and ova mature at different times. As a rule, the testes mature earlier than the ovaries.

Gametogenesis

The process forming gametes is called gametogenesis. The gonads appear only during breeding season. The interstitial cells are transformed into germ cells to undergo gametogenesis which is of two types.

Spermatogenesis

The testes initially appear at the distal end of the body and later reach the budding zone. They are conical protuberances on the body surface and vary in number from a few up to 60. In cases of hermaphroditism, which can even be observed in the *Hydra* that were initially dioecious, the testes are located distally to the oocytes, which develop in the budding zone.

Each testis consists of several cysts with an extended shape, formed by myoepithelial cells. During sperm maturation within the cyst in cell clusters, which represent the progeny of

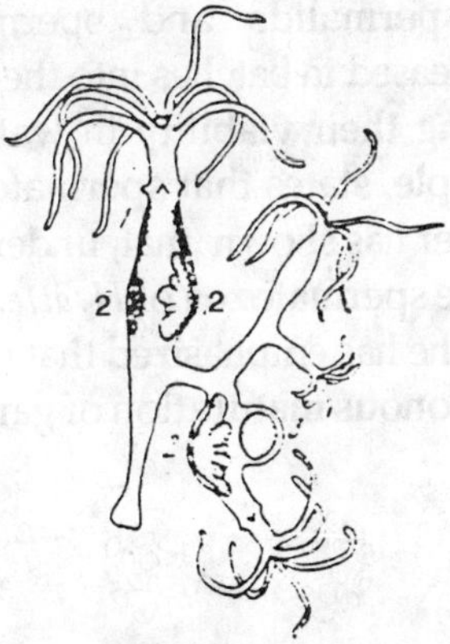

Fig. 2.4. Gametic development in the budding *Hydra*. 1) Testes; 2) oocytes at different stages of development; 3) mature egg.

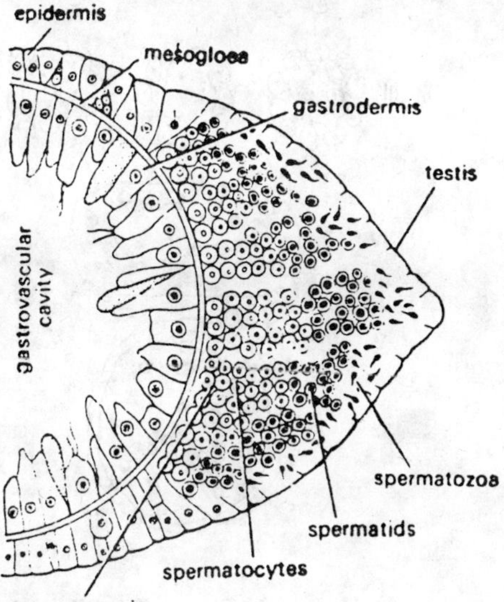

Fig. 2.5. *Hydra*. A testis in T.S. showing stages of spermatogenesis.

one or several interstitial cells, the spermatogonia occupying the basal cyst region can be distinguished; furthermore, primary and secondary spermatocytes can be seen as well as the more distally located spermatids and spermatozoa. Mature spermatozoa are released in batches into the surrounding water. The data concerning their viability in water is controversial; Kanayev, for example, states that spermatozoa stay viable for 1-3 days, while Zihler has shown that, under optimal conditions (9-12°C, pH 7.3), the spermatozoa of *H. attenuata* remain viable for only about 4.5h; he has emphasized that normal insemination requires the synchronous maturation of gametes in both males and females.

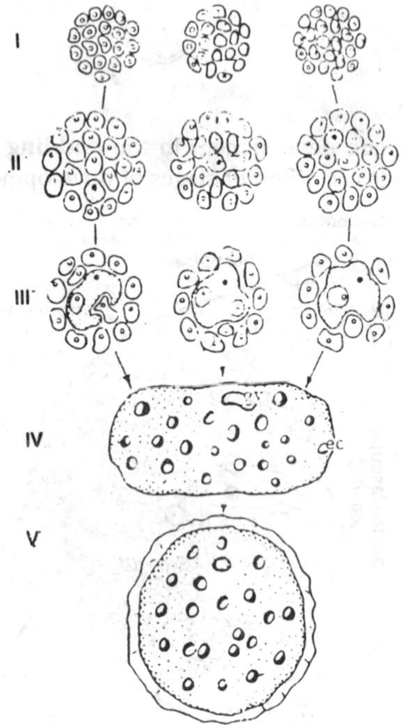

Fig. 2.6. Oogenesis in *Hydra*. 1) Stage of i-cell proliferation in the germinal zone; II) stage of organelle appearance in the cytoplasm of growing oogonia; III) phagocytosis of oogonia II by the grown oocyte; IV) fusion of oocytes; V) egg. gv, germinal vesicle; ec, endocytes.

Oogenesis

The *Hydra* oocytes, which are frequently called "ovaries,"

are gigantic cells which show phagocytic activity. In females or hermaphrodites which have started sexual reproduction, a whitish spot appears in the budding zone; this spot corresponds to a thickening of the epidermal layer formed by the proliferating interstitial cells. The cells within the thickening rapidly increase in size. In histological sections it can be seen that they form grape-like accumulations of syncytially connected oogonia. As in the testis, these accumulations are separated by strands of myoepithelial cells. Each grape-like cluster is formed by the descendants of one or several interstitial cells. One cell within the cluster grows faster than the others and absorbs the surrounding germ cells by phagocytosis. According to a number of authors actively phagocytosing oocytes may fuse with each other.

According to the most recent interpretation only the fast-growing cell of a cluster is the oocyte. All other cells begins to synthesize yolk precursors intensively before they are phagocytized. Hence, they are similar to the ovarian nurse cells of the other species.

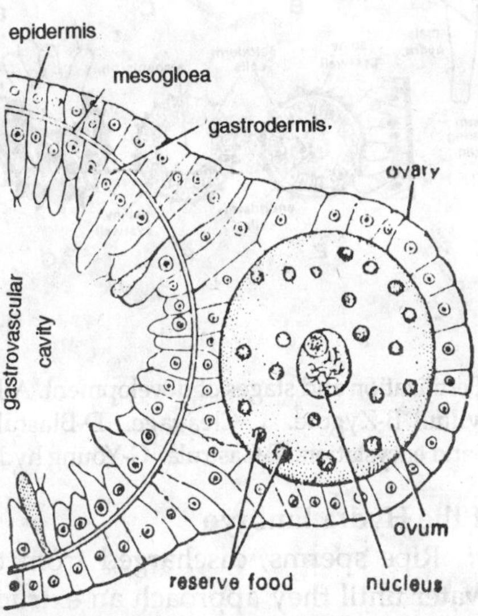

Fig. 2.7. *Hydra.* An ovary in T.S. showing mature ovum.

The oocytes that have stopped growing acquire a spherical shape and become connected with the body surface through a pedicle consisting of epithelial muscle cells surrounding the whole egg and forming a characteristic follicle. In certain species this structure has a diameter of 1.5 mm. The number of eggs in one animal rarely reaches 10; they develop sequentially.

Before maturing the germinal vesicle (nucleus of the oocyte) is displaced to the distal cell wall. Meiotic divisions are followed by "ovulation" involving a rupture in the layer of epidermal cells surrounding the egg which, however, stays connected to its mother through a pedicle until the formation of the embryotheca.

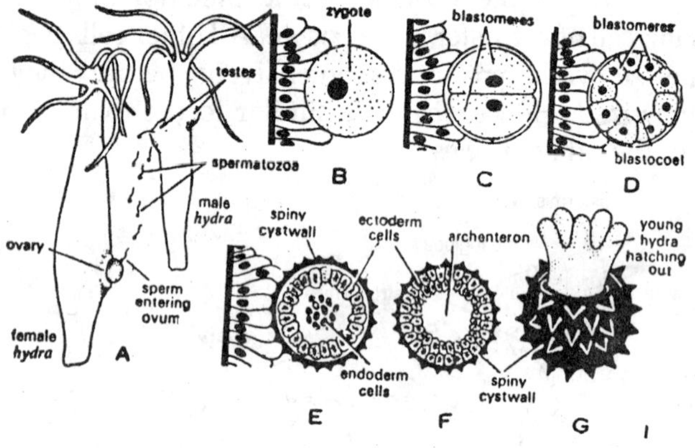

Fig. 2.8. *Hydra*. Fertilization and stages of development. A-Fertilization of an ovum. B-Zygote. C-Cleavage. D-Blastula. E-Early gastrula and encystation. F-Gastrula. G-Young hydra hatching.

Development of the Hydra Embryo

Fertilization : Ripe sperms, discharged from the testes, swim about in water until they approach an extruded ovum and surround it. Several sperms may penetrate the gelatinous covering, but only one enters the ovum and fuses with it completely. This process is called *fertilization* and the fertilized ovum is called the *zygote.* The male and female nuclei unite to form the *fusion* or *zygote nucleus.* The ovum will die and degenerate, unless fertilization occurs within a short time after it is first exposed.

Development : The development begins soon after fertilization while the zygote is still attached to the parent. The *cleavage* is total and equal *(holoblastic)* and results in the formation of a hollow spherical ball, called *blastula,* with a central narrow cavity or *blastocoel.* Its wall is made of a single layer of more or less equal-sized epithelial cells, called *blastomeres.* Some of these cells migrate inward, while-others divide into outer and inner cells by tangential division. By repetition of this process, called *delamination,* the blastocoel is completely filled by cells, so that the hollow blastula is converted into a *solid gastrula.* The inner mass of cells, derived from the blastomeres, forms the *endoderm,* while a single layer of outer cells forms the *ectoderm.* Soon a new cavity, called *gastrocoel or enteron,* appears in the central solid mass of endodermal cells. Meanwhile, the ectodermal cells have secreted a two-layered protective *shell* or *cyst* around the gastrula. The outer layer of cyst wall is thick, horny or chitinoid and spiny, while the inner layer is a thin gelatinous membrane. Different species of *Hydra* can be identified by the specific pattern of their cysts. At this stage, the gastrula usually drops off the parent and either sinks in the mud at the bottom of the pond or adheres by means of its spikes to any solid object it contacts. In *P. oligactis,* the gastrula is shed from the ovary before the formation of the shell, and fastened to aquatic plants by its sticky gelatinous covering.

Hatching : The encysted embryo remains dormant and unchanged for several weeks, until the next spring. As it can withstand drying and freezing it carries the race through droughts and winters. It is also probable that this resting stage serves for dispersal, for it can be carried by currents of wind, or in mud on the feet of animals to other ponds in which water is present.

With the advent of favourable conditions of water and temperature development is resumed. *Interstitial cells* arise in the ectoderm and *mesoderm* is secreted between the two cellular layers. The embryo elongates, and a circlet of *tentacle buds* develops at one end with a *mouth* appearing in their midst. As the embryo increases in size, the outer cyst wall softens to break open and the embryo hatches into a partially differentiated polyp. There is no free larval stage in the development of *Hydra*. This life-history of hydra, thus, has the scheme: polyp —>egg —>polyp.

Regeneration

Regeneration, as already noted, is the ability of an organism to replace its lost or damaged body parts. It usually occurs naturally after an accident, but can be induced artificially by mutilation. The power to regenerate was first reported in 1744 in *Hydra*, by its discoverer *Abraham Trembley*. Regenerative powers also occur in many other coelenterates and other invertebrates, but *Hydra* alone has been a favourite material for experimentations.

A. Trembley (1744) found that if a living animal is cut across into two, three or more pieces, each grows the missing parts and becomes a complete through smaller hydra, provided the pieces are not too small. A fragment of hydra, measuring 1/6 mm, or more in diameter, is capable of regenerating an entire individual. Although regenerating results in an increase in number, it is not reproduction because it is not a normal method of multiplication. Regeneration of hydras is made possible by the amazing generative powers of the totipotent interstitial cells.

One characteristic feature of regenerating piece in hydra is that it retains *polarity*. The end nearer the mouth develops mouth and tentacles, while the end nearer the base forms a new base.

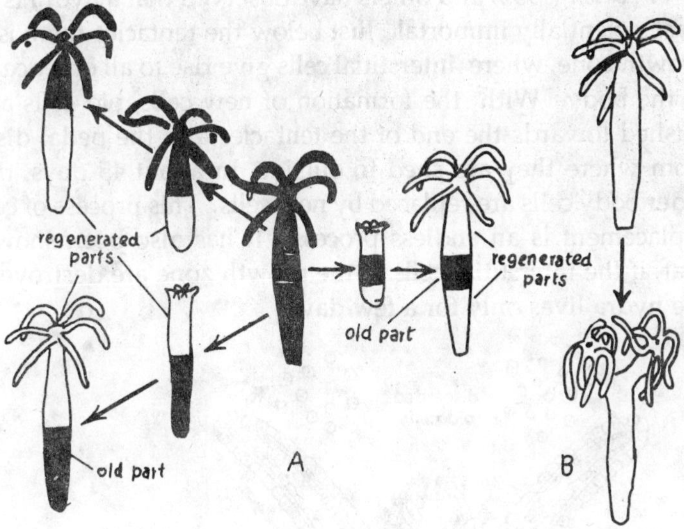

Fig. 2.9. *Hydra*. Regeneration. A—Regeneration of three individuals from one hydra. B—Formation of two-headed hydra by regeneration.

Parts of one hydra may be easily grafted upon another provided they are of the same species. Grafting can be done in various arrangements producing many bizarre effects.

Trembley observed that if the head end of a hydra is split in two and the parts are separated slightly, it results into a Y-shaped specimen, or "two-headed" individual, having two months and two sets of tentacles. Each head may be again split in a similar manner. Thus, *Trembley* succeeded in producing a "seven-headed" hydra. It was the great regenerative power of these animals which won for them the name "Hydra", after a Greek mythical monster which was finally destroyed by Hercules. According to the legend, 'Hydra' had nine heads and no sooner did Hercules cut one off, than two grew in its place.

Immortality in Hydra

P. Brien (1955) and others have observed that a hydra is at least potentially immortal. Just below the tentacles there is a growth zone, where interstitial cells give rise to all other cells of the body. With the formation of new cells, old cells are pushed towards the end of the tentacles and the pedal disc, from where they are shed to outside. In about 45 days, the older body cells are replaced by new cells. This process of cell replacement is an endless process. It has also been shown that, if the interstitial cells of the growth zone are destroyed, the hydra lives only for a few days.

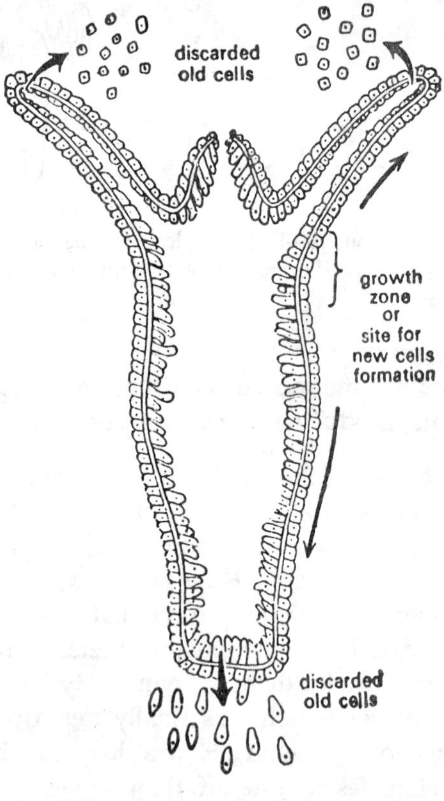

Fig. 2.10. Cell replacement in *Hydra*.

Questions

1. Give an account of histological structure of the body wall of *Hydra*.
2. Describe the modes of locomotion, nutrition and reproduction in *Hydra*.
3. Give the structure of a nematoblast cell of *Hydra*, as revealed by the electron microscope. Add a note on the various types of the nematocysts and their functions.
4. Write short notes on— (i) Locomotion in *Hydra*, (ii) Regeneration in *Hydra*, (iii) Immortality in *Hydra*, (iv) Behaviour of *Hydra*.

OBELIA : A SEA-FUR

Unlike *Hydra*, which is a solitary polyp, most hydrozoans are colonial and many have a complex life-cycle. *Obelia*, commonly called sea-fur, represents both, as it is a colonial hydroid and its life-cycle presents a free-swimming medusa stage. It is, therefore, studied as a typical representative of colonial Hydrozoa. The most commonly occurring species of obelia is O. geniculata.

Habits and Habitat

Obelia is a sedentary, marine and colonial hydroid having cosmopolitan distribution. It is abundant in both the Atlantic and Pacific coastal waters and is usually found up to the depth of about 80 meters. It occurs as asexual and sexual forms. The asexual form occurs as a branched *hydroid colony*, attached to rocks, stones, shells of animals, wooden piling and wharves, fronds of large seaweeds like *Laminaria*, shells of lobsters, etc., as a delicate, whitish or light brown, almost fur-like growth. The sexual form is a bell or umbrella-like free-swimming stage, called *medusa*.

A. Hydroid Colony

Morphology

The colony of *Obelia* is delicate and semi-transparent, having a whitish or light brown tinge. It consists of vertical branching stems, called *hydrocauli* (singular, hydrocaulus), rising 2 to 3 cm. from a root-like *stolon* of the *hydrorhiza*. Both of these are of the thickness of an ordinary sewing thread.

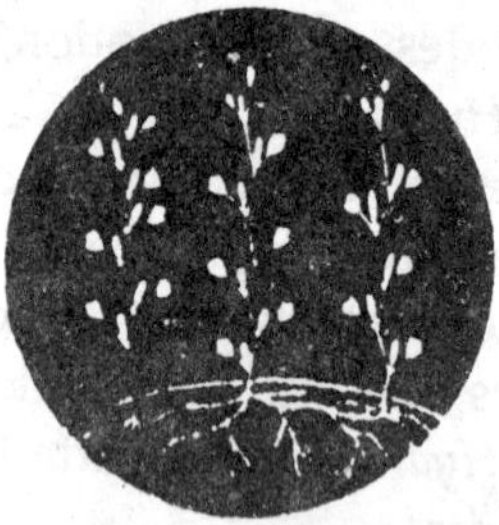

Fig. 2.11. *Obelia*. Colony in natural size.

Each hydrocaulus branches in an alternative manner and the lateral branches may sometimes give off branches of the third order. Each ultimate branch terminates into a zooid, the *polyp* or *hydranth*. In the axils of the older polyps are placed cylindrical zooids or the *blastostyles*. The obelia colony exhibiting two types of zooids, is thus *dimorphic*. When the blastostyles develop saucer-shaped bodies, the *medusae*, the colony then become *trimorphic*.

Microscopical Structure

The branches and zooids of the colony consist of an inner, tubular and living portion, the *coenosarc*, surrounded all round by a tough and non-living horny layer, the *perisarc*.

Reproductive Organs and Development in Obelia

Each medusa of *Obelia* is unisexual i.e., sperms and ova are produced in separate individuals. The *gonads* are situated on the radial canals in the subumbrellar surface. Actually, each gonad arises as an outgrowth from the wall of the radial canal. It has ectoderm, endoderm and mesoglea as well as an extension of the coelenteron. In fact, the *oogonia* and *spermatogonia* both originate in the ectoderm of the hypostome or manubrium from there they move into the endoderm and later they gradually migrate to the gonads in the radial canals.

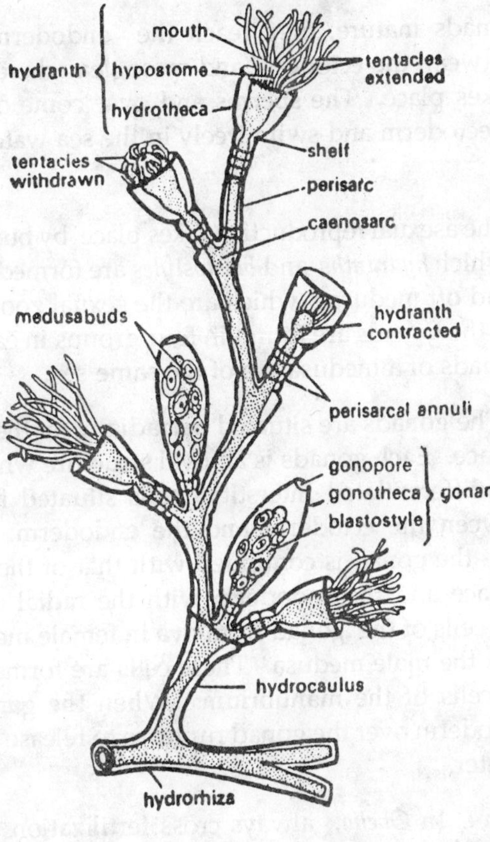

Fig. 2.12. *Obelia*. A part of colony under the microscope.

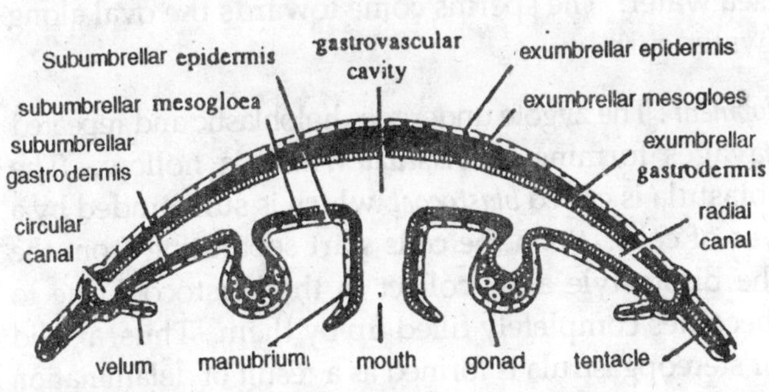

Fig. 2.13. *Obelia*. A vertical section of medusa showing gonads.

Before the gonads mature, they reach the endoderm and come to lie between the ectoderm and mesoglea where their maturation takes place. The sperms and ova come out by rupturing the ectoderm and swim freely in the sea water.

Life History

In *Obelia*, the asexual reproduction takes place by budding as a result of which *hydranths* and *blastostyles* are formed. The blastostyles bud off medusae which are the sexual zooids of *Obelia* because the gonads are found in four groups in each of them. The gonads of a medusa are of the same sex.

Gonads : The gonads are situated perradially on the sub-umbrellar surface. Each gonads is an oval structure which is a group of undifferentiated interstitial cells situated in the mesoglea between the ectoderm and the endoderm. The ectoderm over the gonad is continued with that of the sub-umbrellar surface and the endoderm with the radial canal. The interstitial cells of the gonad from ova in female medusa and sperms in the male medusa. These cells are formed by the ectoderm cells of the manubrium. When the gametes mature, the ectoderm over the gonad ruptures to release them into the sea water.

Fertilization : In *Obelia,* always cross fertilization takes place. The oval are fertilized in the body of the female medusa or in the sea water. The sperms come towards the oval along with the water current.

Development : The zygote undergoes holoblastic and repeated equal cleavages forming a *blastula* which is hollow. The cavity of blastula is called *blastocoel* which is surrounded by a single layer of cells. Soon the cells start separating from the wall of the blastostyle and collect in the blastocoel due to which it becomes completely filled up by them. Thus, a solid gastrula or stereopgastrula is formed as a result of delamination and multipolar immigration of the ectodermal cells. The outer layer of cells of the gastrula forms the ectoderm of the adult while the inner layer of cells forms its endoderm. After sometime, cilia appear in the cells of ectoderm. Simultaneously, a *colenteron* in the centre is formed by the delamination of the endodermal cells. The gastrula in this stage is known as *planula larva.* It comes out by the rupture of the brood pouch and swims freely in the sea water.

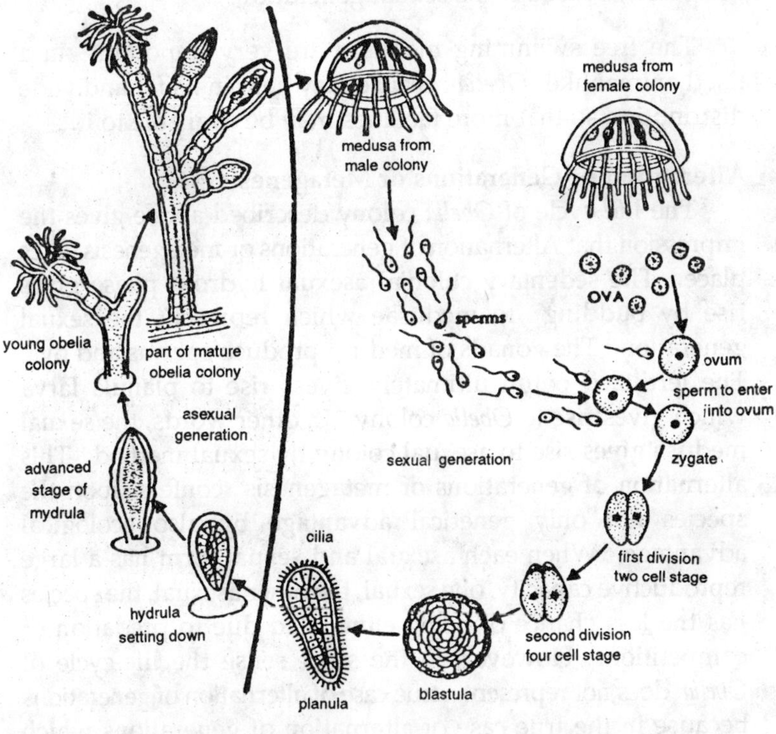

Fig. 2.14.` Life-cycle of *Obelia.*

Metamorphosis : After leading a free swimming life for sometime, the planula larva settles down and attaches itself by its broad anterior end to some object and the process of metamorphosis starts in it. Its anterior end forms the basal disc by which the animal remains attached to the substratum and the posterior end forms the manubrium. A mouth is formed at the free end of the manubrium around which a circlet of tentacles is present. In this way a simple polyp-like structure is formed which is called *hydrula stage*. The basal part of hydrula elongates to form the *hydrohiza* from which vertical and lateral branches arise forming the adult *Obelia* colony. Again the medusa buds arise on the blastostyles of this colony and start the sexual generation.

The free swimming medusae are very important in a fixed animal like *Obelia* because they help in its far and wide distribution so that more facilities may be available to it.

Alternation of Generations or Metagenesis

The life-cycle of *Obelia* colony described above gives the impression that Alternation of generations or metagenesis takes place. The sedentary colonial asexual hydroid phase gives rise by budding to medusae which represent the sexual generation. The gonads of medusa produce sperms and ova. The fertilized ovum ultimately gives rise to planula larva which gives rise to *Obelia* colony. In other words, the sexual medusa gives rise to asexual colony by sexual method. This alternation of generations or metagenesis confers upon the species not only genetical advantage by also ecological advantage. When each asexual and sexual form has a large reproductive capacity, one sexual, the other asexual, the species has the less chance of being eliminated due to predation or competition. However, in the strict sense the life-cycle of *Obelia* does not represent a true case of alternation of generations because in the true case of alternation of generations which occurs in the life-cycle of certain plants such as Ferns, the asexual diploid (2n) phase forms haploid (n) sexual phase by its single cell. But in the case of *Obelia*, both the sedentary asexual colonial polypoid generation and mobile sexual medusoid generation are diploid (only the sperm and 0Va produced in the gonad of medusa having brief existence are haploid.) More over, the medusa (sexual phase) is not derived from a single cell of the asexual phase but is formed by many cells of the blastostyle by budding. Hence, the life-cycle of *Obelia* though apparently seems to represent alternation of generations does not do so in *sensustrictu*.

Reproduction and Development of Aurelia

Aurelia is *dioecious* or unisexual but they are alike in structure i.e., there is no sexual dimorphism. Four gonads (test or ovaries) lie interradially on the floor of the gastric pouches. The gonads are horse-shoe shaped. Their colour is reddish or pinkish. The gonads are endodermin in their origin. The testes of mature male liberat spermatozoa into the genital pouches from where they pass out through the mouth with the out going water. The female gametes or ova are liberated in the genital pouches where they remain till fertilization occur.

Fertilization: Fertilization is effected in the gastro-vascular cavity of female medusa. The spermatozoa from water find their way into the female with the ingoing water current and reach its genital pouch. Here, fertilization takes place. Thus fertilization is *internal*. The zygote or fertilized eggs pass out with the outgoing water current and get lodged in the groovers of the oral arms of the female. Thus oral arms serving as temporary brood chambers and development takes place within these chambers.

Cleavage: The zygote undergoes holoblastic but unequal cleavage to produce a solid, ball-like *morula*. The morula changes into a single-layered hollow, closed sphere, the *blastula* by developing a fluid filled central cavity or blastocole. The blastula undergoes gastrulation by an invagination. Thus a two layered *gastrula* is formed which has an outer layer of ectoderm and inner layer of endoderm enclosing a cavity, the *coelenteron cavity* which has an opening, the *blastopore*. As the invagination progresses, the coelenteron cavity enlarges at the coast of blastocole which finally disappear. The gastrula elongates and its ectoderm develops cilia the blastopore is closed. Thus the embryo becomes a free swimming oval *planula larva.* The planula differs from the planula of Hydrozoa in its method of formation and in having a coelenteron and blastopore. These larvae, can be seen on the oral arms of female *Aurelia.*

Planula larva : They become free from the oral arms of the female and after a short free swimming. They sink at the bottom of sea. Now planula loses its cilia and the larva is fixed to the substratum by its broad aboral end.

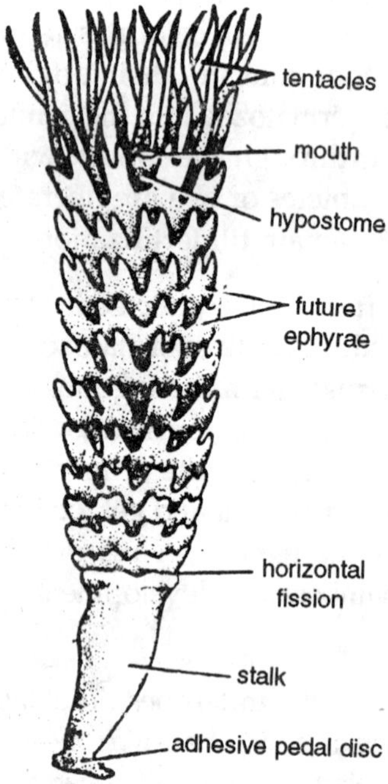

Fig. 2.15. *Aurelia.* A strobila.

The planula metamorphoses into a small trumpet-shaped polyp or *hydratuba* which has a manubrium and the blastopore opens to become the mouth. Later, the tentacles are also formed around the mouth first of all, two *perradial* tentacles as buds. After sometimes two other perradial tentacles are formed longitudinally to the first two are formed and then four *inter-radial* and *eight ad-radial* tentacles arise. The endoderm of coelenteron forms four inter-radial longitudinal ridges called the *gastric ridges* or *mesenteries*. The mouth becomes square and the manubrium forms funnel-like depression called *septal funnel* or *infundibula*. These changes convert the planula into a hydratuba. A stolon arises at the base of hydratuba.

The hydratuba buds off new hydratubae, from its stolon in summer. There hydratubae may separate after summer. Hydratuba stops budding but each continues to feed and generally passes the winter of the first year and buds other hydratuba. In the next winter the hydratuba undergoes a process of transverse fission, strobilation. The dividing hydratuba is called *Scyphistoma* or *Strobila*.

Strobilization of Scyphistoma : In winter and autumn seasons, the body of scyphistoma divides by a series of ring-like transverse constrictions. These constrictions gradually deepan and the body of scyphistoma appears like many saucers placed upside down over each other. The segmented scyphistoma is known as the *strobila* and the process is strobilation and is saucer-like segments as the *ephyrae* or *ephyrulae*.

During the formation of the ephyrae, eight bifid arms, 4 perradial and 4 interradial grow out from the periphery. A rhopalium appears in the terminal notch of each arm. The gastric ridges break down, septal funnals disappear due to expansion of the area occupied by them. The enteron sends a branch in each arm. When fully formed, the ephyrae break from scyphistoma. The scyphistoma loses its tentacles before it casts off the first ephyra. Thus the ephyrae are constricted off one by one from the upper surface of scyphistoma. The successive ephyrae released from strobila have no tentacles. While the ephyrae are being detached, their enteron closes on the proximal side but on the distal side it remains open and its margin grows out to form a short manubrium with a square mouth at its tip. The side becomes subumbella while proximal becomes exumbrella.

Strobilization is of two types: *polydisc* and *monodisc.* In polydisc strobilization several ephyrae are formed. In monodisc strobilization only one ephyra is formed and it takes during unfavourable conditions.

Sometimes scyphostoma does not undergoes strobilization. It gets detached and changes into a single free-swimming ephrya that under goes metamorphosis into a medusa.

Ephyra: The ephyra is a young medusoid form showing tetramerous symmetry. Its margins has four perradial and four interradial befid arms. Each arm is deeply notched distally to form marginal lappets. The groove between them is sensory in nature and has a short tentacle which later becomes tentaculocyst. The larva has a small stomach with gastric filaments (which are derived from mesenteries), a manubrium and a rectangular mouth on the subumbrellar surface. The stomach gives off canals which go into the arms. These canals represent perradial and interradial canals. The adradial canal appears a little later.

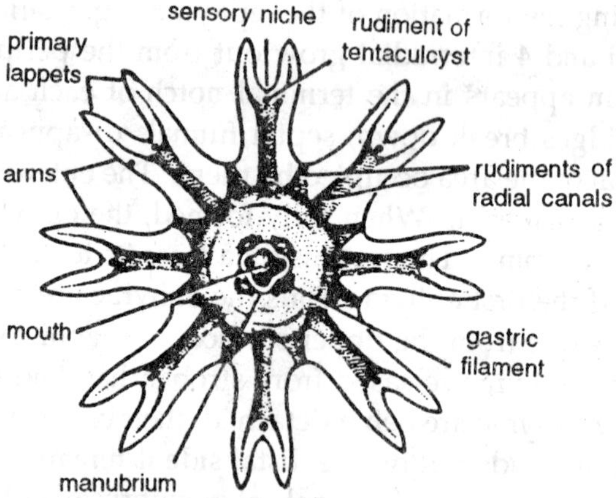

Fig. 2.16. *Aurelia.* A free ephyra.

Metamorphosis of ephyra in to Aurelia: The ephyra swims actively in sea water feeding on minute organisms such as protozoans which are caught by the lappets and passed into the mouth. It grows in size. The mesogloea increases so that the two layers of endoderm together to form an endoderm lamella except the enteron. The adradial region grows more repidly and fills their clefts so that the umbrella becomes circular and saucer-shaped and resembles the adult medusa. Now the four oral arms and numerous marginal tentacles appear so that the ephyra is finally metamorphosed into an adult *Aurelia*.

It has been seen that very large eggs of *Auralia* develop into actinula larva which gives rise to ephyrae larvae directly. The small eggs give rise to *planula larvae* which become hydratubae and then scyphistomae which strobilate ephyrae.

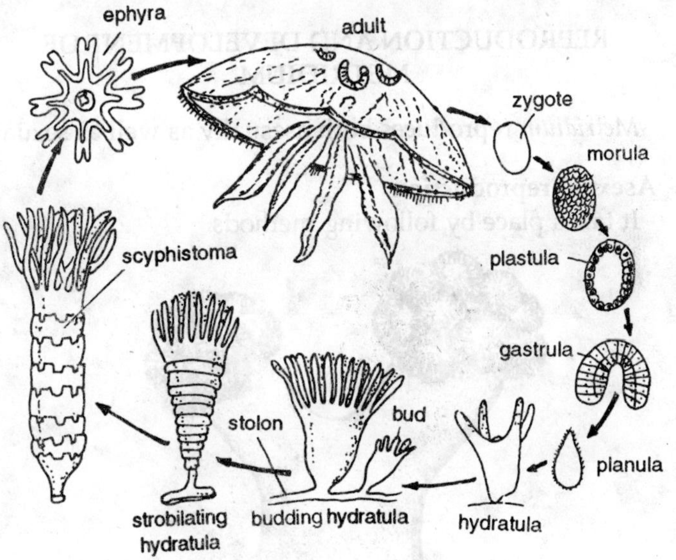

Fig. 2.17. Life - history of *Aurelia*.

Alternation of Generations

The life cycle of *Aurelia* is said to show alternation of generations. The adult medusa represent the sexual phase producing gametes. The zygote produces fixed scyphistoma (polypoid) which represents the asexual phase resembling the polyp of *Obelia* colony. The scyphistoma reproduces asexually by strobelization to form ephyrae which metamorphose into adult *Aurelia* which is a sexual form.

According to some workers, the development of *Auredia* does not show true alternation of generations. The adult *Aurelia* medusa is formed by the metamorphosis of an ephyra larva produced as one of the several transverse segments of the polypoid scyphistoma, whereas in *Obelia* the medusa is developed as a bud on a branched polypoid colony. In *Aurelia* the polypoid phase is greatly reduced in the life-cycle. As the medusa is formed by the metamorphosis of an ephyra and the life-cycle is continuous process so that there is no metagenesis or alternation of generation.

REPRODUCTION AND DEVELOPMENT OF METRIDIUM

Metridium reproduces both asesually as well as sexually.

I. Asexual reproduction

It takes place by following methods.

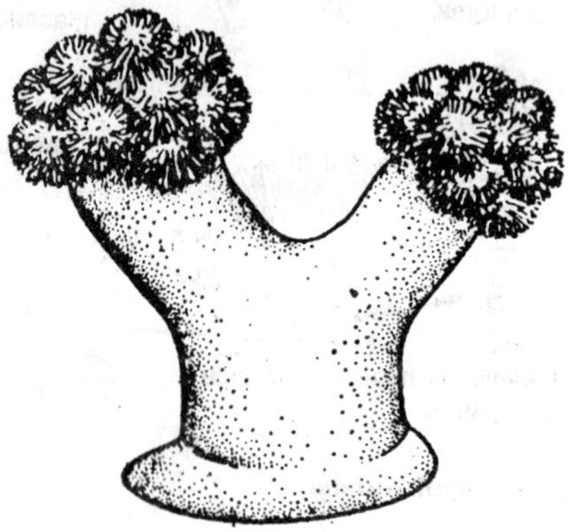

Fig. 2.18. *Metridium* showing longitudinal; binary fission.

(a) *Pedal Laceration :* Pedal laceration is the predominant mode of asexual multiplication. In this process, parts of the pedal disc may firmly adhere to the substratum and get torn off as the anemone creeps along. These pieces in a few days develop into small new anemones at the old site and the parent completes its pedal disc at a new place. Or, alternatively, small lobes arise from the margin of the pedal disc and get constricted off. These are carried by water currents to new places where they give rise to young anemones.

(b) *Longitudinal Binary fission* : The pedal disc elongates in the sagital plane till ruptures along the transverse plane. The cut then rapidly extends upwards through the column and the oral disc.

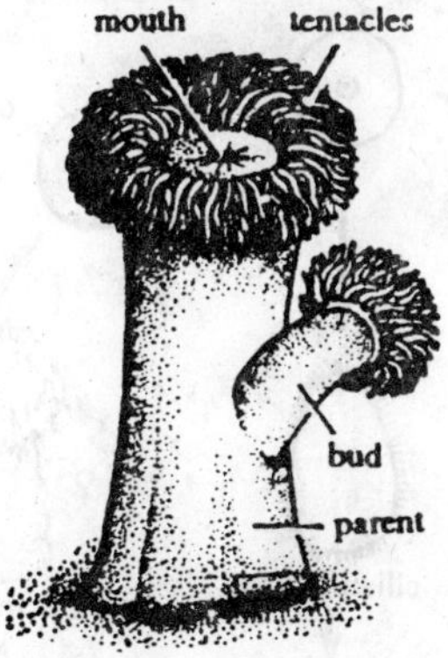

Fig. 2.19. *Metridium* showing budding.

(c) Budding. It involves formation of outgrowths from the column or pedal disc and their transformation into new individuals. The latter ultimately detach from the parent and establish themselves at new places. Budding is of rare occurrence.

II. Sexual reproduction and life cycle

Sea-anemones are dioecious or unisexual. Near the edges of the septa or mesenteries and lying parallel to them, gonads are located. The sex cells are gastrodermal interstitial cells that ripen in the mesoglea. Eggs or spermatozoa produced by gonads are discharged into gastrovascular cavity and finally pass out in sea water through mouth. Fertilization is external and takes place in sea water. Zygote develops after undergoing total, unequalor equal cleavage in a *Coeloblastula.* The endoderm is formed by invagination. The embryo elongates into a ciliated *planula* having in some forms an aboral large hard tuft of cilia. At the oral (posterior) pole the blastopore persists in cases of invagination or soon breaks through. Then blastopore becomes the lower opening of stomodaeum. The planula larva swims for sometime and feeds on micro-organisms. Finally, it settles down at the bottom of the sea and attaches itself by aboral end. The tentacles, septa and mesogloea are formed and the planula larva changes into a minute animal.

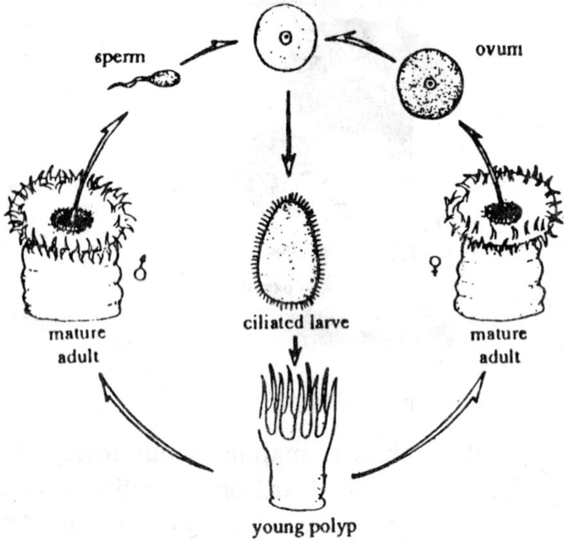

Fig. 2.20. *Metridium.* Life history.

Regeneration : Metridium has a power of regeneration. If the column is cut across, the aboral part regenerates a new oral disc but the oral part fails to develop a pedal disc. In some cases, instead of a new pedal disc, the oral part regenerates a second set of tentacles at its lower aboral surface. Thus, it exhibits a case of *heteromorphosis* or reversed polarity.

3

The Development of Various Helminthes

In helminthes sexual reproduction occurs and asexual reproduction is restricted to the regeneration only. Most of them are *manoceous* or *bisexual* or *hermophodite* with or without cross fertilization. However, some forms are unisexual or *dioceous* with or without clear cut sexual dimorphism. Most of the species are parasitic — Planarian is non-parasite form. Parasites have greatest reproductive capacity. They are usually *digenetic* simply to present the termination of their race. Development is usually indirect i.e., with larval form or forms.

Here we are taking the reproduction and development of some selected types prescribed in various syllabi.

Particularly concentrated on the ventral surface around the mouth, auricle and the body margins. Thus ventral surface is positively thigmotactile while the dorsal surface is negatively thigmotactic.

(v) *Rheoreceptor*: These are sensitive to water current. They are of general occurrence and their sensory processes or bristles project much beyond the level of cilia.

REPRODUCTIVE SYSTEM

Dugesia reproduces with asexually as well as sexually.

(1) Asexual Reproduction

Asexual reproduction occur by the transverse binary fission. The plane of fission usually forms posterior to the pharynx.

The posterior part attaches to substratum by mucus, the anterior part stretches forward until the animal snap into two. Each half regenerates the wanting parts resulting in the formation of two complete planarians. Some planarians fragment into two or more pieces each growing into a complete individual. Asexual reproduction is restricted to some species only. Some species reproduce asexually in one season, and may develop sex organs in the other, but these are some species that reproduce exclusively asexually and rarely develop sex organs.

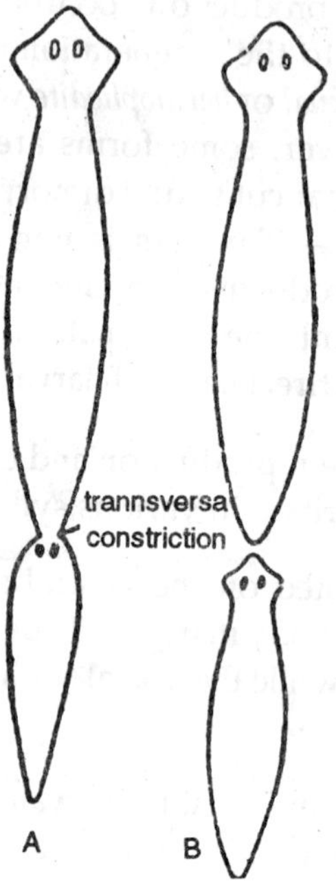

Fig. 3.1. *Dugesia.* Transverse binary fission.

(2) Sexual Reproduction

The reproductive organs are more complicated in these animals. The gonads i.e., testes and or aries are derived from the meroderm and there is a system of tubules and chambers in which fertilization occurs.

The animal reproduces in early summer and the gonads

are developed temporarily in the breeding season. The *Dugesia* is bisexual or hermaphrodite but self-fertilization does not take place and the cross fertilization is the rule. After breeding season the gonads degenerate and disappear and the work reproduces asexually.

Male reproductive system consists of a large number of small spherical testes, which are situated on both the sides of the body. Each testis is connected with minute tubes, called vasa *efferntia.* There form two large ducts called *vasa deferentia.* The two vasa deferentia run parallel to one another towards the posterior end, where these bend mesially towards the penis. Each vas deferens dilated to form the seminal or spermlducal vesicle with thick muscular wall. The sperms are stored here. The penis shows two regions, the anterior large spherical *penis bulb* and the posterior narrow conical *penis papilla* with muscular wall. The lumen of penis is very narrow and called ejaculatory duct. The penis papilla opens into the genital atrium which terminates in the common gonopore, situated midventrally behind the opening of month. The lumen of penis bulb is formed by *bulbar cavities.* The bulb contains numerous uni *cellular glands,* the prostate glands which open into it lumen.

Female reproductive organs : The female reproductive organs consist of a pair *ovaries,* which are small apherical bodies situated laterally just behind the head. From each ovary arises an oviduct which extends posteriorly through the parenchyma. Immediately after its emergence each oviduct presents a dilation, which is known as *seminal receptacle.* The two oviducts unite to form a common vagina which opens into the *genital artium.* The oviducts along their whole length are surrounded by numerous vitelline glands which open into oviducts by short *vitelline ducts.* Because of the close association of oviduct and vitelline glands, these are sometime collectively referred to as the ovovitelline ducts. Opening into the genital chamber is found a muscular sac-like *copulatory sac* or *uterus* or *bursal canal.* The bursal canal opens into a large sac the *bursa conplatrix* which receives penis and temporarily stores the sperms received during copulation. It opens below into female genifal atrium.

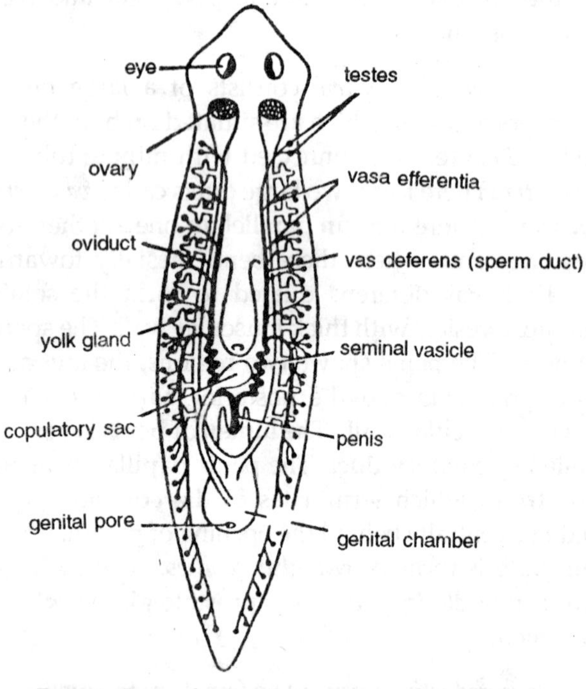

Fig. 3.2. Dugisia Reproductive System.

Copulation

Although *Dugesia* is hermaphrodite, yet *cross-fertilization* is a rule. Copulation takes place between the two worms in which their posterior and ventral surface come together and the penis of each is inserted into the *genital atrium* of the other. In this way sperms are deposited in the *seminal receptacles* of the mating partner. The sperms are stored inside the copulatory sac for sometime and then travel along the oviducts to reach the seminal receptacles. The eggs are deposited in the seminal receptacle and are fertilized. Thus, the fertilization is internal. During breeding season, each worm may copulate many.

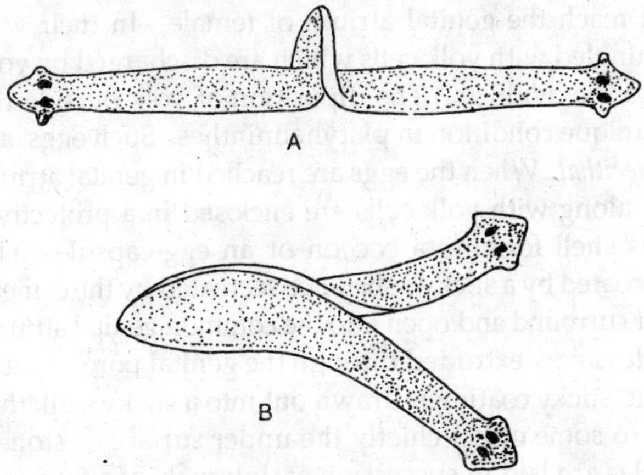

Fig. 3.3. *Dugesia.* Copulation.

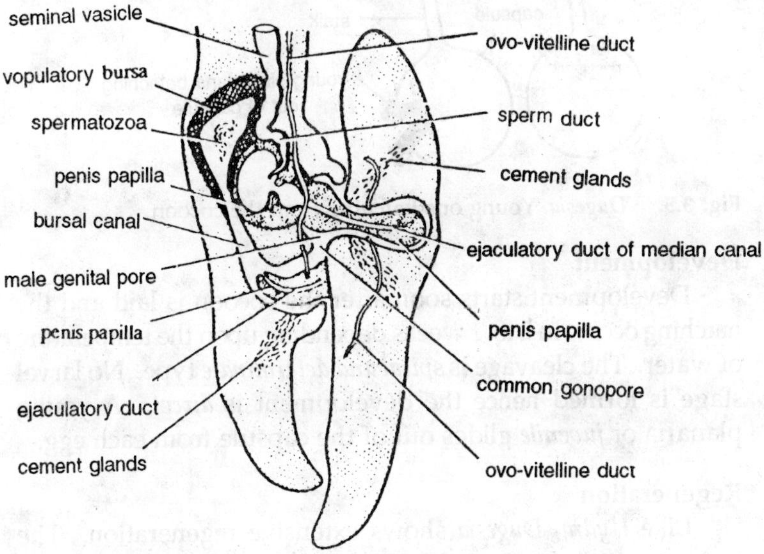

Fig. 3.4. *Dugesia.* A pair of copulating animals in section.

Cocoon Formation

The zygote thus formed, pass backwards through the ovovitelline ducts and reach the genital atrium of ovovitelline ducts and reach the genital atrium of female. In their way they are mingled with yolk cells which are discharged by yolk glands into the oviducts. The diposition of yolk outside the zygote is unique condition in platyhelminthes. Such eggs are called *ectolecithal*. When the eggs are reached in genital atrium 5-10 eggs along with yolk cells are enclosed in a protective, proteneous shell forming a cocoon or an egg-capsule. The capsule is coated by a sticky substance secreated by the cement glands that surround and open into the common genital atrium. The capsule is now extruded through the genital pore. As it is passing out, sticky coating is drawn out into a sticky stalk that fastens it to some object chiefly the under surface of stones. The capsules are laid in succession at intervals of a few day.

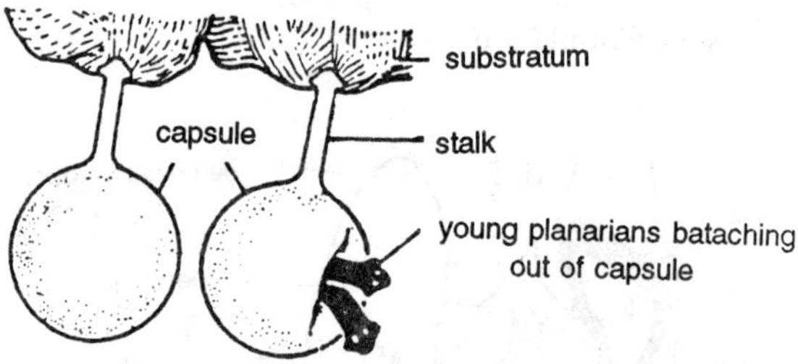

Fig. 3.5. *Dugesia*. Young one hatching from the cocoon.

Development

Development starts soon after the cocoon is laid and the hatching occurs in 2 to [illegible] weeks depending upon the temperature of water. The cleavage is *spiral and determinate* type. No larval stage is formed hence the development is *direct*. A young planaria or *juvenile* glides out of the capsule from each egg.

Regeneration

Like *Hydra, Dugesia* shows extensive regeneration. The lost parts can be easily replaced. If the animal is cut transversely into two halves, the anterior half regenerates a new Tail, while the posterior half regenerates a new head. If the animal is cut into three pieces, the middle piece will regenerates head on the anterior side and a tail on posterior side. The regenerating piece retains the original *linear polarity* of the whole animal. Regeneration involves *epimorphosis* and *morphollaxis*. In epimorphosis the missing parts are formed while in morphollaxis, the original parts fit to function with the regenerated parts in the new individuals.

Power of regeneration is maximum near the head and minimum near the posterior end, this is associated with the metabolic rate. The metabolic rate is highest in the head and becomes progressively lower towards the tail. This variation in metabolic rate and regeneration power is called *axial gradient (Child)*.

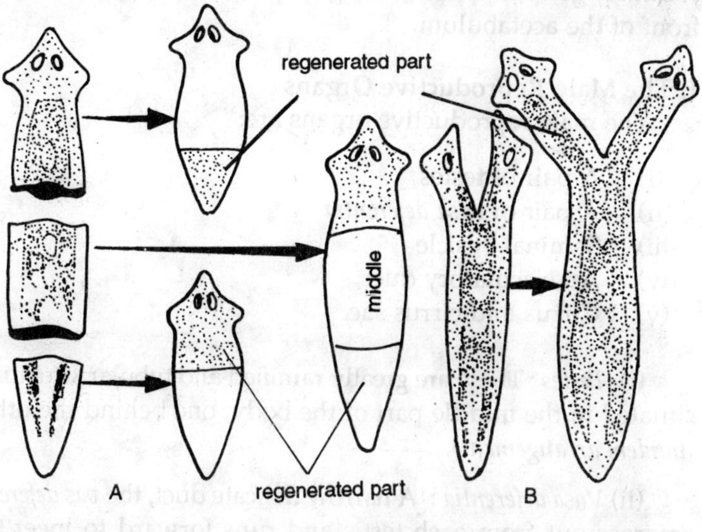

Fig. 3.6. Dugesia Regeneration. A = Transverse, B = Longitudinal.

Regeneration is brought about by the free formative cells or *neoblasts* of the mesenchyme. These cells migrate to the cut surface and by repealed division produce new tissues.

Regeneration is always not perfect. It a sexually mature Dugesia is cut transversely between its pharynx and copulatory apparatus, its reproductive organs degnerate and fail to regenerate, but each help grows into asexual animal.

Dugesia shows a lateral polarity. If the anterior end is cut longitudinally, each part will grow into a complete head. Thus several headed monster results. The phenomenon is called *hetermorphosis.*

REPRODUCTION & DEVELOPMENT OF LIVER FLUXE

Fasciola hepatica is hermaphrodite or disocious. The gonads are well-developed and the male and female genital ducts open into a common chamber, the genital atrium. It is situated anteriorly in the body and opens to the exterior through the common genital aperture or gonopore, located ventrally in front of the acetabulum.

1. The Male Reproductive Organs

The male reproductive organs are:

(i) A pair of testes
(ii) A pair of *vasa deferentia*
(iii) Seminal vesicle
(iv) An ejaculatory duct
(v) Cirrus and cirrus sac.

(i) *Testes* : These are greatly ramified and tubular structures situated in the middle part of the body, one behind the other (*tandem arrangement*).

(ii) *Vasa deferentia* : A narrow delicate duct, the *vas deferens*, emerges out from each testis and runs forward to meet the fellow of opposite side forming the common sperm duct just below the ventral sucker.

(iii) *Seminal vesicle.* The common sperm duct immediately leads into a large muscular pear-shaped sac, the *seminal vesicle*, which lies in front of ventral sucker.

(iv) *Ejaculatory duct* : The ejaculatory duct arises as a fine convoluted tube from the seminal vesicle and runs forward in a zigzag fashion through the cirrus to open into the *genital chamber* through the *male genital pore.*

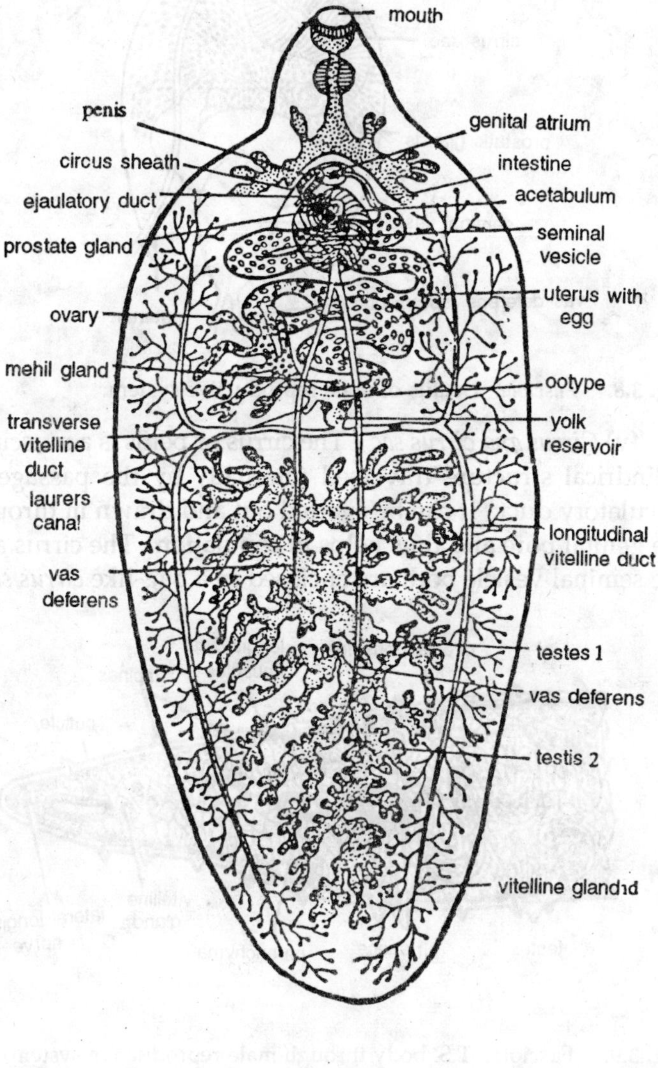

Fig. 3.7. *Fasciola.* Reproductive system.

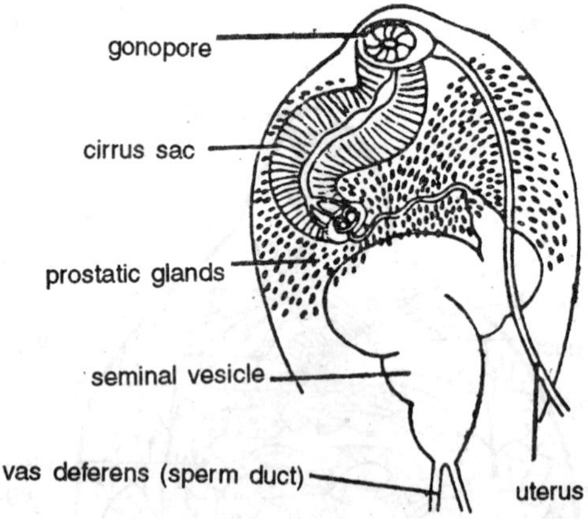

Fig. 3.8. *Fasciola.* Details of male reproductive system.

(v) *Cirrus and cirrus sac* : The cirrus or penis is a muscular cylindrical structure traversed internally by the passage of ejaculatory duct. It can be everted out and drawn in through the genital pore and thus helps in copulation. The cirrus and the seminal vesicle both are enclosed in a bag-like *cirrus sac.*

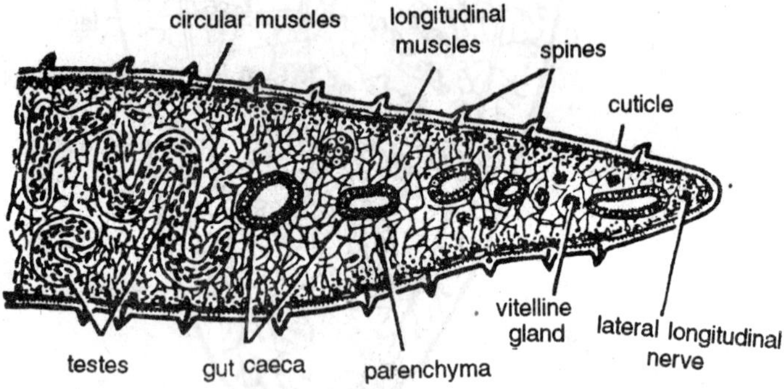

Fig. 3.9. Fasciola. T.S. body through male reproductive system.

(vi) *Prostate glands.* Numerous unicellular prostate glands are found around the ejaculatory duct.

(b) Female Reproductive Organs

The female reproductive organs are:

(i) Ovary
(ii) An oviduct
(iii) Uterus
(iv) Laurer's Canal
(v) Mehli's gland
(vi) Vitelline glands and vitelline ducts

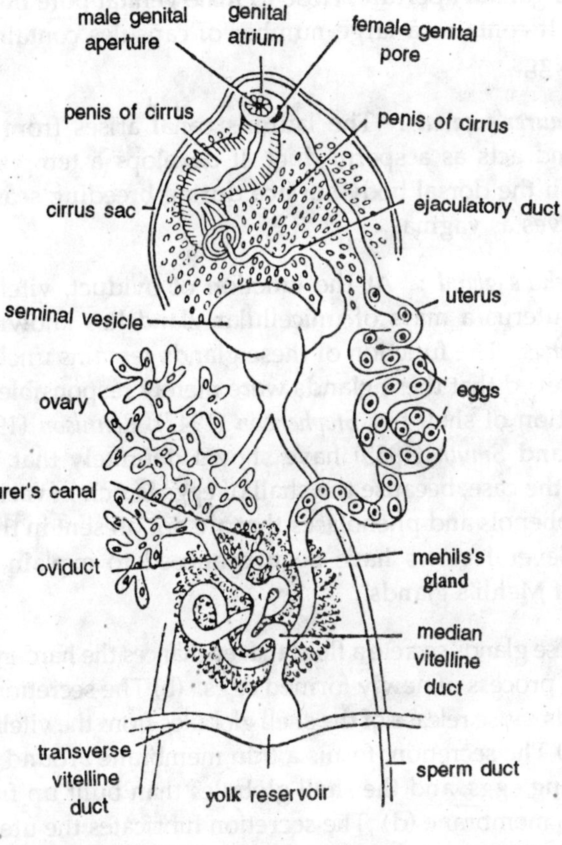

Fig. 3.10. Fasciola. Details of male and female reproductive system.

(i) *Ovary* : The ovary is a large, highly branched tubular structure which is situated on the right side in front of testes in the mesenchyme. It occupies the anterior one-third of the body.

(ii) *Oviduct* : From the inner side of ovary arises a short, narrow and convoluted duct, the *oviduct.* It runs down ward to join the uterus.

(iii) *Uterus :* It is a long wide and highly convoluted tube that extends upto the genital atrium, opening into it through the female genital aperture, close to male genital pore on the left-side. It contains a large number of capsules containing fertilized eggs.

(iv) *Laurer's canal* : The laurer's canal arises from the oviduct and acts as a sperm duct. It develops a temporary opening on the dorsal body surface during breeding season. It may serves as vagina.

(v) *Mehli's gland* : At the junction of oviduct, vitelline duct and uterus a mass of unicellular gland lies known as *Mehli's glands.* The function of these glands remains unclear. It was believed that these glands were merely responsible for the formation of shell but *Stephenson* (1947), *Rennison* (1953) and *Johri* and *Smyth* (1955) have shown definitely that this cannot be the case, because the shall of eggs is rich in various proteins, phenols and phenolases that are not present in these glands. Several views have been proposed to explain the function of Mehli's glands:

(a) These glands secrete a fluid that enchances the hardening or tanning process of newly formed eggs. (b) The secretion of these glands cause release of the shell globules from the vitelline glands. (c) The secretion forms a thin membrane around the cells forming eggs, and the shall globules than built up from within this membrane (d) The secretion lubricates the uterus facilitating passage of the eggs. (e) The secretion activates sperms which are passed down. All or some of these hypothesis may be valid.

(vi) *Vitelline glands and vitelline ducts* : The vitelline glands or the yolk glands are in the form of numerous minute vesicles or follicles arranged on either side of the body from one end to the other. A fine duct arises from each vesicle and after uniting with similar ducts from other vesicles opens into the *lateral longitudinal ducts.* The two longitudinal ducts are connected in the middle line by a *transverse vitelline duct.* A medium duct arises mesially from the transverse vitelline duct and joins the oviduct.

The functions of vitelline glands is now know. In addition to contributing yolk material for incorporation within the egg, the vitelline glands also secrete large globules known as *shell globules* — which envelope the developing egg and eventually coalesce and become hardened to form the shall. This hardening process involves the tanning of protein present within the coalesced globules by quinone. Thus vitelline glands contribute both yolk and shell material.

	Phenolase		
Phenol	------------------>	quinone --------->	Tanned protein
Protein	------------------>	protein ↗	

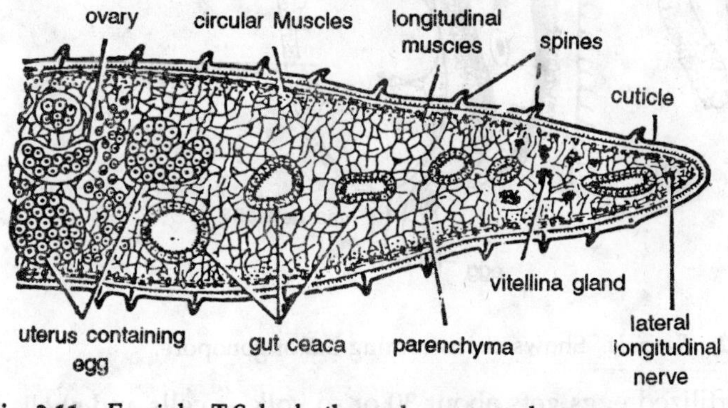

Fig. 3.11. Fasciola. T.S. body through ovary and uterus.

Ootype : In some cases of the oviduct after receiving Laurer's canal and a common vitelline duct swells to form ootype. In *F. hepatica* according to *Stephenson* (1947) the ootype is absent.

Life History

Copulation and fertilization : Cross — fertilization takes place by insertion of cirrus of one fluke into the Laurer's canal of another fluke. Transference of biflagellate sperms directly into the oviduct is possible. Secretion of prostate glands and as mentioned above, the secretion of Mehli's gland keep the sperms active for fertilization. Laurer's canal become invisible after copulation. Fertilization takes place in oviduct. Self-fertilization is also seen here the sperms enter the uterus of same fluke through female genital pore and pass down to reach oviduct where fertilization takes place.

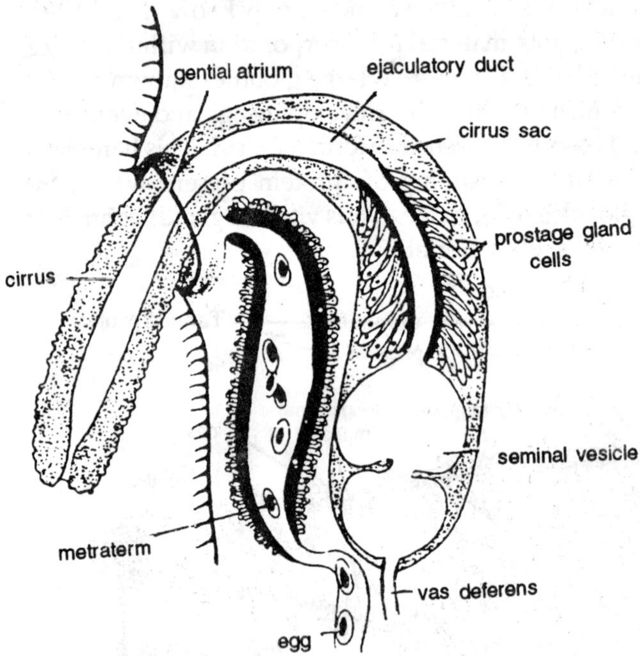

Fig. 3.12. *Fasciola.* Shows cirrus coming out of gonopore.

Fertilized eggs gets about 30 or so yolk — cells and shell globules from vitelline glands by the way of various vitelline ducts. The shell is formed with a lid or *operculum* on one side. *Rowen* (1956, 57) demonstrated that there is a viscous granular cushion at the opercular end of the egg, situated immediately beneath the operculum. The zygote and a mass of yolk — cells enveloped by the shell constitute the capsule. When complete the *capsule* passes into the uterus.

Capsule formation occurs throughout the year and one fluke may produce about 50,000 capsules. A single fluke may produce a total of 2,00,000 capsules in about 11 years.

The capsules differ in shape and sizc in different animals. Hence, with the sized and shape of capsules the parent flukes are recognized.

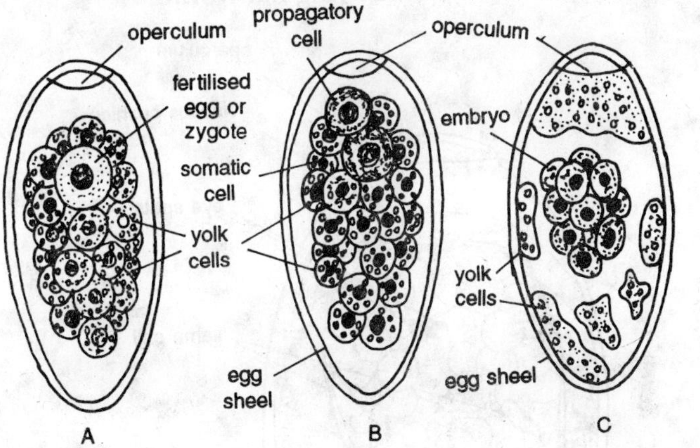

Fig. 3.13. *Fasciola*. Early stages of development. A = Fertilized egg, B= Two cell-stage, C = Many cell stage.

Development : Cleavage starts immediately after fertilization. The fertilized egg divides into a small granular *propagative cell* and a large ectodermal *somatic cell.* Somatic cell divides and gives rise to the body-wall of the larva. Propagative cell later on divides into a cell similar to *somatic cell,* which takes part in the development of mesoderm, endoderm, and a cell called *germ cell* afterwards reaches the posterior part of the body and forms germ-ball consisting of 14 to 20 germ cells.

Capsules then escape through the gonopore and are liberated into the bile capillaries from where they reach the intestine of the host and are passed out with the faeces.

The development remains arrested while the embryos remain in the faces, they may survive in wet faecal matter for several months. If washed free, the development of the embryo proceeds. The optimum temperature for development ranges from 10 - 30°C. At 30°C the encapsulated embryo differentiates into *miracidium larva* within eight days. The availability of oxygen is also a factor. The fully formed miracidium is a more or less conical little animal covered with ciliated epithelium. It does not look at all alive an adult fluke but does show resemblance to some free-living flat worm.

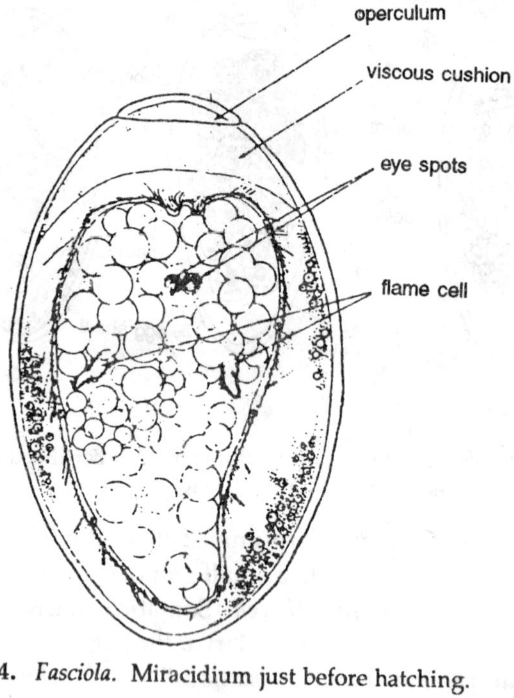

Fig. 3.14. *Fasciola.* Miracidium just before hatching.

Miracidium larva : As the miracidium is formed the hatching enzyme is secreted in response to exposure to light. The enzyme dissolves the cementing material by which the operculum is attached, thus releasing the operculum. Expension of the granular cushion, accompanied by exosmosis of salts and other materials from within the egg pushes off the operculum. The hatching occurs in water and free-swimming miracidium penetrates the intermediate host.

Structure of miracidium : It is an elongated, about 0.07 mm long creature with a broad anterior end and a narrow posterior end. Anteriorly there is a *conical* or *apical papilla* of terebratorium. It is cliated except in the origin of apical papilla. However, anterior cilia are slightly longer.

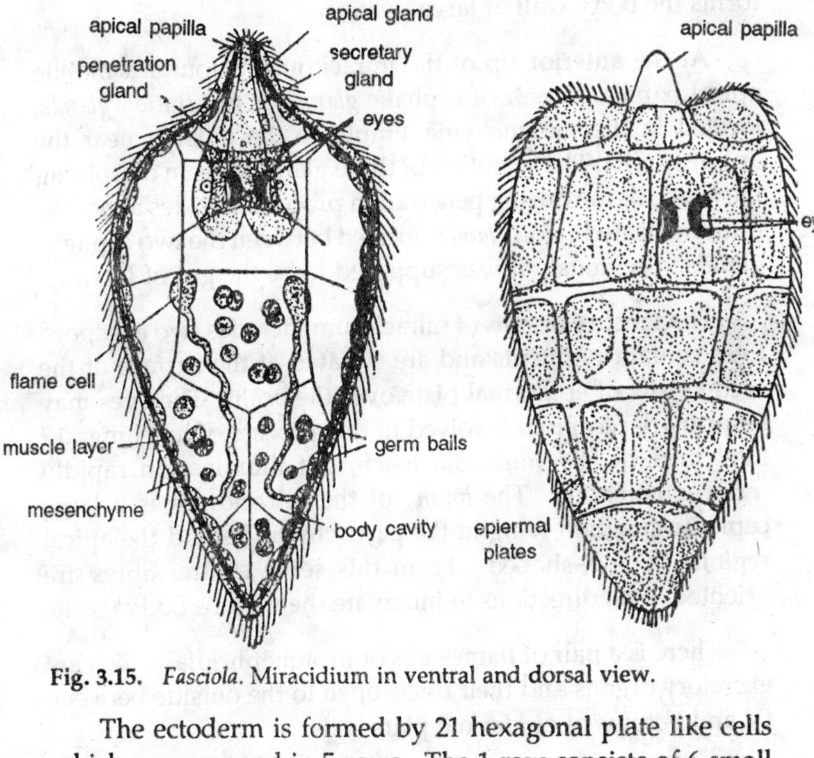

Fig. 3.15. *Fasciola*. Miracidium in ventral and dorsal view.

The ectoderm is formed by 21 hexagonal plate like cells which are arranged in 5 rows. The 1 row consists of 6 small plates. The II row contains again 6 plates, III row has 3 plate cells, 4 plates in IV row and 2 plates in V row. The arrangement of the ciliated plates is quite constant and is of considerable value in determining phylogenetic relationships. In *F. hepatica* the cells of first row arranged are two dorsal, two ventral and two lateral in position. The cells of second row are three dorsal and three ventral. In the third row the cells are one dorsal and two venterolateral. The fourth row has two right and two left cells and the fifth row has one left and one right cells.

The ectoderm is followed by musculature consisting of outer circular and inner longitudinal muscle fibres. The musculature is followed by a layer cells forming the sub-epithelium. The ectoderm musculature and sub-epithelium forms the body wall of larva.

At the anterior tip of the miracidium is found a mobile apical papilla. A pair of cephalic *glands* or *penetration glands*, located in the mesenchyme, empty to the outside near the apical papilla. The secretion of these glands aids in dissolving the host's tissue during penetration process. Besides a sac-like multinucleated *apical gland* is located between the two cephelic glands. Previously it was supposed to be the gut of larva.

The sensory organs of miracidium includes two eyespots. Eyes contain pigments and are situated at the surface of the second row of epidermal plate over the brain. The eyes may be primary receptors involved in the process of hatching. 17 kept in dark few miracidia hatch, but illumination rapidly induces hatching. The *brain* of the miracidium is a large cephalic ganglion lying in the parenchyma behind the apical region. It is X-shaped. From this serve centre, fibres are oriented in all directions to innervate the various body tissue.

There is a pair of flame-cells or protonephridia, which are excretory organs and their ducts open to the outside between IV and V rows of epidermal plate cells.

During the differentiation of the miracidium within the egg shall, certain germinal cells become trapped in the parenchyma. These develop into germ balls by increasing in number and size and becoming enveloped, eventually giving rise to the next larval generation. Such germ balls are usually located in the posterior region of the body.

Miracidium swims about in water or moisture film, where cattle usually pass out their faeces. It dies in 30 hours unless it is able to reach intermediate host, which is always an amphibious snail of the genus *Limnaea*. Intermediate host in case of *Fasciola indica is Limnaea* acuminatisata while in the case of *Fasciola hepatica* intermediate host is Limnea truncatula. When larva enters the tissue of the intermediate host with the help of apical papilla, it reaches soft parts, such as mantle, pulmonary sac and hepato-pancreas. Penetration into the intermediate host is helped by the secretion of the penetrant glands or other glands. On reaching the intermediate host, ciliated epidermal cells are cast off, while eyes and brain are lost. Larva swells up and changes in shape and size, so as to become a sac-like structure, called sporocyst. Germ-balls of miracidium do not disintegrate.

Sporocyst : The sporocyst is an elongated, hollow, sac-like larva. Its body wall consists of non-cellular cuticle followed by musculature and sub-epithelium. The brood chamber is hollow space in the centre of the sporocyst within which are found the germ balls that eventually give rise to the next generation. Sporocyst contains no alimentary canal, nervous system or reproductive systems. The protonephridia are present opening laterally by pores.

The sporocyst moves about by muscular contractions in the host tissues, absorbing nourishment from them and causing them a considerable damage. Its germ balls develop into the next larval forms, the redial. Each sporocyst produces 5-8 redial. Some of the germinal cells are again set aside within the redial for the formation of the next larval forms.

Redia larva : The rediae emerge from the sporocyst by the

rupture of body ball of sporocyst. Each radia is an elongated and cylindrical form, measuring 1.3-1.6 mm in length. It has a circular ridge, called collar near the anterior end and a pair of projections termed *lappets* near the posterior end. The body-wall consists of the usual cuticle, musculature and subepithelium. Unlike the sporocyst, the redia possesses a mouth located at the anterior terminal of the elongated body. The mouth leads into a muscular pharynx, which in term leads a blind-sac, the gut. Numerous unicellular pharyngeal glands open into the pharynx.

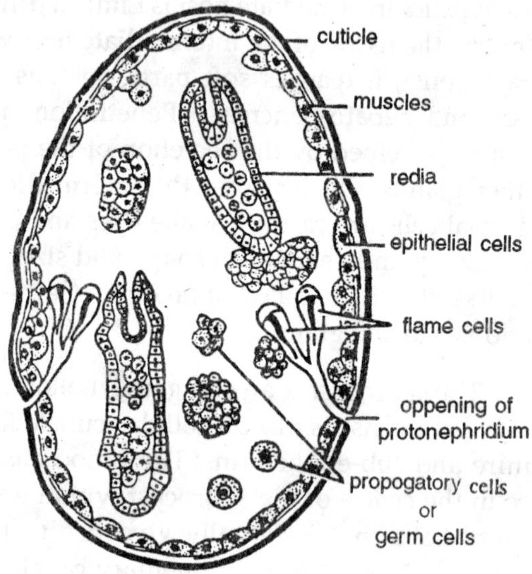

Fig. 3.16. *Fasciola.* Sporocyst.

Protonephridia divide further and form a much branched system. All the flame cells of one side open out through a common excretory duct. A birth pore is present, lateral to and in proximity of the mouth. Within the brood chamber are found germ balls.

Redia is a very active little creature. It moves about by the contractions of the body and lappets. It nourishes itself by sucking fluid and cells of the host tissues with its muscular pharynx. It soon migrates into the digestive glands of the snail. During summer months when the food is sufficiently available, mother redial mature in 12-18 days, the germ balls give rise to a second generation of daughter redial. Germ balls of daughter redial in the winter develop into next larval stage, the cercariae.

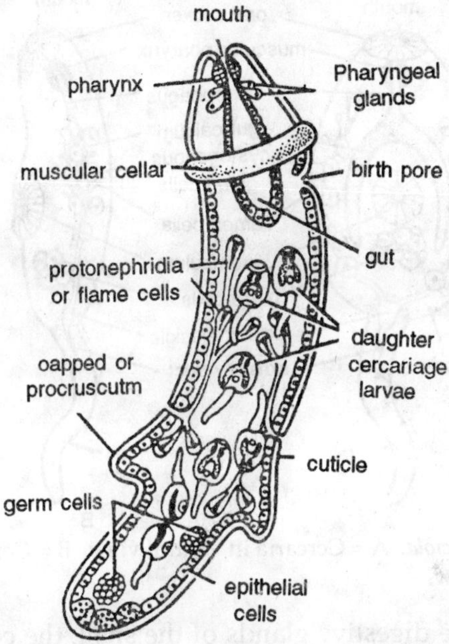

Fig. 3.17. *Fasciola.* Redia.

Cercaria larva : Cercariae leave the mother or redial through their birth pores. From each redia 14-20 cercaria are produced. Cercaria is about 0.25-0.35 mm long and is an oval larva with a tail. The body wall consists of cuticle, musculature and subepithelium. On the anterior side, there is a mouth with a sucker, a sectorial pharynx, long oesphagus and a bifid intestine. A ventral sucker or acetabulum is present on the ventral side. A large number of dark, brown cells, called *cystogenous glands* are found beneath the cuticle. There is a pair of excretory ducts, each with several flame cells. These open near the hind end of the larva into a bladder that leads out through a pair of nephridiopores. A nervous system similar to that of adult has also appeared. Rudiments of productive organs develop from the germ cells near the hind-end.

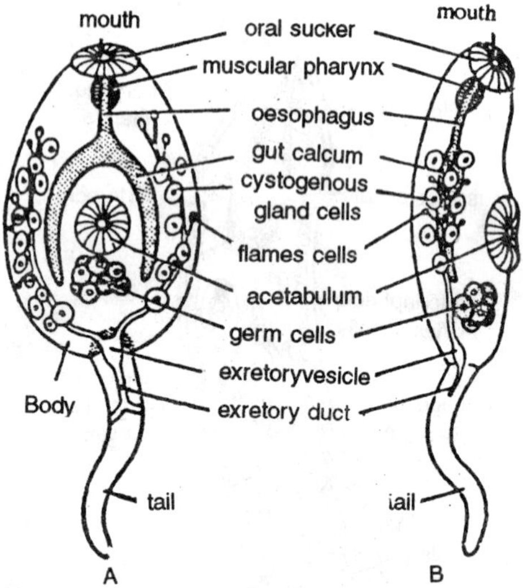

Fig. 3.18. *Fasciola.* A = Cercaria in ventral view. B = Cercaria in lateral view.

From the digestive glands of the snail, the cercariae pass into the pulmonary sac and then escape into the surrounding water. Cercaria swims about 5 minutes to an hour in summer. Afterwards, it settles down on some green leaves of water plants. Tail is cast off and the body becomes rounded and forms around itself a thick brownish cyst wall from the cystogenous glands. The encysted cercaria is now called metacercaria.

Factors Affecting Development in Snail

The rate of development in the snail host is a function of the environmental temperature. *Kendell* reported a curious effect of temperature on the development of *Fasciola.* In snails maintained at laboratory temperature, daughter redial were not produced, the first generation of redial giving rise to cercariae. On the other hand, snails exposed daily to a period of low temperature i.e. 4-5°C, daughter redial were produced from redial.

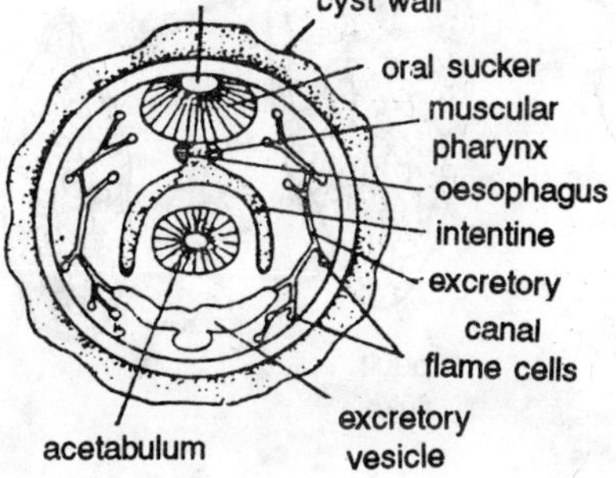

Fig. 3.19. *Fasciola.* Metacercaria.

The state of snail nutrition also affects development. The effect of food availability of development in the snail host is probably the explanation for the observation that fewer larval develop in small snails than in large ones, *Metacercaria.* It is a juvenile fluke having a rounded form with a diameter of about 0.2 mm. It resembles cercaria larva but differs from later in the absence of tail, cystogenous glands and presence of thick cyst wall. Excretory bladder opens directly through a single pore.

The metacercaria, encysted on vegetation, may live for as long as a year at low temperature, or for as short time as two or three weeks at 25°C.

Inspection to Final Host

The final excystment in the host requires an elevated temperature i.e. 38°C, low redox potential, a high carbon dioxide tension, and the presence of bile. The factors necessary for the excystment of *Fasciola* matacercariae seem to serve as stimuli to activate the worms, but the mechanism is not clearly understood. When metacercaria reaches the intestine of host,

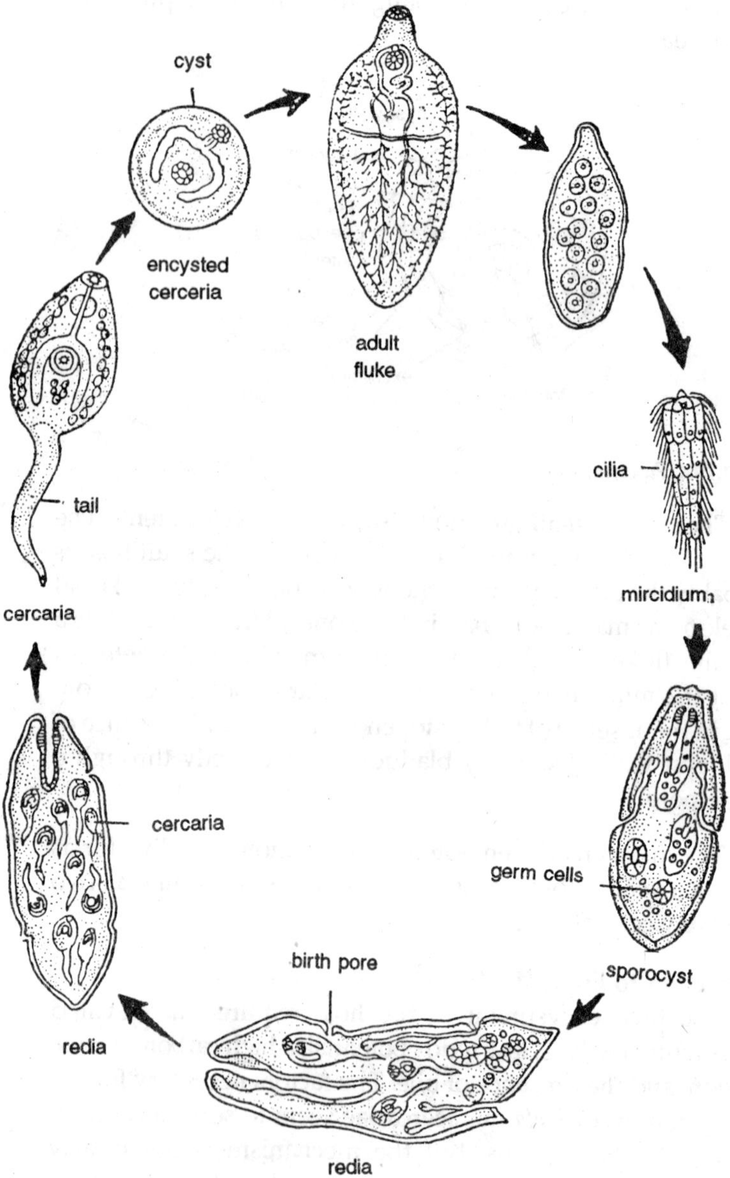

Fig. 3.20. *Fasciola.* Life-cycle.

the cyst is digested and young fluke is liberated. The young worm penetrate the wall of gut and passes into the peritoneal cavity. In the body cavity the young worms seem to move about rather aimlessly until the encounter the capsule of liver. On contact with the liver, the flukes burrow into it, developing liver parenchyma as they move. It grows in size 5-6 weeks and finally settles in the bile duct. It attains sexual maturity and begins to lay eggs 11-13 weeks after ingestion by the mammalian host.

REPRODUCTION & DEVELOPMENT OF SCHISTOSOMA

Reproductive System

Sexes are separate. Males are broader and enclose the females in the gynecophoric canals.

1. *Male Reproductive system :* The male reproductive system consists of 4-7 testes which are spherical and are situated anteriorly. Each testis gives rise a minute duct known as vas efferens. All the *vasa efferentia* unite to form a common spermduct or *vas deferens*, which after a short distance dilates into a structure, the *seminal vesicle* just before opens into a genital aperture situated just below the acetabulum or ventral sucker.

2. *Female reproductive system :* Female reproductive system consists of a single ovary. From the posterior portion of *ovary* arises a coiled tube known as *oviduct*. The oviduct bends upwards and runs afte•a short distance to open into a ootype. The ootype in turn gives off a wide coiled tube, the *uterus*. The uterus is thin and long tube running between the two branches of intestine and opens into a female genital pore near the acetabulum. The *ootype* is surrounded by Mehli's gland. *Vitelline glands* are situated in the posterior region of the body. From these glands a vitelline duct arises that runs along the oviduct and joins the latter before it opens into ootype. In *S. haematobium* seminal recepticles is absent.

Life-history

Copulation and fertilization : The male and female adults in the definitive host move against the blood flow into small venules until they can move no further. Fertilization occurs when spermatozoa from the male are introduced into the female. Fertilization takes place in the oriduct. The fertilized eggs are encapsulated by the secretion of vitelline glands which also form yolk cells. In *S. haematobium* the uterus contains only 10-100 capsules while uterus of female *S. mansoni* contains only 14 capsules at a time.

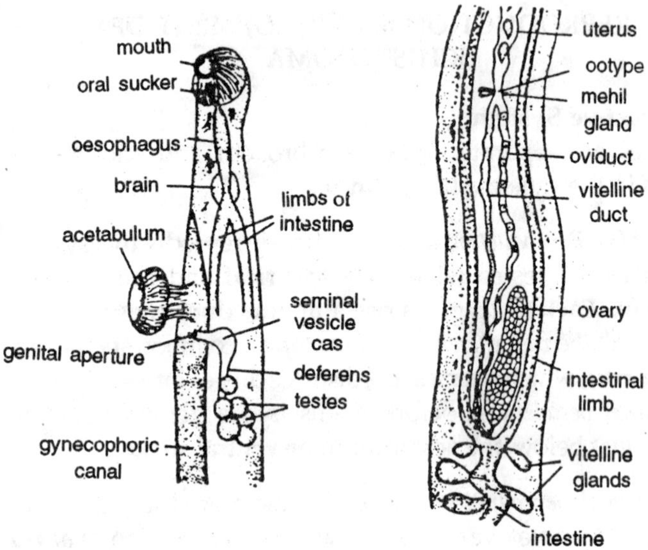

Fig. 3.21. A=Antomy of male *S. mansoni*. B=Anatomy of female *S. haematobium*.

Eggs laying : The fully formed eggs or capsules measures 114-175 m x 45-68 u in *S. mansoni*. In *S. japonicum* 70-100 u x50-65 u and in *S. haematobium* 112-170 u x 40-70 u. The capsules are nonoperculate and bear a large posterio-lateral spine in S. mansoni. In *S. haematobium*, the spine is terminal and in *S. japonicum* the spine is small and subterminal. In *S. mansoni* such eggs are laid in the mesenteric vessels and are carried by blood to the proximity of large intestine where they gradually penetrate the intestinal wall, become included in the lumen, and are passed out in feces. In *S. haematobium* the capsules are passed into the urinary bladder and are expelled in urine.

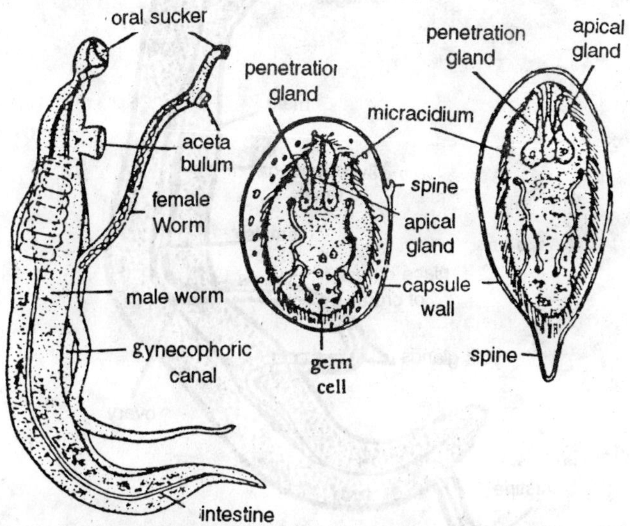

Fig. 3.22. A=*Schistosoma japonicum*. B=Capsule of *S. japonicum*. C=Capsule of *S. haematobium*.

Female lays a single egg at a time. *Pesigan et al.* (1958) estimated that a female *S. japonicum* lays about 1200 eggs per day.

Miracidium larva : Eggs do not hatch in the urinary bladder or intestine of the host. The eggs of *S. japonicum* takes 9-10 days to develop mature miracidia while passing through the tissue and will live for 10 or 12 days longer if not expelled out, but may die in the tissue. Miracidium measures about 130 u in length and 60 u in breadth. The larva contains ciliated epithelium. Two pairs of penetration glands are present, one pair anterior in position other posterior. Meracidium also possesses a primitive gut, nerve centre and two pairs of flame cells. Germ balls are usually located in the posterior portion of the body.

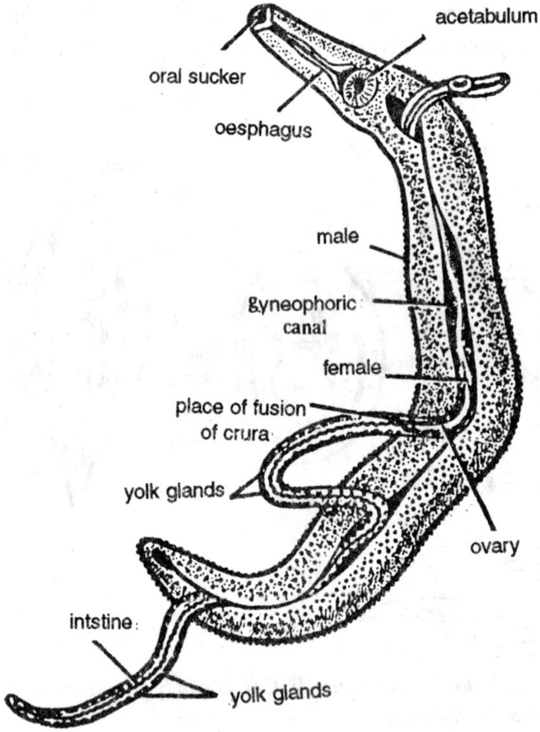

Fig. 3.23. *Schistoma.* Male and Female.

When the eggs come in contact with water, at 28°C the enclosed miracidia break out the shells by cracking them. The free-swimming miracidia finds and penetrates a pulmonate snail and transforms into mother sporocyst which is located in the foot region of the host. Free swimming miracidium survive for upto 24 hours. It dies if not find its intermediate host within this single day of free life.

Sporocyst : In *S. mansoni* the sporocyst grows 1 mm in about 14 days. The sporocysts start producing daughter sporocysts which burst free of the mother sporocyst. These

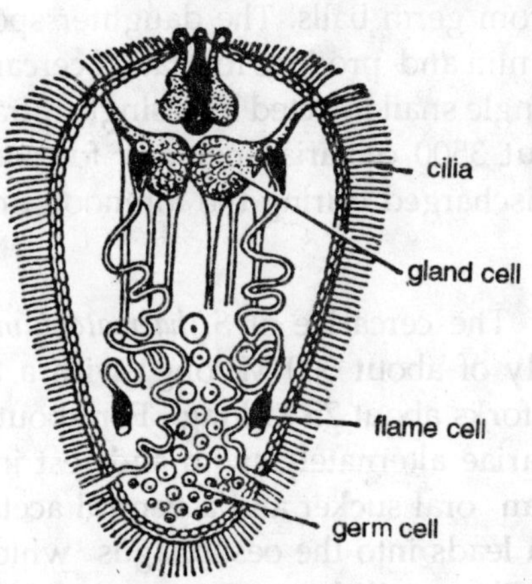

Fig. 3.24. *Schistosoma.* Miracidium.

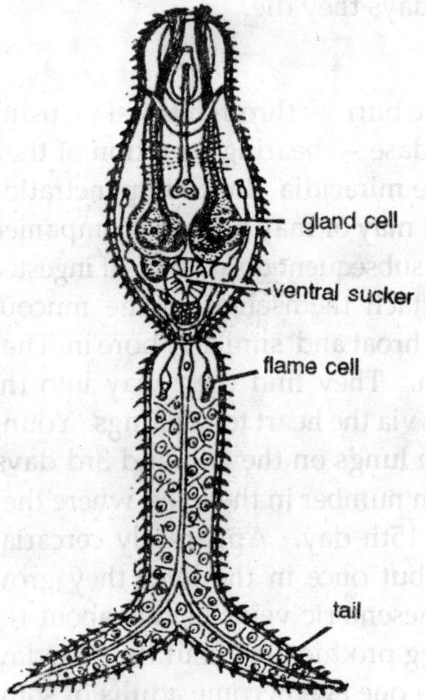

Fig. 3.25. *Schistosoma.* Cerceria.

are differentiated from germ balls. The daughter sporocysts reach a length of 1.5 mm and produce forktailed cercaria from the germ balls. A single snail infected by a single miracidium may discharge about 3500 cercariae per day for sometime. The cercariae are discharged during the forenoon on sunny days.

Cercaria larva : The cercariae of *S. haematobium* and *s. mansoni* have a body of about 300 M long with a tailstem slightly longer and forks about 75 M long. For about two or three days the cercariae alternately swim and rest in water. Cercaria possesses an oral sucker and a ventral acetabulum are present. Mouth leads into the oesophagus which leads into a trifid intestine. Pharynx is absent. Five pairs of penetration glands are present - 2 pairs of anterior oxyphilic glands and 3 pairs of posterior basophilic glands. If the cercariae fail to reach final host within two days they die.

Infection to the Final Host

If successful the cercariae burrow through the skin, using the histolytic and hyaluronidase — bearing secretion of their penetration glands just as the miracidia do. Skin penetration requires several minutes and may or may not be accompanied by a prickling sensation and subsequent dermatitis. If ingested with water the cercariae attach themselves to the mucous membranes of the mouth or throat and similarly bore in. They leave their tail behind them. They find their way into the blood system and are carried via the heart to the lungs. Young *S. mansoni* accumulate in the lungs on the 2nd and 3rd days. By the 6th day they appear in number in the liver, where they are well established by the 15th day. Apparently cercariae feed only on portal blood, but once in the liver they grow rapidly. Migration to the mesenteric veins begins about the 23rd day, and mating and egg production about the 40th day. All the larval produced from one egg become adults of some sex. No female worms mature in a host, if the male worms are absent.

REPRODUCTION & DEVELOPMENT OF CHINESE LIVER FLUKE

Reproductive System

Opisthorchis is hermaphrodite trematode possesses well-defined reproductive organs and a common genital pore.

Male Reproductive System

Testes are posterior, branched and lying one behind the other. From the main body of each testis arises a thin duct the *vas efferens.* This duct proceeds forward to a point slightly anterior to the ovary where it joins its mate from the other testis to form a vas deferens vas deferens enlarges into a sperm-filled seminal vesicle. A narrow ejaculatory duct connects the vesicle to the genital atrium which opens at the ventral genital pore. A cirrus sac, prostate and cirrus all are absent.

Female Reproductive System

A single branched ovary is present in the median line slightly anterior to an oblong sperm-filled seminal recepticle. The *seminal recepticle* lies just ahead of the anterior testis. Anteriorly the seminal recepticle gives off a short duct that joins Laurer's canal and which terminally joins a short *oviduct. Laurer's canal* curves posteriorly ground the seminal recepticle and also ruins abruptly dorsally to open by a small aperture, the *opening of laurer's canal.* This opening lies between the anterior testis and seminal recepticle near the middle. The oviduct, after receiving the seminal recepticle, proceeds forward a short distance and receives a *common vitelline duct.* Vitelline glands are small and numerous located in the middle third of the body. Two lateral vitelline duct arises from the middle of transverse vitelline duct. The oviduct receiving all these ducts (i.e., median vitelline duct, Laurer's canal and seminal recepticle) then penetrates an *ootype* which is surrounded by *Mehli's* gland. The Mehlis' glands are loosely organized and composed of small acinus shaped cells. A very conspicuous uterus arises from the ootye anteriorly and proceeds an on irregularly coiled duct to the genital atrium. The uterus becomes larger in diameter anteriorly because of the massing of eggs.

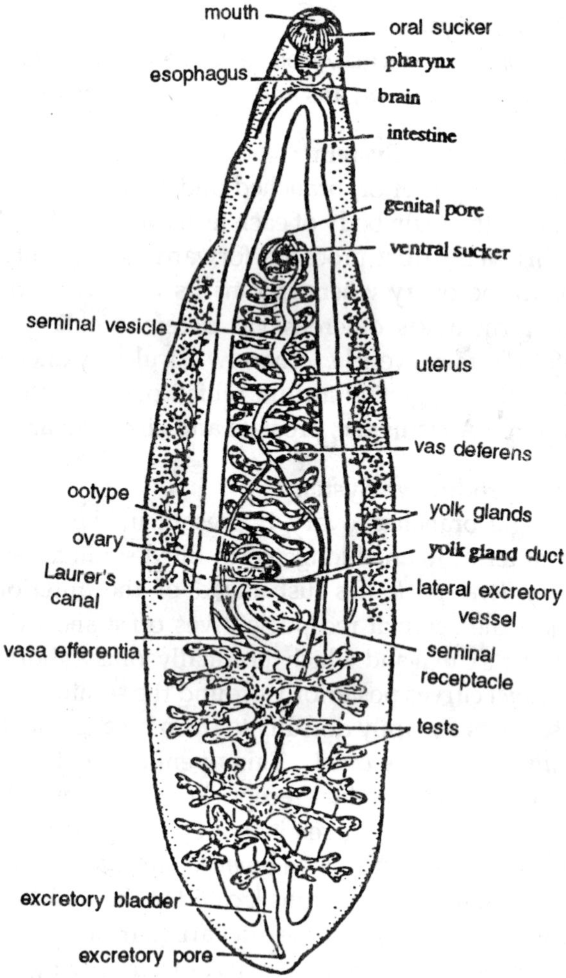

Fig. 3.26. *Opisthorchis sinensis.*

Fertilization

Mutual cross - fertilization occurs after copulation between two flukes. The ova liberated from ovary are fertilized by the sperms from the seminal recepticle. In the ootype the fertilized egg yolk are enclosed in a thin walled, oval, operculate shall which is often provided with a comma-shaped posterior appendage. The egg gradually acquires a darker brown colour and proceeds with cleavage until a fully formed miracidium is formed. The embryonated capsules, pass to out side through the genital pore and are carried by the bile to the host's intestine, they are finally eleminated with the faces.

Life-history

Man is the primary host besides snail and fish are two intermediate hosts.

Miracidium and Sporocyst

As stated above the eggs when laid contain fully mature miracidia. Each ovoid miracidium bears long, motile cilia on an outer ectodermal sheath. The apical region bearing no cilia. In the apical region there is no apical spine and an apical gland opening. On the dorsal side there is a single excretory pore. An apical histolytic gland and a large, ventral, sausage-shaped secretory gland with a narrow duct opening besides the spine. About 25 germ balls are present, that fills the miracidial body space. Two dorsally placed flame bulbs and their ducts comprise the excretory system. Nervous system is situated ventral to the apical gland.

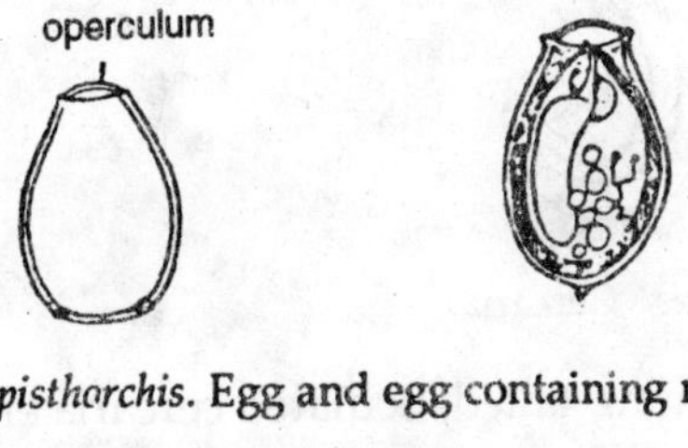

Fig. 3.27. *Opisthorchis.* Egg and egg containing miracidium.

Miracidium do not hatch free in nature, rather they are released from the eggs only after ingestion by appropriate snail host. The miracidium than actively penetrates the gut wall into the peri-intestinal blood sinus. Now it becomes a thin walled sporocyst. Sporocyst produces the redial generation which migrates to the hipatic blood sinuses.

Redia : A mature redia is elongated in appearance. Its anterior end is rounded whereas posterior end is some what pointed. The larva internally lined with sub-epithelium. A birth pore is absent. An anterior pharynx and a brownish sacculate gut extending about half-way posteriorly are present. Several germ balls are present which develop synchronously into cercariae. About 6-8 cercariae are liberated by the ruptuse of redial wall.

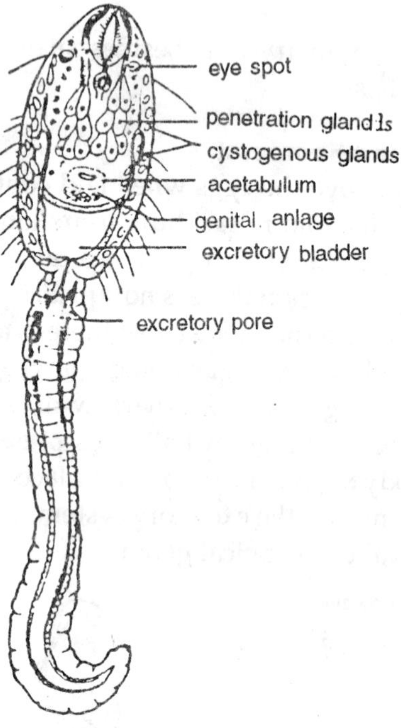

Fig. 3.28. *Opisthorchis.* Cercaria.

Cercaria : The long-tailed, oculated cercaria emerges from the snail host into the water and swims vigorously in the search of another host i.e., a fish. The tail is provided with muscle-fibres and bears thin, lateral, transparent elytra. Two bean-shaped eye-spots are located near the anterior end of body. An oral-sucker, pharynx and acetabulum are present but oesophagus and gut caeca are not formed yet. Seven pairs of unicellular penetration glands occupy. The middle of the body. Their ducts open near the mouth. Dorsal and ventral peripheral mesenchymal cells are packed with cystogenous glands. These glands help in the formation of cyst of metacercaria. Excretory system comprises a triangular excretory bladder with anterio-lateral collecting ducts and a bifurcated posterior caudal excretory tubule.

Metacercaria : On coming into contact with the fish, member of Cyprinidae family, the cercaria attached to the surface, discards its tail, and creeps under a scale into the fish. A cyst wall is formed by the material secreted by cystogenous glands. The cyst is thin, tough and spherical. Organ formation continues e.g., the gut caeca etc.

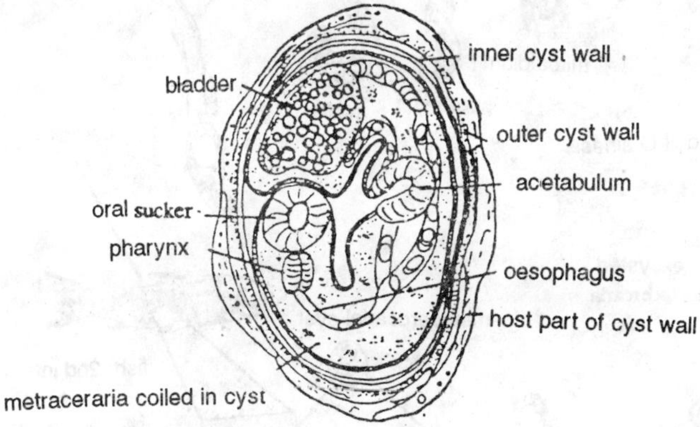

Fig. 3.29. *Opisthorchis.* Metacercaria.

Inspection to the primary host : Primary host i.e., man ingests uncooked fish containing metacercaria. Metacercarial cyst walls dissolve in the duodenum or intestine and juvenile flukes migrate up the common bile-duct into the small bile-ducts of liver. Egg laying begins 3-4 weeks after infection.

Effects on the Host

Symptoms are classified as mild, modrate or severe, depending upon the burden of worm, length of infection and health of host. Light infections may be asymptomatic. Heavy infections are characterized by chills and fever, followed by liver enlargement and tenderness, sometimes with splenomegaly (Spleen enlargement). Chronic heavy infections cause cholecystitis (inflammation of gall bladder) and hepatitis (inflammation of liver). Modrate infection causes diarrhoea, edema and some hepatomegaly (enlargement of liver). The disease is termed *opisthorchiasis* or *clonorchiasis*.

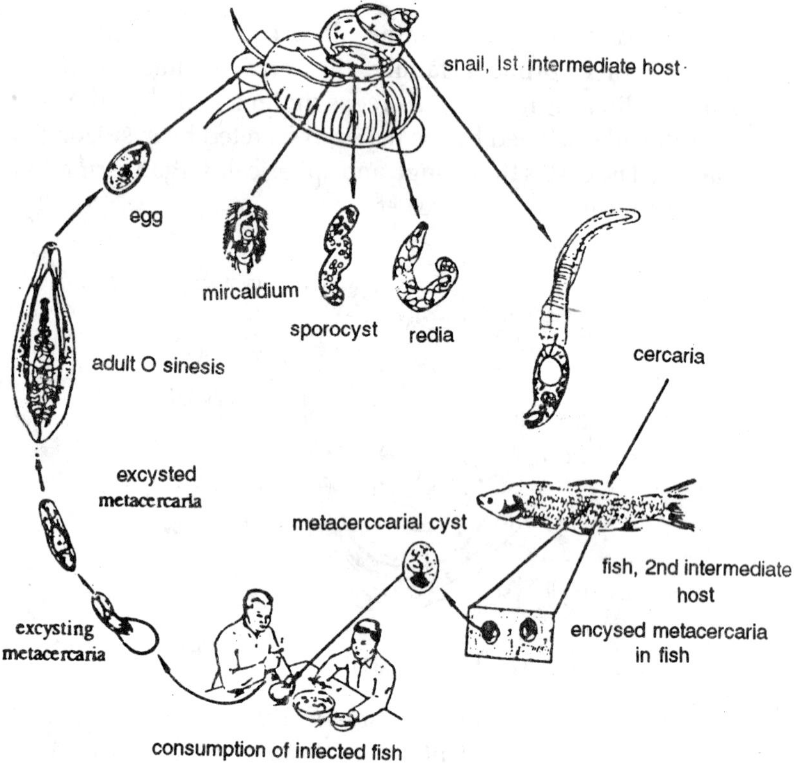

Fig. 3.30. *Opisthorchis.* Life cycle.

REPRODUCTION & DEVELOPMENT OF TAENIA SOLIUM

Reproductive System

Taemia is a *hermaphrodite.* Reproductive organs are usually found in the proglottids after 200th partition. Some anterior proglottids are functionally male. Middle proglottids are both made as well as female. Proglottids at the posterior end are filled with enlarged *uterus*.

Structure of the proglottids in the middle region is as follows:

I. The male *reproductive system* consist of the following organs:

1. Testes
2. Efferent ducts
3. Vas deferens
4. Cirrus or penis
5. Cirrus sac.

1. *Testes :* They are numerous small, rounded bodies situated in the medullary past of the parenchyma. There number is about 15-200 and are distributed throughout close to the dorsal surface of the proglottid. Some workers are of the openion that the numerous rounded follicle, in fact, constitute a single and highly sub-divided testis. However, the number of testes is of texanomic value in tapeworms.

2. *Efferent ducts* : From each testis arises a fine ductule or capillary known as vas efferens. All the *vasa-efferentia* unite to form a vas deferens.

3. *Vas deferens :* It is a long and tube, arising from the mid-plant of the proglottid. It runs transversely to the left or right to open into cirrus. In some cestodes the vas deferens enlarge to form the seminal vesicle for the storage of sperms. But in *Taenia* no such enlargement is seen.

4. *Cirrus or penis*. The distal end of vas deferens in modified into a protusible *cirrus*. The cirrus is armed with bristles, hooks or spines. It is eversible through a male genital pore into a cup like *genital atrium* of the genital papilla.

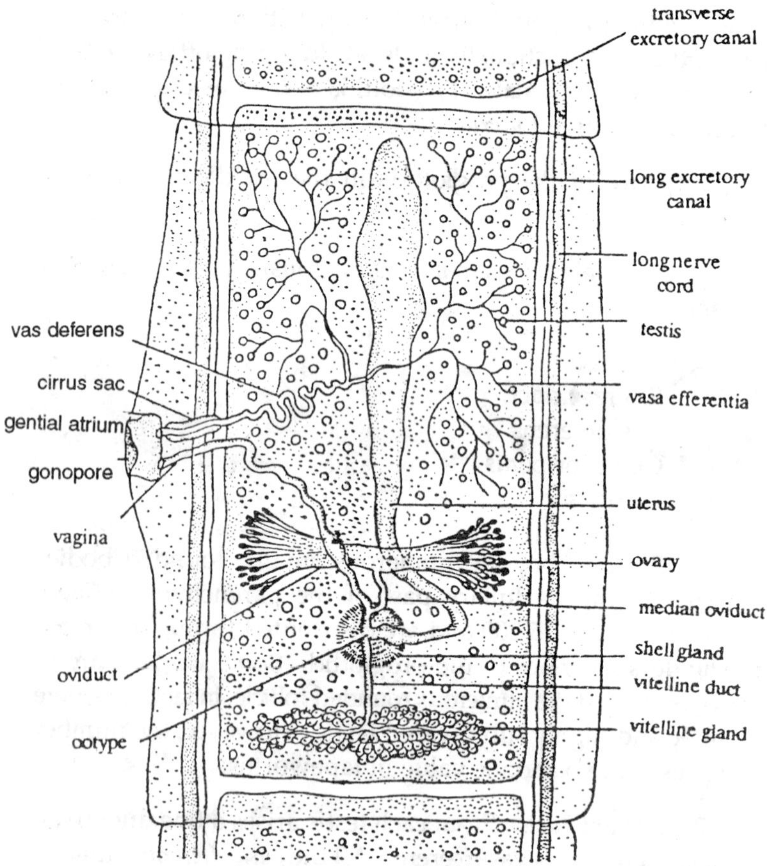

Fig. 3.31. *Taenia solium.* Mature proglottid showing reproductive system.

5. *Cirrus sac:* Cirrus is surrounded by a kind of muscular sac known as cirrus sac. Its function is to erect penis during copulation.

II. The female reproductive system comprises:

1. Ovary
2. Oviduct
3. Ootype
4. Uterus

5. Vagina
6. Vitelline gland
7. Mehli's gland

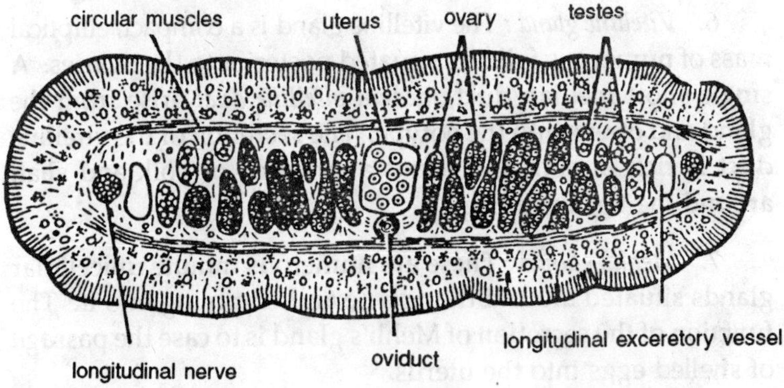

Fig. 3.32. *Taenia solium.* Diagrammatic T.S. of a mature proglottid.

1. *Ovary :* A simple bilobed ovary lies ventrally in the posterior past of the proglottid. Each lobe of ovary is dorsoventrally flattened and consists of a number of radially-arranged follicles. The right and left lobes are connected medially by a transverse bridge, the *ovarian isthmus.*

2. *Oviduct :* The oviduct arises from the middle of the bridge. It is a short but wide duct and joins the ootype distally.

3. *Ootype :* It is a small rounded chamber developed at the junction of oviduct with the vitelline duct. It is surrounded by numerous unicellular *Mehlis glands,* the cells of which form an amorphous complex.

4. *Uterus :* A blind sac-like or tube-like uterus arises from the ootype and runs forward in the segment. In gravid proglottids the uterus enlarges and gets branched to oocupy the whole of the proglottid. It remains filled with the fertilized ova or developing embryos.

5. *Vagina :* The narrow tubular vagina arises from the receptaculum seminis and runs obliquely outwards to open into the genital chamber by female reproductive opening.

The seminal receptacle or receptaculum seminis is a small wide tube connected on one hand with the ootype by a narrow *spermatic duct* and on the other with vagina.

6. *Vitelline gland :* The vitelline gland is a compact, elliptical mass of numerous follicles situated posterior to the ovaries. A single short median duct, *median vitelline duct* arise from the gland and opens into the ootype. The secretion of the vitelline duct is rich in yolk and forms a covering of yolk and shall around the fertilized egg.

7. *Mehlis glands :* These are numerous minute unicellular glands situated around the ootype and emptying into it. The function of the secretion of Mehli's gland is to case the passage of shelled eggs into the uterus.

Life-Cycle

Fertilization : Fertilization occurs by introducing the cirrus into the vagina of the same proglottid and sperms are stored in the receptaculum seminis, which later on fuse with eggs in uterus. Thus in *Taenia* self-fertilization takes place. Sometimes, the anterior mature proglottids, devoid of female genitalia, can enter into copulation only with the posterior mature proglottids with fully developed female genital organs. Fertilization, following copulation between two different proglottids, is sometimes termed *cross-fertilization* to distinguish it from that occurring between gametes of the same proglottid (self-fertilization).

Thus the eggs are fertilized in the oviduct and get surrounded with yolk and egg-shall in the ootype. These materials are provided by vitelline gland. The capsulated eggs enter the uterus and are collected there. The uterus enlarges in size, gets branched and occupies the whole space. The eggs are small measuring 40 μ (micron) in diameter.

Cleavage : The zygote or egg cell undergoes cleavage when the capsule is in the uterus. The first unequal division results in a larger megamere and a smaller *embryonic cell.* The megamere divides further and froms several similar megameres, while the embyoonic cell divides repeatedly producing two types of embryonic cells, larger mesomeres and smaller micromeres. Thus, three types of cells result from the zygote; small micromeres, medium mesomeres and large megameres.

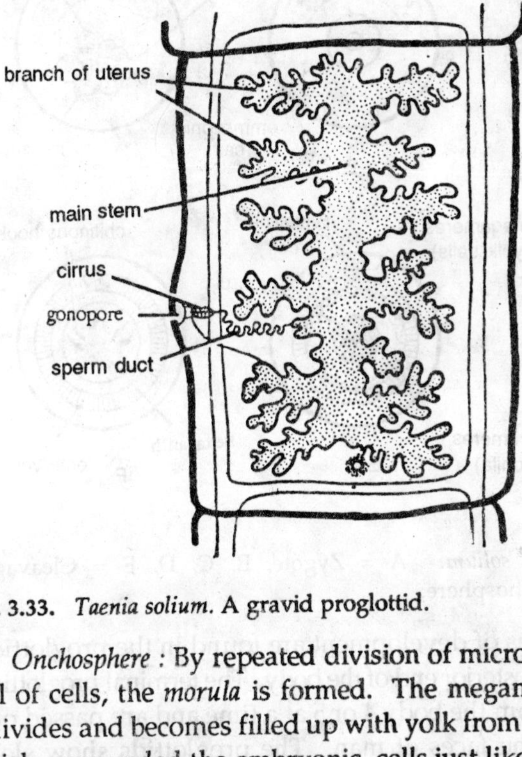

Fig. 3.33. *Taenia solium.* A gravid proglottid.

Onchosphere : By repeated division of micromeres a solid ball of cells, the *morula* is formed. The megameres divides, redivides and becomes filled up with yolk from the yolk cells, which surrounded the embryonic cells just like an envelope. It provides nourishment to the embryonic cells and finally disentigrates. The mesomeres form the inner thick striated membrane known as *embryophore.* Beneath embryophore lies basement membrane. The, morula, develops three pairs of chitinous hooks. These *hooks* are secreted by the differentiated cells known as *onchoblasts.* Now the six-hooked embryo is called *hexacanth.* It possesses a pair of large penetration glands and is surrounded by two membranes, the hexacanth membrances. The hexacanth, together with all the membrances surrounding it is known as *Onchosphere.* The onchosphere loses the shell.

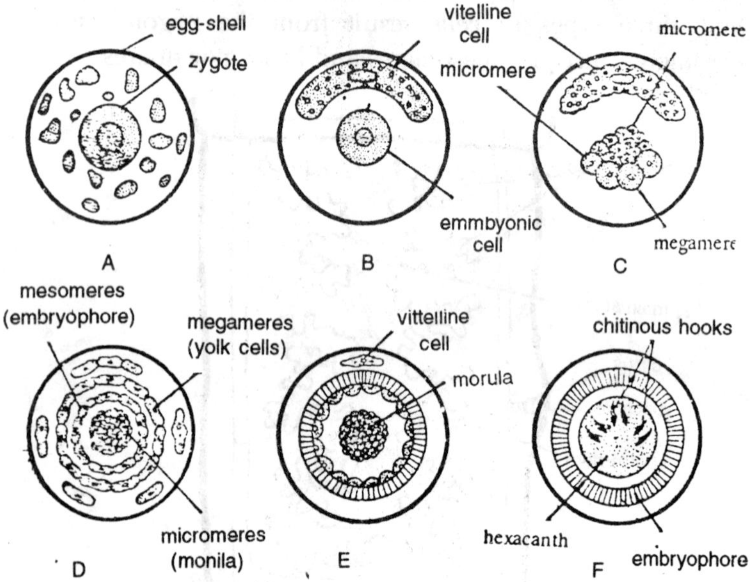

Fig. 3.34. *Taenia solium.* A = Zygote. B, C, D, E = Cleavage. F=Onchosphere.

These stages of development are found in the proglottids at the extreme posterior end of the body. The terminal proglottids are liberated from the body 4 or 5 at a time and are passed out quietly with the faces of man. The proglottids show slow movements of contraction and later on perish. *Embryos* remain active for a considerable time. They are capable of further development only if they are taken in by a secondary host, which is a pig.

Infeciion to secondary host. The pig is the secondary host which regularly feeds of the faces of man. Dogs, monkeys and sheeps are also known to get infection.

Inside the alimentary canal of a pig, proglottids and

thousands of egg's shells, nearly 30,000 to 40,000, are digested and *hexacanths* are set free. Later on, hexacanths bore their way with the help of hooks and penetration glands, through the wall of the intestine and reach intestinal blood vessels. The process requires only 10 minutes.

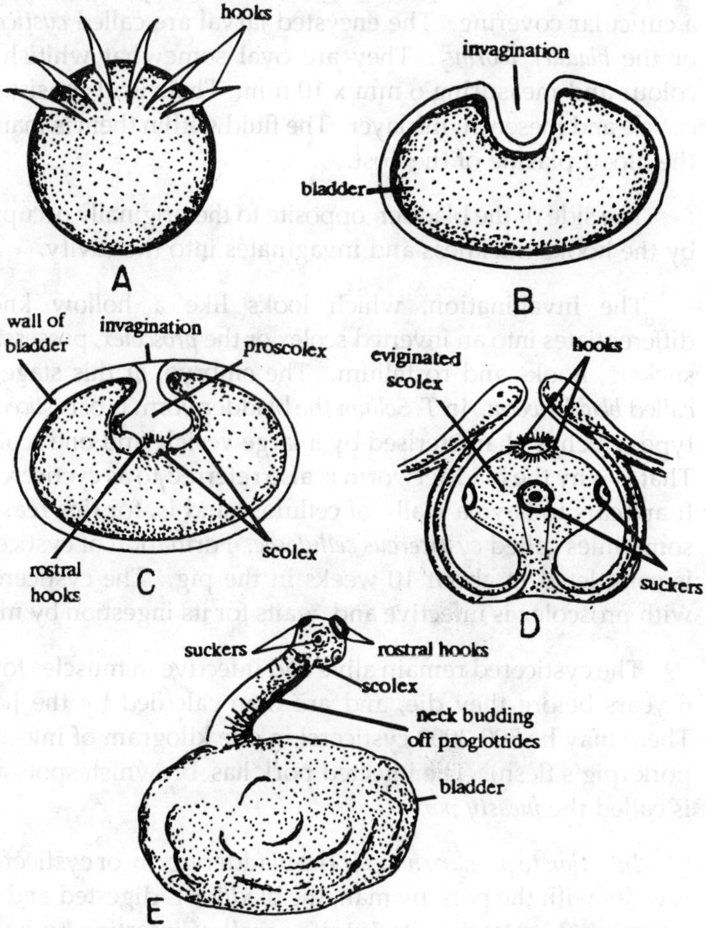

Fig. 3.35. *Taenia solium*. A = hexaconth, B = bladder worm with invagination, C = bladder with proscolex, D = bladder worm with evaginated scolex, E = Cysticercus.

Migration : Travelling through the heart, these enter the

muscles of a various parts in the body. The usual site where the hexacanths get encysted arc the voluntary muscles of tongue, heart, liver and shoulder.

In the muscles, hexacanths lose their hooks, grow in size and develop a fluid-filled cavity in them. They now encyst in a cuticular covering. The encysted larval are called *cysticerci* or the *bladder worms.* They are oval somewhat whitish in colour and measuring 6 mm x 10 mm. Their wall consists of cuticle and mesenchyme layer. The fluid within them is mainly the blood plasme of the host.

The side of the bladder opposite to the originally occupied by the hooks thickness and invaginates into the cavity.

The invagination, which looks like a hollow knob, differentiates into an inverted scolex or the *proscolex,* possession suckers, hooks and rostellum. The embryo, at this stage, is called *bladderworm.* In *T. Solium* the bladderworm is of *cysticercus* type which is characterised by a large vesicle and one scolex. That is why the bladderworm is also referred to as cycticercus. It appears to have a wall of cellulose and is, for this reason, sometimes called *cysticercus cellulosae.* Formation of cysticerci is completed in about 10 weeks in the pig. The cysticercus with proscolex is infective and waits for its ingestion by man.

The cysticered remain alive and infective in muscles for 5-6 years before they die, and are than calcified by the host. There may be 500-3000 cysticerci in one kilogram of infective pork (pig's flesh). The infected pork has brownish spots and is called the *measly pork.*

Infection to primary host : If the bladderworm or cysticercus is eaten with the pork by man, the bladder is digested and the young scolex attaches itself to the wall of intestine by hooks and suckers. A region of proliferation of neck is formed, which develops proglottids of the adult tapeworm. While anchoring itself to the wall of small intestine in matures in two or three months.

Pathogenesis

Both aduts as well as cysticercus stages are harmful:

1. *Adult worms :* The scolex of adult worm remains embedded in the mucous lining of alimentary canal for the attachment. The injury of musoca may lead to bacterial infection. There may be abdominal discomfort, diarrhoca, frequent hunger pangs, nervousness, insomnia, nausea nad epileptic etc. These symptons are due to the presence of adult worm are collectively known as *Taeniasis.*

2. *Cysticercus larvae :* Sometimes, man may accidentally ingest eggs or proglottids with contaminated food. Serious effects may result from cycticercosis in humans. Although they may be found in almost any organ, cysticerci are commonly located in striated muscles, liver, eye and the central nervous system, where they most often are in the meninges and ventricles of brain (*Marquez-Monter,* 1971). In an aberrant from of *T. solium*, apparently found only in the central nervous system, a cysticercus has daughter cysticerci budding from the bladder wall and forming an invasive, proliferating parasite. This type of metacestode is referred to as a *racemose cysticercus.* No viable scolices have been found in recemose cysticerci (*Beaver* and *Roton,* 1981).

Living cysticerci usually elicit little inflammatory response. They damage the host by forming space - occupying leisons, that is, their growth puts pressure on the adjacent tissue and interfers with its function or causes necrosis due to reduced blood circulation. In the brain ventricle the organisms cause *hydrocephalus* because the cerebrospinal fluid cannot be properly severe inflammatory reactions.

REPRODUCTION & DEVELOPMENT OF ECHINOCOCCUS

Reproductive System

Echinococcus is haermaphrodite : The male and female

reproductive organs are present in the second proglottid or mature proglottid.

Male Reproductive System : In the middle region of proglottid, encircling the other reproductive organs, are about 20-30 spherical structure known as *testes.* From each testis a fine duct arises known as *vas efferens.* In the postmedial portion of the proglottid the vasa efferentia join to the base of a vas deferens. Vas deferens is an extremely convoluted, sperm filled tube, which passes lateral to an enlarged, coiled, marginal *cirrus* or *penis.* The cirrus lies in a bulbous *cirrus pouch* or sac. The cirrus, during copulation is protruded through a lateral *genital atrium* and *genital pore.*

Female Reproductive System : Opening into the genital atrium closely behind the cirrus is a female copulatory tube, the *vagina.* The vaginal opening is guarded by a muscular sphincter. Beyond the sphincter, the vagina passes downwards and enlarges at its inner end to form a sperm filled sac, the *seminal recepticle.* On the posterior side of the proglottid there is a bilobed *ovary* with a narrow, central *isthmus* from which an oviduct arises posteriorly. The oviduct joints a duct from the seminal receptical, and this combined fertilization duct continuous posteriorly into a median compact *ootype.* The ootype is surrounded by unicellular *Mehlis' gland.* In the posterior most region there lies a bilobed *vitelline gland.* The two lobes of vitelline glands are oriented dorsoventrally from the vitelline gland a median *vitelline duct* arises that joins the ootype. From the ootype a tube-like uterus arises which extends to a point near the anterior border of the proglottid.

Eggs : Eggs, resembling those of taenoid, are monoperculate and measuring about 30x38 u. These are ectolecithal i.e., the zygote is surrounded by the yolk cells provided by vitelline gland and shall material. After an egg shell is laid down the eggs pass from the ootype into the uterus. When the uterus of a gravid proglottid bursts, the eggs are liberated and passed out along with faeces of dogs. These eggs gain enterance into the intermediate host (man or other mammal) either through water or in forage. Each egg contains a fully developed *onchosphere.* Human infection develops from too intimate an association with dogs. Infection of children is commonly established when dogs are permitted to lick their months. Eggs can be ingested from contaminated fingers.

The eggs, upon being swallowed by a mammalian intermediate host, hatches in duodenum as a typical six-hooked onchosphere. The matching occurs in slight alkaline medium. Activation of embryos take place in either human bile or intestinal juice of ruminants, they are not activated in bile or intestinal juice of dogs and cats. The escaping onchospheres penetrate the intestinal wall, enter the mesenteric venules, and become lodged in the capillary beds of various organs. The hooks are now shed. After three hours of egestion, the embryo reaches its final place of development. Among moose, deer etc., they reach lungs where as in man and other mammals, they reach in the liver, lungs, brain, kidneys, spleen etc., but liver being the most common site. Implantation and growth occur here, Onchosphere now develops into *hydatid cyst.* The cyst formation is a slow process, with the host laying down as envelope of connective tissue around the parasite. In about 4-5 months, the cyst reaches approximately 10 mm, in diameter. It goes on increasing in size and eventually becomes as big as the orange or even sometimes a football.

The various types of larval cysts, formed in the life history of *Echinococcus are:*

1. *Unilocular hydatid cyst :* The hydatid cyst is composed of three layers. The inner most layer is called *endocyst.* It is thin syncytium forming the germinal layer. The middle layer is called *ectocyst.* It is made up of cuticle which is amorphous and laminated. The outer layer is *pericyst.* It is composed of connective tissue and fibres secreted by the host.

The cyst produces daughter cysts upon invagination of its wall. These daughter cysts project into the lumen of mother cyst. The cavities are filled with a sterile fluid known as *hydatid fluid*. Within some daughter cysts, tertiary or grand-daughter cysts, form by invagination. These are endogenous or exogenous cysts or brood capsules.

The endogenous cysts fall free into the hydatid fluid of mother cyst white the exogenous cysts, after detaching, migrate to other part of body, where they continue to grow further. The daughter cysts have a fibrous cyst wall internally lined with endocyst.

Rounded and hollow brood capsules are also continuously budded off from the endocyst of mother cyst. These capsules remain attach for sometimes by means of a narrow pedicle.

From the wall of these brood capsules, minute scolices are formed. Scolex formation may occur from the invagination of walls either of daughter or grand-daughter cyst.

The anterior end is inturned to form a central canal into which the four suckers open and into the expanded base of which the hooklets project. The body of each scolex is filled mesenchymal calls aggregated to form a subcuticular layer and to form a compact cup-shaped mass below the hooklets. The syncytium of the endocyst retains its embryonic potentialities, and new brood capsules may be produced indefinitely. Hydatids thus may expand to enormous size are contains nearly 50 litres of fluid as well as thousands of scoleces. The pressure so produce may severely damage adjacent host organs.

When the infective brood capsule is ingested by the final host, each scolex quickly evaginate and attach to the duodenal mucosa. For the first four to five days in the host they undergo little growth. By the 14th day the scolex has grown in size, and first proglottid has appeared. The genital anlage and a second proglottid appear by the 18th day. Growth continues and fully developed shelled embryos are formed by about the 35th day, and the life cycle is begun again.

2. *Multilocular cyst :* It is characterized by many small bladders embedded in a common enclosing membrane. these are found in cattles. It is considered to be the result of differential reaction inside the host.

3. *Alveolar hydatid in man :* It is characterized by a malignant type of growth with jelly-filled proliferating vesicles embedded in a common dense stroma. The very thin laminated membrane does not restrict the germinal membrane that grows out into surrounding host tissues. It is commonly sterile and it undergoes degeneration and calcification.

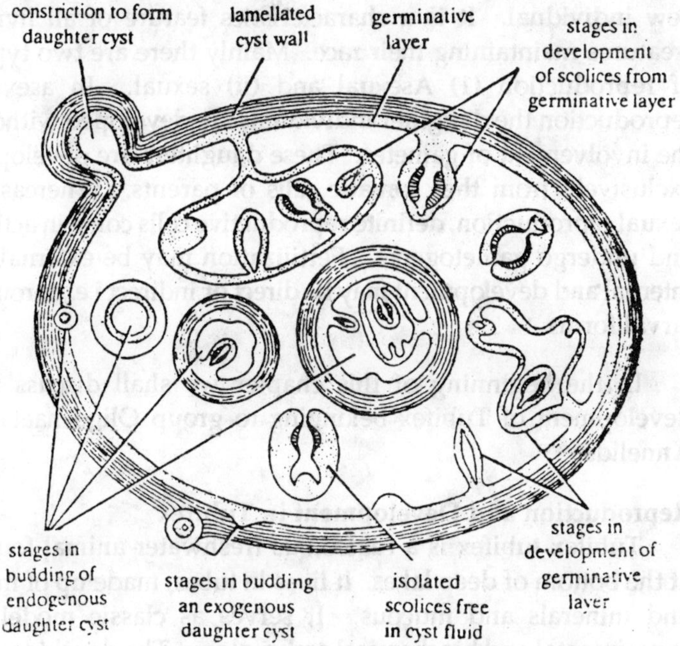

Fig. 3.36. *Echinococcus granulosus.* Hydatid cyst.

4. *Multivesicular :* It is characterized by many adjoining and connected bladders, each having its own separate germinal membrane.

4

Reproduction and Development in Some Annelids

Reproduction is the process by which organisms give rise to new individual. It is a characteristics feature of all living creatures maintaining their race. Mainly there are two types of reproduction (1) Asexual and (ii) sexual. In asexual reproduction the daughter individuals are developed without the involvement of gametes. These daughters are developed exclusively from the somatic cells of parents. Whereas in sexual reproduction, definite reproductive cells come in action and undergo gametogenes. Fertilization may be external or internal and development may be direct or indirect i.e., through larval forms.

In the beginning of this chapter we shall discuss the development of Tubifex belonging to group Oligochaeta of Annelids.

Reproduction and Development in Tubifex

Tubifex tubifex is a tubicolous freshwater animal found at the bottom of deep lakes. It lives in tubes, made up of mud and minerals and mucous. It serves as classic model in experimental and biochemical embryology. The development of this worm is an impressive example of ooplasmic segregation which occurs prior to the first cleavage as a formation of pole plasms rich in mitochondria and endoplasmic reticulum. Owing to the strictly determined heteroquadrant spiral cleavage, the material of these plasms is distributed over definite individual blastomeres. Therefore *Tubifex* has for a long time been used for studying the problem of the early localization of anlage, using methods such as centrifugation, blastomere isolation and cytochemistry. As different blastomeres may easily be identified during early cleavage, the biochemical characteristics of certain cells or groups of cells may be studied in relation to subsequent differentiation.

The simplicity with which worms can be kept under laboratory conditions, plus the opportunity of obtaining eggs virtually all year round, add to the advantages of *Tubifex* as a laboratory experimental model. Furthermore, *Tubifex* eggs may be cultured throughout development outside the cocoon in a simple salt solution. The egg membranes of *Tubifex* are permeable to a wide range of substances, including, specifically, mitochondrial and mitotic poisons. This provides an opportunity for experimental studies of mitosis. Early *Tubifex* embryos have been used for studying the effect of magnetic and electromagnetic fields on the chromosomal apparatus and chronology of mitoses. Methods of studying the structure of *Tubifex* eggs using the transmission and scanning electron microscope have been developed, as well as methods for isolating meiotic spindles *Tubifex* has been extensively used by Japanese scientists to study the function of microfilaments in oocyte maturation, ooplasmic segregation, and early cleavage. Cytoskeleton poisons, ionophoresis, and protein synthesis inhibitors are employed as tools in these studies. Along with methods enabling nucleate and non-nucleate egg fragments to be obtained, these studies open up the way to the elucidation of mechanisms underlying the intrinsic asynchrony of the cleavage divisions characteristic of *Tubifex*. For example, evidence has been collected for the existence of the rhythmic pattern of actin polymerization/deploymerization in the egg cortex; this pattern does not depend on the nucleus and determines the rhythm of oocyte maturation and cleavage divisions. Although no systematic studies of macromolecular synthesis in *Tubifex* development are known, it should be mentioned that techniques have been developed which allow protein synthesis on semithin sections to be studied using autoradiography.

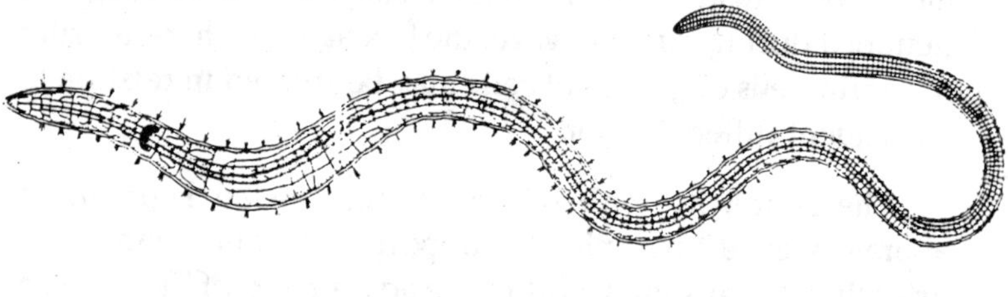

Fig. 4.1. General view of *Tubifex tubifex.*

Adult worms may be used for studying regeneration. Since sludgeworms are generally resistant to oxygen deficit, the biochemistry of their energy metabolism and the physiology of their respiration provide another developing area of research.

The drawbacks of *Tubifex* as an experimental animal include the small number of eggs in the cocoon, the not-very-high synchrony of their development, the rather high proportion of spontaneous developmental abnormalities, and the fact that they can be easily infected by fungi and other pathogens.

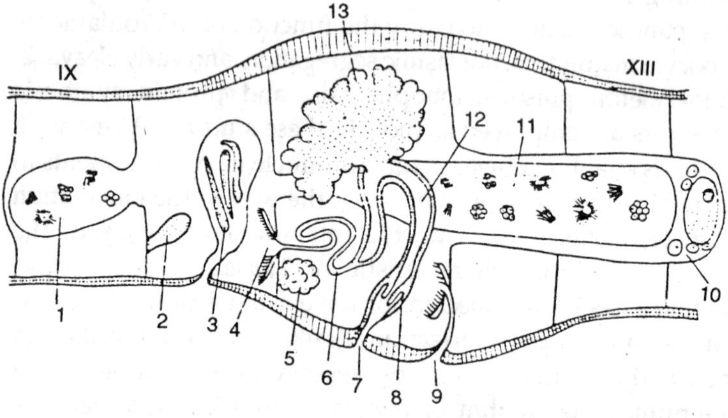

Fig. 4.2. Diagram of the reproductive system in *Tubifex tubifex.* 1) Anterior seminal sac; 2) testis; 3) spermatheca containing spermatozeugmata; 4) seminal funnel; 5) ovary; 6) seminal

duct; 7) male genital pore; 8) penis in the penial sac; 9) female genital pore; 10) egg sac; 11) posterior seminal sac; 12) atrium; 13) prostate. Roman numerals refer to segment numbers.

Taxonomy and Distribution

The genus *Tubifex* includes several species living of which *T. tubifex Mull. (= T. rivulorum Lam.)* is the most common and can be found almost worldwide. It is impossible to identify freshwater oligochaetes precisely without a detailed examination of the reproductive system and bristles. *T. tubifex* almost invariably comprises the only species of this genus found in material commercially sold in zoological shops as fish food. The salient features for the identification of *T. tubifex* are given below. Penial bristles are absent. The penis is not included in the chitin tube, and the distal notch of the ventral bristles is no shorter than the proximal one. Long dorsal hairlike bristles are present, the length of these not exceeding twice the body width. Live worms have a light-red pigmentation of various shades ranging from yellowish to brownish. Their length is 20-100 mm and the number of segments varies from 36to 120. Clitellum is found in segments 11 and 12th.

Another species, *Tubifex hattai* Nomura, is used in embryological studies in Japan. The early development of this species is better known in many respects than that of *T. tubifex*.

The taxonomic position of *Tubifex* is as under:

Phylum	=	Annelids
Class	=	Oligochaeta
Order	=	Archi oligochaeta
Genus	=	Tubifex

Ecology and Culture

Tubifex occurs in predominantly freshwater environments such as slow rivers, streams, ditches, and stagnant reservoirs. It prefers shallow places with slimy, sometimes sandy beds.

The species is particularly abundant in polluted reservoirs. It is resistant to various chemicals and is tolerant of low or fluctuating oxygen levels. *Tubifex* feeds by passing mud through the intestines. No specialized respiratory organs are present; respiration takes place through the skin and is helped by rhythmic movements of the body's caudal section. The head region and approximately half the body are submerged in the river bed; particles of the substratum together with mucous secretions from the skin from a tube around the worm.

Worms obtained commercially or caught using a metal sieve can, if necessary, be kept wet for several days in a cool place. The thickness of the worm layer should not exceed 1 cm and the worms should be rinsed once a day. If the layer of worms is thicker, the animals are rapidly poisoned by the products of their own metabolism. If a constant culture is desired, the worms should be washed several times, and the longest specimens, possessing a marked spindle-like thickening in the anterior third of the body (where gametes are formed), should be selected; the large oocytes can often be seen through the transparent wall of the body.

River mud passed through a mesh, or alternatively a mixture of mud and sand, should be boiled with water for an hour and then placed into medium-sized dishes (15 to 20 cm diameter); the thickness of the mud layer should be 3-4 mm, with water carefully added until its level is 3-4 cm above the surface of the mud. About 100-200 worms are placed into each dish, which is then loosely covered by glass. If the substrate consists only of mud, no additional feeding is required for 1-2 months; after this period the mud should be changed.

Japanese scientists cultivate the worms on a 1.5 mm-thick layer of sterilized river sand. To feed the animals, a small quantity of wet yeast is added every 3 days. There is no need to aerate the water, but Inase believes that aeration aids reproduction. The water may be changed once a week or at longer intervals. Dead worms should be removed. A film of bacterial growth on the water surface is one indication that the water should be changed. The worms are best kept at 18-20°C; higher temperatures are less desirable than lower ones. Oviposition usually begins within several days of the worms being placed into cultivation vessels.

Experiments aimed at elucidating the optimal culture conditions for *T. tubifex* have been performed. It is recommended that the animals be kept at a temperature below 24°C and at a population density no greater than 1500 animals per square meter; they can be fed on lettuce.

Reproduction

Reproduction in *T. tubifex* takes place during most of the year, although recent data indicate that reproduction is interrupted during the summer months (June to September). Oogenesis and spermatogenesis begin before the end of September, while oviposition takes place between mid-January and May.

The structure of the reproductive system, oogenesis, and the formation of the cocoon of the Japanese species have been described in great detail, and there is a detailed account of the structure of *T. tubifex* spermathecae. These worms are hermaphrodite, but self-fertilization is avoided in the course of copulation as clusters of spermatozoa are deposited into spermathecae of opposite partners. Oogenesis of the nutrimentary type. During the period of sexual reproduction the worms possess a girdle or "saddle" — a modified region of the skin epithelium which secretes material necessary for forming the cocoon envelope. A protein containing liquid is secreted under the envelope and the cocoon is gradually pushed toward the head region. When it passes over segment XI, it picks up oocytes; the spermatozeugmata enter the cocoon as it passes over segment X. Finally, the cocoon passes over the head and is shed. Mating and oviposition may be separated by a considerable time interval.

It has been established that *Tubifex* is capable of parthenogenetic reproduction, which takes place under conditions of arrested spermatogenesis.

Under the above culture conditions, the whitish cocoons are deposited directly onto the mud surface where they can be easily detected. When the worms are cultured on sand the collection of cocoons is conveniently performed every 3-4 h. The likelihood of quick oviposition may be increased if one initially selects worms with 6-8 mature eggs in the ovisacs located in the first segments immediately behind the girdle. Freshly laid cocoons may be identified by the relatively low contamination of the external membrane by mud particles.

According to our observations, a single worm can lay 4-5 cocoons in 1 month. According to Kosiorek, one worm lays about 14 cocoons on average in 100 days, and the complete life cycle of *T. tubifex* (from zygote to the first oviposition by mature worms) at 24°C lasts 50-57 days, while embryogenesis (up to hatching from the cocoon) lasts 10-12 days. The first cocoons are laid by worms at least 40 mm in length.

Gametes and Fertilization

When the eggs are laid into the cocoon at metaphase I, they are not yet fertilized. The cocoon is roughly elliptical in shape and has one tubular appendage at each end. These appendages later serve for the emergence of hatched juvenile worms. The thin elastic transparent membrane of the cocoon is covered by an opaque sticky layer of external mucus with adhering particles of mud and detritus. Between 1 and 20 eggs may be found inside the cocoon, whose length varies from 1.2 to 2 mm. The cocoon is filled with a protein-containing fluid.

The *tubifex* egg belongs to the isolecithal type. The equatorial diameter is 0.2-0.5 mm. The egg is opaque, yellowish, or rose-white, and contains a large amount of yolk. The mature eggs in the ovisacs are surrounded by a tightly apposed single-layered vitelline membrane consisting of a loose fibrillar network. The plasma membrane contains numerous microvilli. Cortical granules or their analogs cannot be detected in the egg cortex. In addition to yolk granules and lipid droplets the ooplasm contains mitochondria, ribosomes, endoplasmic reticulum, and multivesicular bodies.

Immediately after deposition the egg is somewhat irregular in shape but gradually becomes ellipsoid, the short axis coinciding with the animal-vegetal axis. A small region located at the animal pole during this stage is relatively poor in yolk. The vitelline membrane is in close contact with the egg surface. The *Tubifex* egg is protected from the environment by not only the vitelline membrane, but also by protein-containing fluid, the cocoon envelope and the external layer of mud.

The cocoon envelope is resistant to a variety of enzymes and appears to consist of scleroproteins. The envelope is permeable to inorganic ions, water, and mitochondrial poisons such as dinitrophenol and dyes; colloids of the gum arabic type cannot, however, that both the cocoon membrane and the protein-containing fluid of the cocoon protect the egg to some extent from the above substances. The vitelline membrane is also permeable to colchicine, theophyline, various quinones and narcotics, cytochalasine B, cycloheximide, ionophore A 23187, and ^{3}H-amino acids. Transfer through the membrane proceeds faster at higher temperatures.

The spermatozoa are deposited in the spermatheca during mating and enter the cocoon in the form of spermatozeugmata, long cylindrical arrow-shaped complexes of spermatozoa attaining 1.2 mm in length. After deposition the spermatozeugmata disintegrate and the spermatozoa disperse in the protein-containing fluid. It has been established that a spermatozeugma consists of an axial cylinder and a cortical layer. The spermatozoa typical of the Oligochaeta are located in the axial cylinder. Each consists of an acrosome, a long cylindrical nucleus 0.1.-0.3 μm in diameter, a middle section, and a flagellum. The cortical layer is formed by helically arranged aberrant spermatozoa glued to each other by their caudal parts (tails). The tail ends are surrounded by a layer of mucopolysaccharides and are capable of wave-like movements. The mucopolysaccharide layer appears to play a protective role.

Insemination in the Japanese species, *T. hattai,* takes place roughly 30 mm after oviposition. On spermatozoon enters the egg close to the vegetal pole where the fertilization cone is formed. This is where the vitelline membrane begins to death from the egg membrane and forms the perivitelline space; its initial width is approximately 0.25 μm, but increases to 0.5-1.0 μm 40-50 min after oviposition. The vitelline membrane acquires a three-layered structure, the surface microvilli gradually disappear, and microfilaments appear in the cortex for the first time. The cortical layer becomes 0.5 μm thick on average, although later its thickness may undergo local variations. It does not contain any membranous organelles. Golgi elements and sac-like vesicular structures appear for the first time after fertilization. The so-called meiotic deformations appear during the course of the maturation divisions (see the description of developmental stages). Cortical microfilaments are involved in these deforming movements. The microfilaments also play a part in ooplasmic segregation and they appear to be closely connected with the meiotic deformations, segregational cytoplasm movements, and the cytokinesis of cleavage divisions.

The formation of pole plasms, one of the most remarkable features of *Tubifex* early development, proceeds in two stages. Initially the cytoplasm consists predominantly of Golgi complexes, multivesicular bodies, and mitochondria but is devoid of yolk, while lipid droplets move away from the egg center to the periphery, forming a subcortical layer under the entire egg surface. Thereafter, this layer moves parallel to the cortex toward the animal and vegetal poles and forms lens-shaped accumulations at the poles by the metaphase of the first cleavage division. Usually, the animal pole plasm protrudes markedly from the egg surface and the organelles become compacted in the pole plasms. By this time the organelles are connected to each other by microfilaments. Mechanisms underlying the cytoplasmic movements at these two stages of segregation appear to be different.

Very occasionally the author has been able to perform artificial insemination of *Tubifex* eggs. This was achieved as follows. The ovisacs with oocytes are usually located near the spindle-like thickening of the worm body and are easily discernible through the translucent body wall. The non-anesthetized worm is placed into a solution (composition described below), and the part of the body with oocytes is cut from the head and tail regions using a surgical knife. The body wall and ovisac membrane are then cut laterally from each side and the oocytes are carefully pressed out into the salt solution. Only the largest oocytes showing germinal vesicle breakdown are selected. The oocytes with a germinal vesicle may be easily identified because the vesicle is located near the surface of the egg.

The spermathecae are a pair of small oval sacs located on the ventral side of the body at the border between the spindle-like thickening and the head region. They usually contain easily discernible spermatozeugmata, appearing as several whitish cylindrical bodies. The body region with the spermathecae is cut away from the worm in a salt solution, and the spermathecae are excised and suspended together with the spermatozeugmata in the salt solution. A suspension of generally immobile spermatozoa obtained from 2-4 spermathecae is then added to the oocytes along with 2-5 ml of the salt solution. A protein-containing fluid is needed to render the spermatozoa mobile.

To conclude this section, a procedure for the fixation of early *Tubifex* embryos for transmission electron microscopy is described.

The eggs taken from the cocoons are passed through three changes (of 2 h each) of 3% glutaraldehyde solution in 0.5 M phosphate buffer (pH 7.4) containing 0.2% tannin at room temperature. They are then placed overnight in a 0.05 M sucrose solution made in 0.05 M phosphate buffer at 0-4°C; the egg is then cut with a safety razor into several parts and post-fixed for 1.5 h in a 1% O_sO_4 solution made in a 0.5 M phosphate buffer in a cold room. The eggs are then dehydrated in alcohol, the vitelline membrane is removed with watchmaker's forceps, and the eggs are embedded in Epon 812. For general cytological purposes, semithin Epon sections may be stained with 1% toluidine blue.

Culture of Embryos

To observe the development in the cocoons the external mucus should be carefully removed with needles. The cocoon wall is not quite transparent, and strong dark-field illumination is needed for accurate staging. According to Penners, the cocoons should be cultured in soft aquarium water to avoid the deposition of calcium salts on the walls; aeration should be adequate and the water changed daily. It should also be pointed out that those cocoons whose mucus has been removed are often attacked by fungi. Cocoons with a moderate number of loosely located embryos are the most suitable for observation.

It often happens that not all embryos in a cocoon develop normally. Some embryos may stop developing while various abnormalities, including twin abnormalities, form in others. The proportion of normal embryos at hatching averages 74%. Apparently, the fluid present in the cocoon is not assimilated by the embryo and serves rather as mechanical protection and salt reservoir.

Salt solutions suitable for culturing an embryo throughout its developmental period have been described. The composition of such a solution for *T. tubifex* is as follows :

$MgSO_4$	20 mM	50 parts

$CaCl_2$	20 mM	130 parts
NaCl	20 mM	16 parts
KCl	20 mM	4 parts

Embryos of the species common in Japan, *Tubifex hattai*, are cultured in a solution containing 5 mM NaCl, 1.4 mM KCl, 8.1 mM $CaCl_2$, 2 mM $MgSO_4$, pH 7.1. The success of the culture depends predominantly on the care with which an embryo is removed from the cocoon. It should be remembered that large *Tubifex* eggs are extremely sensitive to mechanical treatments, and this frequently results in the plasma membrane being disorganized. As a result the perivitelline space acquires a characteristic bluish opalescence. This happens when this space is overfilled with mitochondria which comprise the main bulk of the active subcortical plasm. In these cases, the vitelline membrane undergoes a uniform detachment from the egg surface at a distance of 50-60 μm. This is a reliable indication that the egg is dying. Just as often mechanical trauma results in an abnormal cleavage, but here the rhythmicity of divisions is less affected. The eggs are relatively less sensitive between cell divisions and in the more advanced stages of development (24 blastomeres and later). The stage with prominent pole plasms is very suitable for taking the fertilized egg out of the cocoon prior to cleavage. The cocoon is placed onto the paraffin-coated bottom of a cuvette filled with a salt solution. Then, using a pair of sharp forceps, the cocoon wall is incised in two places and the region between the incisions removed; alternatively one of the cocoon tips (about one-third of the overall size) is carefully dissected together with several eggs, using an ophthalmological surgical knife. Afterward the cocoon is turned tip down (the hole should be sufficiently large) and the eggs are allowed to sink to the bottom. Touching the eggs with instrument is not recommended. The removal of the eggs from the cocoon is performed under a dissceting microscope.

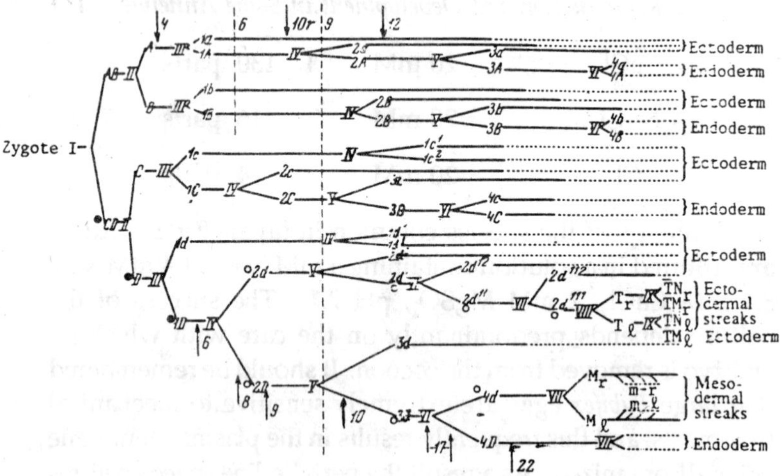

Fig. 4.3. Diagram illustrating cleavage in ***Tubifex tubifex***. M_l, M_r, left and right mesoblasts; T_1, T_r, left and right primary ectoteloblasts; TN1, TNr, left and right neuroteloblasts; TM_l, TM_r, left and right myoteloblasts; m-1, m-r, elements of the mesodermal streaks. Roman numerals refer to division numbers of different cell lines, Arabic numerals with arrows indicate the number of blastomeres.

Successful culturing requires working with a dim light; bright light has an adverse effect on the isolated embryos. When high light intensity is required for examining the embryos under a microscope, heat filters should be used. Another problem is associated with the possible contamination of the cultures by fungi, bacteria, and infusorians: the latter are the most dangerous. They can be avoided if, prior to the extraction of embryos, the cocoons are thoroughly washed on cheesecloth, using first water and then a sterile salt solution. Subsequent surgical intervention should be performed using sterilized instruments. If, despite these measures, contamination still occurs, the culture solution should be changed at frequent intervals. The embryos are suitably cultured in small beakers containing 3-5 embryos per 5.10 ml of solution. Under favourable conditions of culture about 74% or 85% of normal embryos survive. Henzen pointed out that there was no significant difference in the time embryos take to develop inside or outside the cocoons. Even so, it is not practical to work with isolated embryos unless there are particular reasons for doing so. Isolating embryos is not suitable for experiments aimed at elucidating the long-term results of the action of various substances which may pass through the cocoon envelope. It would be beneficial to take another cocoon at the same developmental stage as a control, especially since high synchrony in *T. tubifex* development is rather rare; usually no more than 60-70% of embryos develop synchronously at the early stages.

The temperature range of normal *T. tubifex* development is around 10-25°C. Bearing in mind that drastic temperature changes (between 5-10°C) result in various developmental abnormalities, such temperature fluctuations should be avoided.

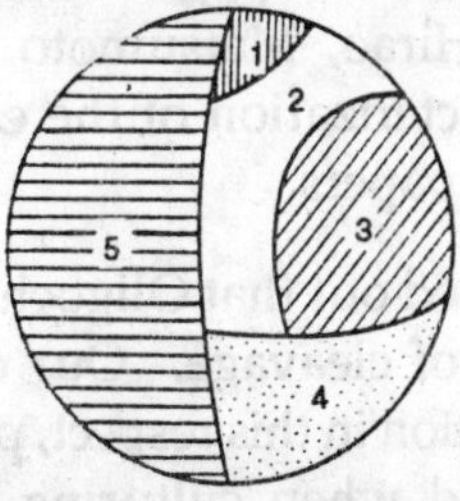

Fig. 4.4. Fate map of *Tubifex* blastula, view from the left, 1) Head and ectoderm; 2) epidermal ectoderm; 3) ectodermal embryonic streaks; 4) mesoderm; 5) midgut. Animal pole is on the right.

Normal Development

Tubifex development belongs to the mosaic type. Penners studied in detail the cleavage and fate of the derivatives of different blastomeres. The main organ systems in *Tubifex* are formed of the following cells. The midgut and esophagus are formed from the yolk-rich derivatives of 3A-3C and 4D macromeres. The head ectoderm originates predominantly from the micromeres of the first quartet. The entire circulatory system, the connective tissue envelopes of the inner organs, the longitudinal muscles and nephridia, all originate from the coelomic mesoderm, derived from progeny of the 4d micromere.

The same is true for the dissepiments and peritoneum of the coelom and probably for the gonads as well. The second and third quartets of micromeres give rise to a considerable part of the epidermis. The $2d^{111}$ blastomere is a source of two ectodermal streaks; these are the precursors of a part of the epidermis, the bristle sacs, ring muscles (from rows M_1 and M_2), and the entire nervous system (from N rows) including the ventral chain and the pharyngeal ring.

The description of stages presented below has been compiled for the first time by the author. To distinguish the stages up to four blastomeres we have used the classification of Lehmann and Hadorn; the later stages have been distinguished and described according to Penners and Meyer and partially to Lehmann and Hadorn. A detailed description of the embryo structure as determined by microscopy may be found in the aforementioned papers of Penners, as well as in the publications of Hirao, Matsumoto and Shimizu. The submicroscopic characterization of the early stages has been given in a number of papers.

It should be pointed out that Oligochaeta in general show a marked variability of cleavage. Our observation point to *Tubifex* being no exception in this respect, particularly beginning from stages 10-12 and when culturing outside the cocoon. Therefore, people studying *Tubifex* development may find deviations from the description of stages as regards the order of appearance of individual blastomeres and their size and relative arrangement.

There appears to be no information in the above-mentioned studies about the time of onset of maternal and paternal genome function in the development of *Tubifex* or about the timing of macromolecular syntheses.

The papers cited contain controversial information on the completion time of endoderm epiboly by the epidermis and embryonic streaks. Here we take it that epiboly is completed by stage 28.

The stages described are the number of each drawing corresponds to a stage. The table of normal development lists, in a summary form, the diagnostic features of different stages as well as their timing. The data of Lehmann and Hadorn are used for temperatures of 13, 18, and 23°C; these authors observed the development of *Tubifex* embryos in salt solutions (the temperatures were maintained at a constant + 1° c). Unfortunately, the authors did not specify the intervals and methods of observation. Their data regarding the early stages raise some doubts.

For stages 1-2 the Henzen data at 20 ± 1°C has been used. It refers to embryos developing in salt solutions and the dating refers to the beginning of a stage for the most advanced embryos in the observed series.

We have observed the development of *Tubifex* embryos inside the cocoons at stages 3-22 at 21 ± 0.5°C. The observation interval at stages 6-21 was 30 min; the beginning of a given stage (appearance of cleavage furrows) has been recorded in the most advanced embryos. At the early developmental stages asynchrony in the development of different embryos appears to be more prominent owing to the shorter duration of each stage. At stages 24-35 the external diagnostic features of stages seem to be rather ambiguous. The emergence of juvenile worms from the cocoon (stage 35) could take as long as several days. Table 4.1 shows the mean time of the onset of stages 6-22, averaged over 3-4 cocoons; the data for one cocoon is given for stages 3-5.

Lehmann and Hadorn have established that the duration of different stages in *Tubifex* development changes proportionally at different temperatures. Therefore, relative values may be used to characterize the temporal aspects of *Tubifex* development just as they are for other species. However, since synchronous cleavage divisions are lacking during *Tubifex* development, the duration of one synchronous cleavage division (τ_0) cannot be used as a criterion to characterize development.

Detailed Description of Stages

Stage 1 : One-Cell Embryo 1/1. Laying and insemination: Cocoon formation takes place at metaphase I. Mitochondria are uniformly distributed under the cortical layer, they are also present, however, in the deeper parts of the cytoplasm.

Insemination in the Japanese species *T. hattai* takes place roughly 30 min after the cocoon is formed. The spermatozoon enters the oocyte in the region located close to the vegetal pole where the fertilization cone is formed.

1/2. Extrusion of the first polar body: During anaphase I, amoeboid protrusions or protuberances appear on the egg surface. These protuberances are radially symmetrical to the animal-vegetal axis. They seem rather chaotic during the first

---

Fig. 4.5-4.8.

Normal developmental stages of *Tubifex tubifex.* Sketch numbers correspond to stage numbers with abbreviations referring to the direction of the view, an, animal view; veg, vegetal view; ps, view from behind; lef, view from the left; r, view from the right; ven, ventral view. Other abbreviations; an, animal pole; veg, vegetal pole; 1(2) pb, first (second) polar body; protub., protuberances; pr, pronuclei; vm, vitelline membrane (shown only at 1/1); fc, fertilization cone; app and vpp, animal and vegetal pole plasm; M_1 and Mr, left and right mesoteloblasts; T_1 an T_r, left and right primary ectoteloblasts; m1 and mr, elements of mesodermal streaks; TN_1 and TN_r, left and right neuroteloblasts; en, endoderm; ec, ectoderm outside the streaks; TM_1+_2, TM_3, TM_1, TM_2, secondary and definitive myoteloblasts; Mes, mesoblast position; S_{ec} (1, r), ectodermal streak (left and right); m.ep., median epidermis; 1-24, segment numbers; at and pt, anterior and posterior tip of the embryo; a. bl., anterior edge of the blastopore; ph, pharynx; ph. gl., suprapharyngeal ganglion; sm, mesodermal streak; rm, ring muscle cell nuclei; T. ectoteloblasts; st, stomodeum; dv, dorsal blood vessel; b, som., border of somites closure on the dorsal side; bs, bristle sacs; nephr, nephridium; gon, gonoblasts; mo, mouth; M_1, M_2, M_3, muscle elemnts of ectodermal streaks; n, neural elements of the streaks (for designation of blastomeres see the description of stages).

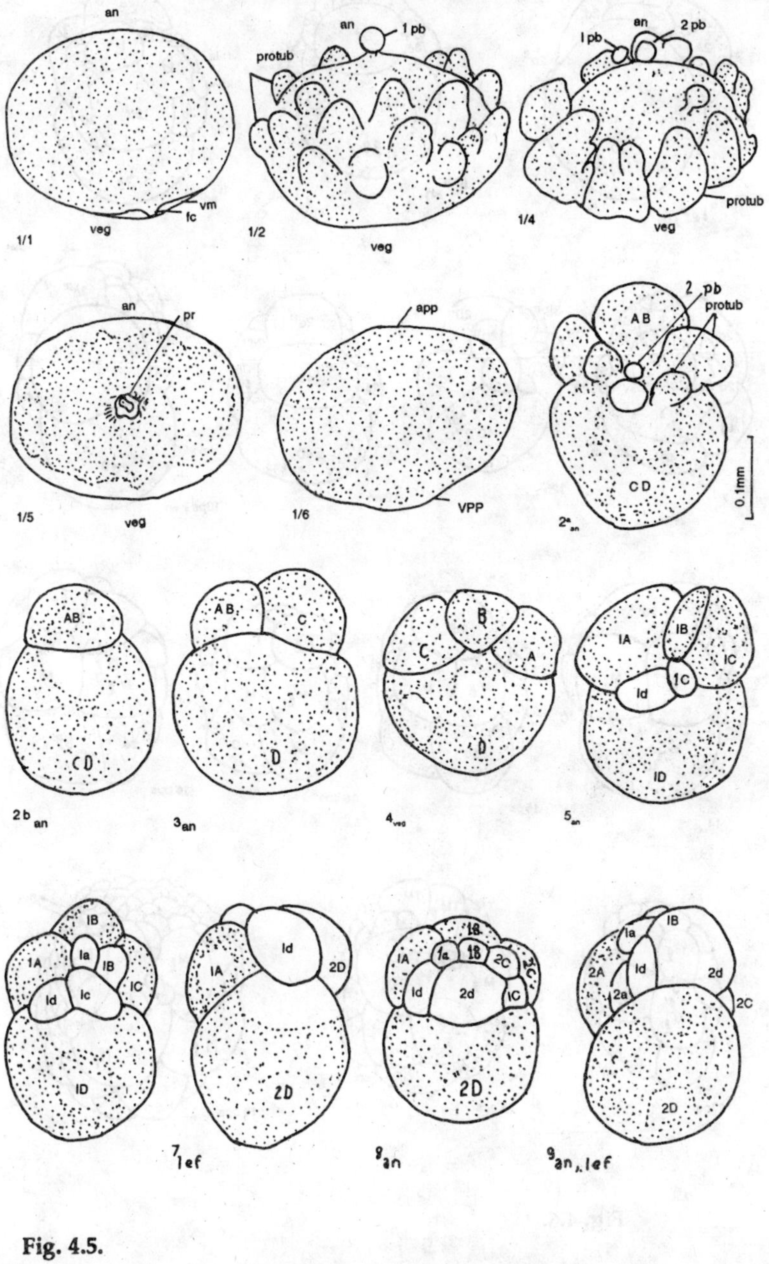
an
vm
fc
veg
1/1
an
1 pb
protub
1/2
veg
an
2 pb
l pb
protub
1/4
veg
an
pr
1/5
veg
app
VPP
1/6
2 pb
protub
AB
CD
0.1mm
AB
CD
2 b an
AB
C
D
3an
B
C
A
D
4veg
IA
IB
IC
1C
Id
ID
5an
IB
1A
Ia
IB
IC
Id
Ic
ID
7lef
Id
IA
2D
2D
8an
1B
1A
1a
1B
2C
2C
Id
2d
IC
2D
9an,lef
IB
Ia
2A
Id
2d
2a
2C
2D

Fig. 4.5.

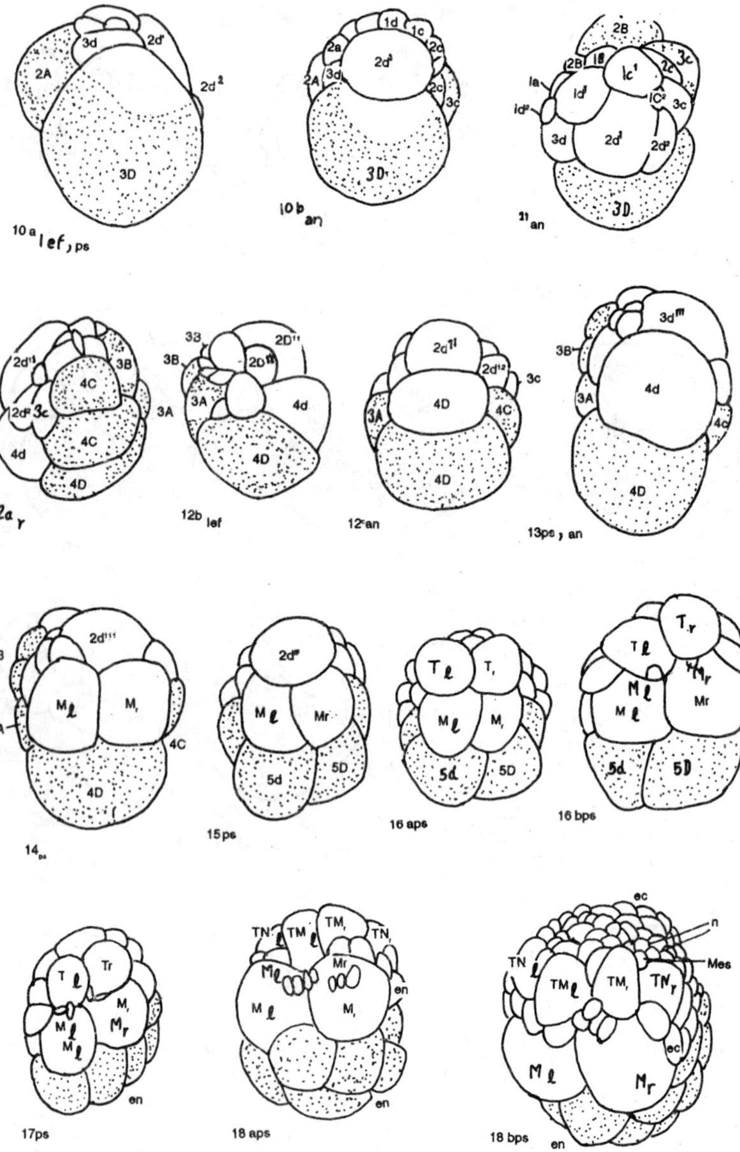
10 a lef, ps
10 b an
11 an
12a r
12b lef
12c an
13ps, an
14ps
15ps
16 aps
16 bps
17ps
18 aps
18 bps

Fig. 4.6.

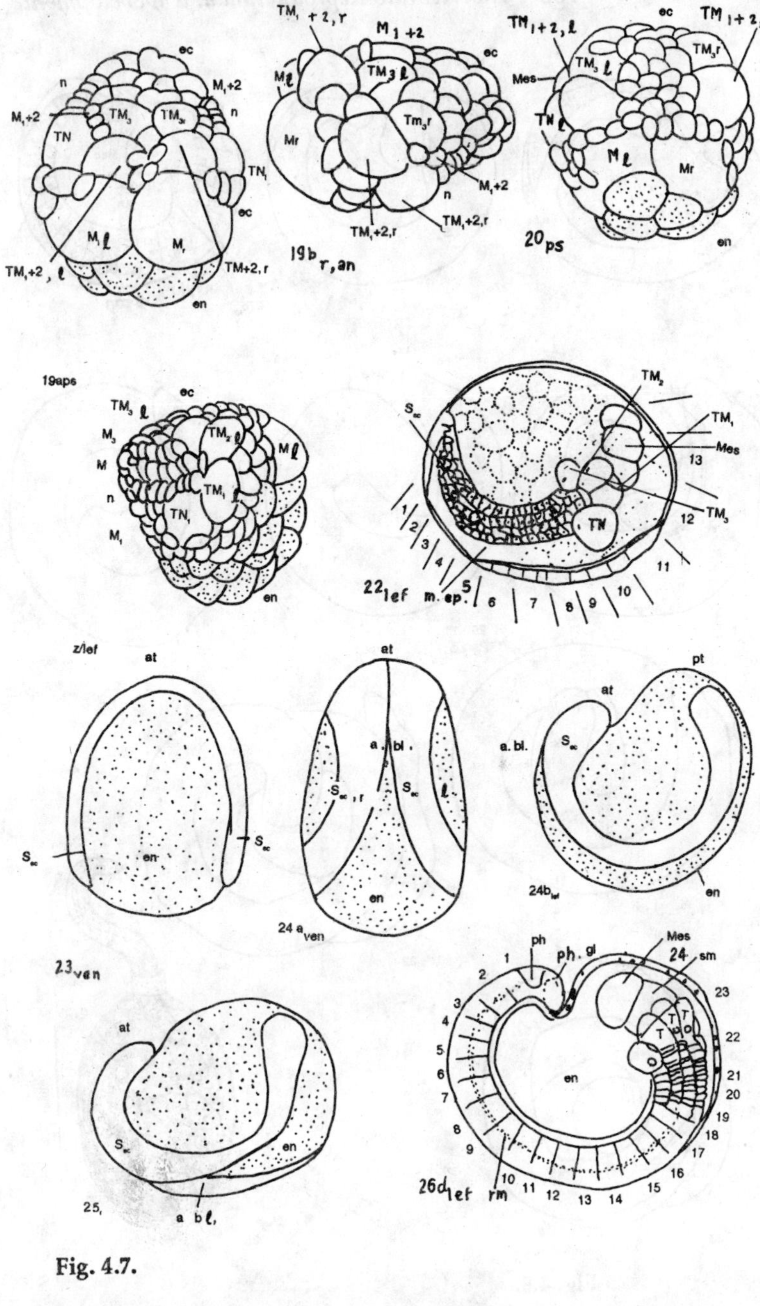
ec
n
M₁+2
TM₃
TM₃r
TN
TN₁
M₁ℓ
M₁
TM₁+2, ℓ
TM+2, r
en
TM₁ + 2, r
M 1 +2
Mℓ
TM₃ ℓ
Mr
Tm₃r
M₁+2
TM₁+2,r
19b r, an
TM 1+2, ℓ
TM 1+2, r
TM₃r
Mes
TN ℓ
M ℓ
20ps
19aps
TM₃ ℓ
M₃
M
TM₂ ℓ
TM₁ ℓ
S_ec
TM₂
TM₁
TM₃
TN
22lef
m.ep.
z/lef
at
en
a. bl.
S_ec, r
S_ec
ℓ
24 a ven
pt
24b lef
23 ven
ph
ph. gl
Mes
24
sm
T
rm
26d lef
25,
a bℓ,

Fig. 4.7.

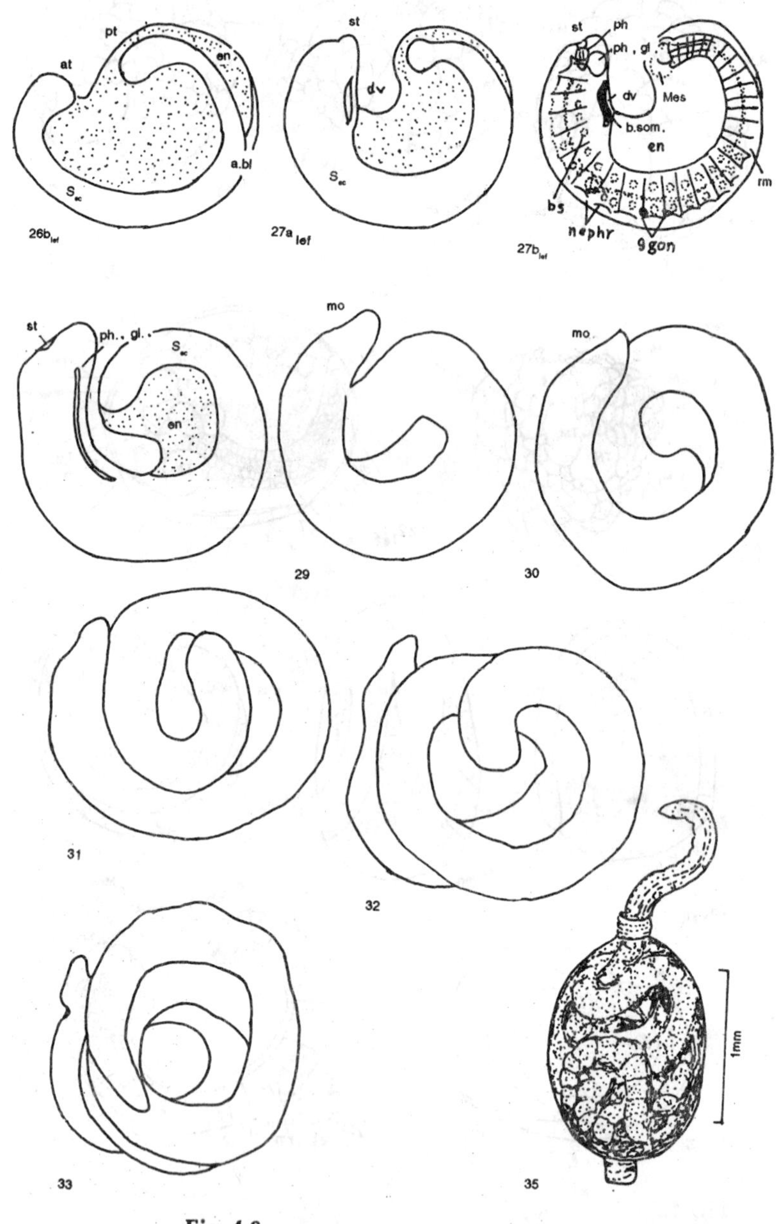
at
pt
en
a.bl
Sec
26b
st
dv
Sec
27a lef
st
ph
ph , gl
dv
Mes
b.som.
en
rm
bs
nephr
gon
27b
st
ph., gl.
Sec
en
mo
29
mo
30
31
32
33
1mm
35

Fig. 4.8.

TABLE 4.1

Normal Development of *Tubifex tubifex*

Stage No.	*Time after egg-laying (hours and minutes) at the following temperatures :*					*Diagnostic features*	
	13ºC	*18ºC*	*20ºC*	*21ºC*	*23ºC*	*External features*	*Internal features*
1/1	00.00	00.00	00.00	00.00	00.00	Perivitelline space is not seen under the light microscope	Metaphase of the 1 maturation
1/2	—	01.00	00.30	—	—	Perivitelline space is noticeable Amoeboid protuberances on the egg surface. One polar body at the animal pole.	
1/3	06.00	02.00	01.45	—	02.00	Ellipsoid shape without any irregularities	Metaphase of the II maturation division
1/4	—	0.400	03.55	—	—	Two polar bodies. Amoeboid protuberances	
1/5	—	—	04.25	—	—	Undistorted shape. Pole plasms are not visible	Fusion of pronuclei
1/6	12.20	0.600	05.50	—	05.20	Pole plasms are well seen	Prophase and metaphase of the first cleavage division
2	—	08.00	07.30	00	00	Two blastomeres	
3	23.00	—	—	11.15	09.10	Three blastomeres	
4	—	—	—	14.25	—	Four blastomeres	
5	—	—	—	16.05	—	Appearance of 1d and 1c	
6	—	—	—	17.25	—	Appearance of 1a and 1b (segregation) of the first quartet	

7	36.20	24.00	—	22.10	14.00	Appearance of 2d	
8	—	—	—	26.00	—	Appearance of 2c	
9	—	—	—	—	—	Appearance of 2a and 2b (segregation of the second quartet)	
10	—	—	—	—	—	Appearance of 3d, 2d2, 2d1 (14 cells)	
11	—	—	—	27.10	—	Division of 1d and 1c. Appearance of 3c (17 cells)	
12	—	28.00	—	28.30	—	Appearance of 4d. Division of 2d1, 3C, 2A, and 2B (22 cells)	
13	—	—	—	31.15	—	Formation of 2d (24 cells)	
14	54.00	36.20	—	33.25	23.30	Two mesoblasts. (First manifestation of bilateral cleavage)	
15	—	—	—	34.30	—	Division of 4D	
16	—	—	—	36.10	—	Division of 2d[111]	
17	—	—	—	42.45	—	4 entoblasts under the mesoblasts	
18	—	—	—	47.00	—	4 ectoteloblasts over the mesoblasts	
19	—	—	—	54.20	—	Continuous row consisting of 6 ectoteloblasts over mesoblasts	
20	—	—	—	61.00	—	3 ectoblasts are present on each side over mesoblasts. Embryonic streaks appeared. Beginning of epiboly	
21	—	72.00	—	62.10	—	Embryonic streaks are located equatorially	Mesoblasts migrated under the ectoderm. 4-10 somites
22	—	—	—	71.15	—	The streaks cover the endoderm over two-thirds of the embryo length along the animal-vegetal axis	13-17 somites. The rudiment of stomodeum

23	—	—	—	—	—	The streaks reach the ventral surface of the embryo	
24	168.00	108.00	—	—	74.0	The streaks bend to the dorsal side of the head tip. They come into contact on the ventral side of the head tip	Appearance of the ventral chain ganglia
25	—	—	—	—	—	On the ventral side the streaks are closed over half the embryo length.	Formation of the pharynx and of the suprapharyngeal ganglia.
						Muscle movements in response to a prick	Beginning of somite differ-emitation into muscles
26	180.00	120.00	00	00	88.00	The streaks are closed over two-thirds of the embryo length. Spontaneous	24-27 somites, Appearance of nephridia Gonoblasta
27	192.30	132.00	—	—	96.00	The head region is well developed. On the dorsal side the streaks are closed up to segment IV	Formation of bristle sacs. The dorsal blood vessel is functional
28	—	—	—	—	—	End of epiboly. The streaks on the ventral side fuse completely; on the dorsal side they fuse up to segment X	Nephridia are multicellular. 34-36 somites
29	—	—	—	—	—	Uniform cylndrical shape	
30	264.00	168.00	—	—	120.00	The first coil is formed	
31	—	—	—	—	—	1.5 coils	
32	348.00	216.00	—	—	156.00	2 coils	
33	—	—	—	—	—	3 coils	
34	—	—	—	—	—	Emergence from the vitelline membrane	
35	552.00	360.00	—	—	264.00	Hatching. Long hairlike bristles are present	

meiotic division and appear predominantly in the animal hemispheres. At this stage the vitelline membrane separates from the egg surface, yielding a well-defined perivitelline space (the membrane is not stretched but is folded at this stage). Microfilaments are only in the animal hemisphere. The meiotic spindle contains centrioles, as it also does in the next cleavage division.

1/3. Acquisition of a stable shape: After the extrusion of the first polar body the protuberances merge and disappear. By metaphase II the egg acquires the stable shape of a rotational ellipsoid, attaining maximal diameter in the equatorial plane.

1/4. Extrusion of the second polar body: The second series of protuberances appears during the anaphase of the second meiotic division; they are more ordered than previously and are separated from the egg surface by rather deep false furrows or even narrow circular constrictions in the equatorial zone. The cortical layer has maximal thickness (0.8-0.8 μm) in the region of these furrows and contains numerous microfilaments which can also be found in other regions of the animal and vegetal hemisphere. Numerous mitochondria are accumulated near the bases of the protuberances.

The polar bodies of *Tubifex* are rather loosely attached to the egg surface and, because the vitelline membrane is separated from the egg surface, they can only rarely be found near the animal pole after the first cleavage divisions. Therefore, they cannot be used to identify polarity.

1/5. Formation of the zygote: Soon after the disappearance of the second series of protuberances, the male and female pronuclei migrate toward the center of the egg where they fuse. Both pronuclei still possess a karyomeric structure at the beginning of the fusion. The spindle, possessing asters of different sizes, begins to form in the center of the nucleus during the prophase of the first cleavage division. This period corresponds to the first stage of pole plasm formation, namely to the appearance of a subcortical layer on the cytoplasmic periphery resulting from the centrifugal movement of the membranous organelles.

1/6. Formation of pole plasms: The accumulation of large quantities of mitochondria and cytoplasmic reticular elements, which resemble two hemispherical convex lenses with their flat surfaces turned toward the cytoplasm, takes place at both egg poles by metaphase of the first cleavage division. These lens-like accumulations appear semitransparent and bluish in colour.

TABLE 4.2

Duration of Stage 1 Periods for *T. Hattai* at 20° C.

Period	*Time, min*
Oviposition	0
Beginning of the first meiotic deformation	40-60
End of the first meiotic deformation. Extrusion of the first polar body	120
Beginning of the second meiotic deformation	170
End of the second meiotic deformation Extrusion of the second polar body	250-270
End of phase I of pole plasm formation (until the prophase of the first cleavage division)	300
End of phase II of pole plasm formation (until the metaphase of the first cleavage division)	330
Completion of pole plasm formation (metaphase of the first cleavage division	360
Beginning of the first cleavage division	570-600

The egg shape is rather stable in this state. Lipid droplets are concentrated on the boundary separating the pole plasms from the yolk. According to Inase the animal pole plasm later goes into the 2d blastomere and the vegetal pole plasm into the 4d blastomere. This contradicts statements from earlier authors.

The second stage of the pole plasm formation (when the subcortical cytoplasm "flows" to the poles) is divided by Shimizu into two phases. During these phases the volume of the subcortical layer does not expand, but its distribution does become extremely non-uniform. The first phase corresponds to a transition from prophase to metaphase of the first cleavage division. The lens-shaped pole plasms are actually formed during the second phase. Protuberances and microvilli appear again on the animal pole; blebbing appears to occur in this place.

Table 4.2 Supplements Table 4.1. It shows the duration of stage 1 periods for *T. hattai* at 20°C.

Stage 2 : Two Blastomeres. During anaphase a large heteropolar spindle is displaced by a smaller aster to the lateral egg surface. Surface activity is served which may accompany the division of large blastomeres containing the pole plasm up to the formation of the 4d mesoblast. This results in the formation of unequal AB and CD blastomeres; the bulk of pole plasm material remains in the CD blastomere. A small cavity may appear between the two blastomeres.

Stage 3 : Three Blastomeres. During metaphase the CD blastomere begins to undergo an asymmetrical deformation. Eventually blastomere C segregates on the right side of AB (when viewed from the animal pole). C is equal in size to, or slightly larger than, AB. The bulk of the pole plasm material remains in D.

Stage 4 : Four Blastomeres. AB divides to yield A (located on the left) and B (sometimes somewhat smaller in size). All three first cleavage divisions are thus meridional.

Stage 5 : Six Blastomeres. Two micromeres, 1d and 1c (the former is usually somewhat larger) are segregated synchronously in the dexiotropic direction. They contain little yolk.

Stage 6 : Eight Blastomeres. The formation of the first

micromere quarter, later to yield the main bulk of the ectodermal part of the head is completed by the segregation of 1a and 1b.

Stage 7 : Nine Blastomeres. A large 2d blastomere containing abundant pole plasm is segregated at this stage. This is the first somatoblast, the descendants of which later yield ectodermal teloblastic streaks. Despite laeotropic segregation, 2d is located virtually in the center of the animal hemisphere.

Stage 8 : Ten Blastomeres. After the formation of the 2d blastomere, the 1C blastomere laeotropically segregates a small 2c micromere.

Stage 9 : Twelve Blastomeres. The formation of the second micromere quartet is completed by a laeotropic division of 1A and 1B macromeres. This stage may be recognized by the position of the 2a micromere.

Stage 10 : Fourteen Blastomeres : The following blastomeres undergo cleavage : 2D segregates the rather large 3d micromere to the left of 2d; and 2d micromere buds the $2d^2$ micromere toward the C quadrant, the size of the latter micromere being slightly less than that of its sister $2d^1$.

State 11 : Seventeen Blastomeres. The micromeres of the first quarter divide. 1d yields a smaller $1d^2$ and a larger $1d^1$ and 1c yields a smaller $1c^1$ and a larger $1c^2$. Directions of these divisions are more or less symmetrical to the B-D axis, the future anteroposterior axis of the embryo. The micromere 3c undergoes dexiotropic segregation.

Stage 12 : Twenty-two Blastomeres. The second 4d somatoblast is formed at this stage. This is a gigantic segregating under $2d^{11}$ which, in turn, forms at this stage as a result of the $2d^1$ division, which buds a small $2d^{12}$ micromere to the left. 4d contains virtually all the remaining pole plasm but differs from $2d^{11}$ in that it has far more yolk. Therefore, its consistency may sometimes be confused with yellow-rose macromeres. In addition, 3C divides into the two yolk-rich endodermal cells

4C and 4c with cells 2A and 2B completing the formation of the third quarter budding the small 3a and 3b blastomeres.

Stage 13 : Twenty-four Blastomeres. The $2d^{111}$ blastomere, which is an immediate precursor of ectodermal teloblasts, forms at this stage. It retains the median position of the $2d^{11}$ blastomere. One of the macromeres, either 1A and 1B, divides.

Stage 14 : Two Mesoblasts. The second 4d somatoblast divides to yield the two identical mesoblasts M_l and M_r (left and right). Cleavage becomes distinctly bilateral.

Stage 15 : Division of the 4D. The 4D macromere divides to yield two identical blastomeres. This division may sometimes come after stage 16.

Stage 16 : Two Primary Ectoteloblasts, Two Mesoblasts. The $2d^{111}$ cell divides to yield two identical ectoteloblasts (Tl and T_r). Teloblastic activity of the mesoblasts begins somewhat later, the mesoblasts budding small cells from their anterior edges. The beginning of this activity may, however, be delayed.

Stage 17 : Second Division of 5d and 5D Macromeres. The 5d and 5D macromeres divide once more, forming four endoblasts under the mesoblasts; during this period each endoblast buds yet another small cell (macromeres after the segregation of 4d are called endoblasts). Each of the primary ectoteloblasts at this stage yields one small cell in the posterior direction that is toward the animal pole.

Stage 18 : Four Ectoteloblasts.

Each ectoteloblast divides in the direction parallel to the B-D axis to yield a neuroteloblast located externally and a more internal myoteloblast. According to our observations, the left and right teloblasts frequently divide asynchronously at an interval of up to 1 h. By this time the mesoblasts may segregate up to 4-5 small cells by budding. Each myoteloblast can yield two cells of an epidermal nature by budding in the posterior direction. Cells of quartets, 1, 2 and 3, which by this time have multiplied, start creeping up on the endomeres and mesoblasts. Ectodermal streaks formed from cells budded by the ectoteloblasts in the forward direction begin between stages 18 and 19.

The next stage begins after five or six cells have been segregated by each myoteloblast.

Stage 19 : Six Ectoteloblasts. The myoteloblasts divide normally with respect to the B-D axis (again frequently asynchronously) to yield myoteloblasts, 3 (TM_3) and 1 + 2 (TM_{1+2}). Initially the teloblast triplets are packed together. This is not an abnormality as Penners suggested. TM_{1+2} buds a row of cells which continue the TM row, while TM_3 begins a new row.

Stage 20 : Beginning of Epibolic Gastrulation. Owing to intensive proliferation, ectodermal cells wedge in between the ectoteloblast triplets, displacing the ectoteloblasts to the lateral sides of the embryo, and then creep up on the mesoblasts. From this time onward the whitish embryonic streaks can be clearly seen among the semitransparent epidermal cells. Cells comprising the streaks, however, cannot be seen, and it is difficult to discern the ectoteloblasts.

Stage 21 : Eight Ectoteloblasts; Equatorial Position of the Ectodermal Streaks. After segregating five or six cells, the TM_{1+2} myoteloblasts divide in a direction perpendicular to the animal vegetal axis; therefore on each side of the embryo embryo four ectoteloblasts, TM, TM_1, TM_2, and TM_3, can be found above the other; TM_1 begins to bud a new row. Cell proliferation in the streaks begins closer to the head region where the streaks touch one another. The mesoblasts are almost completely covered by the ectoderm. The segmentation of mesoderm begins at this stages (between 4 and 10 somites are present). Each somite on one side of the body originates from a single cell budded by the mesoblast. Mesoblasts at this stage are located close to the X segment.

Stage 22 : Thirteen to Seventeen Somites : The embryonic streaks obviously bent in the vegetal direction as well as the epidermis cover the endoderm by roughly two-thirds along the animal-vegetal axis. The ectoteloblasts can no longer be seen in vital observation. In contrast to Penners and Ivanov, Meyer believes that, at this stage, the embryo possesses the first three mesodermal segments which later fuse together. The rudiment of the stomodeum becomes discernible close to the site where the streaks touch each other on the head tip.

Stage 23 : Emergence of the Embryonic Streaks onto the Ventral Surface. The embryonic streaks emerge onto the ventral surface of the embryo but are not yet fused together. The embryo is still egg shaped.

Stage 24 : Beginning of Streak Fusion and Elongation of the Embryo. The embryonic streaks on the head region start bending toward the dorsal side of the embryo and, at the same time, come into contact on the ventral side of the head region. The embryo begins to acquire an elongated and bent shape (the dorsal side is concave). From this stage onward, during the course of streak fusion, differentiation proceeds gradually in the anteroposterior direction. The differentiation of the mesodermal streaks proceeds alongside. At stage 24 the neural part of the ectodermal streaks is multicellular and partially covered by epidermis from the outside. Pairs of rudiments corresponding to the ventral chain ganglia have already appeared.

Stage 25 : Streaks on the Ventral Side are Closed Approximately Halfway. The rudiment of the pharynx appears for the first time as a dense ectodermal mass with the suprapharyngeal ganglion appearing alongside. The muscle part of the ectodermal streak is now multilayered, the external one yielding part of the epidermis while the internal one (predominantly rows m_1 and m_2) yields the ring muscle system. The external parts of the somites begin to differentiate into muscle elements of the longitudinal muscle system.

Stage 26 : Streaks on the Ventral Side are Closed Together for Roughly Two-Thirds of Their Length. Streak fusion approximately reaches segment XVI. The head region of the embryo makes small spontaneous movements. Pairs of rudiments corresponding to the dorsal blood vessels, and protruding up to segment X, appear in the upper part of the mesodermal streaks. Between 24 and 27 somites are present in this stage. Pairs of nephridia rudiments appear in segments VII and VIII; they are probably all of mesodermal origin. One or two gonoblasts appear in the region of segments X and XI; these gonad anlage exist in a two-cell state until hatching.

Stage 27 : Segregation of the Head Region : Movements are obvious. The streaks are closed on the dorsal surface of the anterior part of the body approximately up to segment IV. On the dorsoventral surface they are separated over a very short distance. The stomodeal invagination is quite distinct. The ventral blood vessels is formed in the anterior part of the body. Ectodermal bristle sacs consisting of a follicle and 5-7 large cells of the bristle gland have formed. The ventral ganglia are clearly delimited. The dorsal blood vessel begins to function. Somites have a small lumen. Precursors of the future epithelial cells of the midgut appear inside the endoderm.

Stage 28 : Thirty-four to Thirty-six Somites, the End of Epiboly. The streaks on the posterior end are completely closed : in the dorsal region their closing is advanced up to segment X. According to Meyer the mesoderm of the three first segments loses its segmented character. Nephridia are multicellular. Bristle anlage are connected with muscles. Ectoteloblasts are present but are small. Ganglia and longitudinal connectives have differentiated within the central chain approximately to segment XXII. The length of the embryo in its extended state is about 0.7 mm.

Stage 29 : Acquisition of a Wormlike Shape. The embryo acquires a uniformly cylindrical shape and the characteristic posterior thickening disappears. From now until hatching the number of segments stays at 36 (in rare cases, 38).

Stages 30-33 : Coiling of the Body. When the embryo is still inside the vitelline membrane, body elongation and a decrease in cross section result in helical coiling of the body. Different stages are characterized by a different number of turns. At Stage 30 : the first coil of the body is complete. The length of the embryo is about 2.2 mm. *Stage 31* : about 1.5 coils. *Stage 32* : about 2 coils. *Stage 33* : about 3 coils.

Stage 34 : The juvenile worm emerges from the vitelline membrane, unfolds, and creeps around inside the cocoon.

Stage 35 : Hatching. Juvenile Stage. The juvenile worm emerges from the cocoon through an orifice, head forward. Meyer points out that the worm still does not have long hairlike bristles, but we managed to observe some. By the time of hatching, nephridia also appear in other segments (beginning with segment XIII). The contractile perivisceral vessel is prominent and the midgut possesses a lumen. The oral cavity establishes a connection with the pharynx only after hatching. The length of the hatched juvenile worms is between 3.00 and 5.50 mm (mean length 4.50 mm).

REPRODUCTION AND DEVELOPMENT OF NEANTHES

Reproductive System

Neanthes is unisexual and the male and female may exhibit slight sexual dimorphism. The gonads i.e., ovaries and testes are temporary organs, appearing only during breeding season. The breading season is summer, these gonads disappear after the liberation of gametes. In male *N. dumerilli* there is only one pair of testes lying in each segment between the 19th and 25th segments in one of the segments. But in *N. virens* the testes extend in many sagments around blood vessels. The gonoducts are absent.

During breeding season the cells which detach from the coelomic epithelium in male constituting the spermetogonium

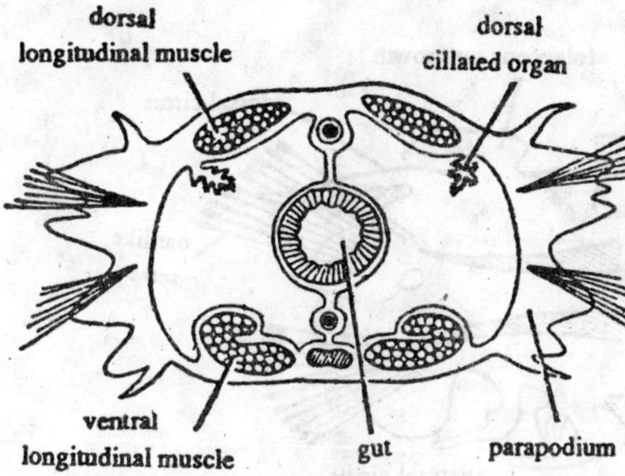

Fig. 4.9. Neanthes. T.S. of a segment showing ciliated organs

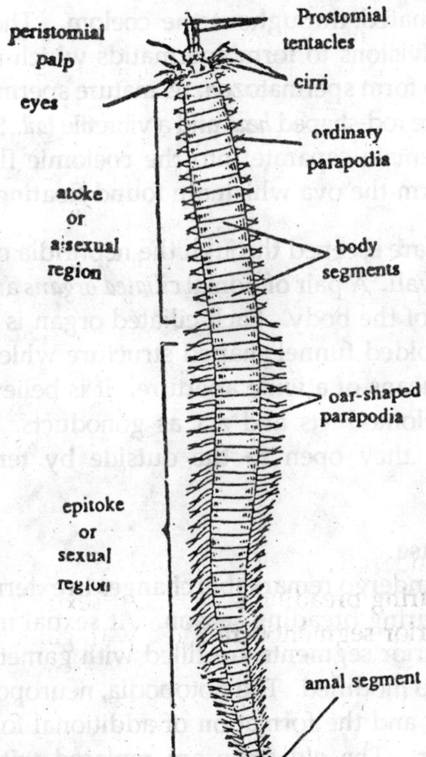

Fig. 4.10. Heteronereis.

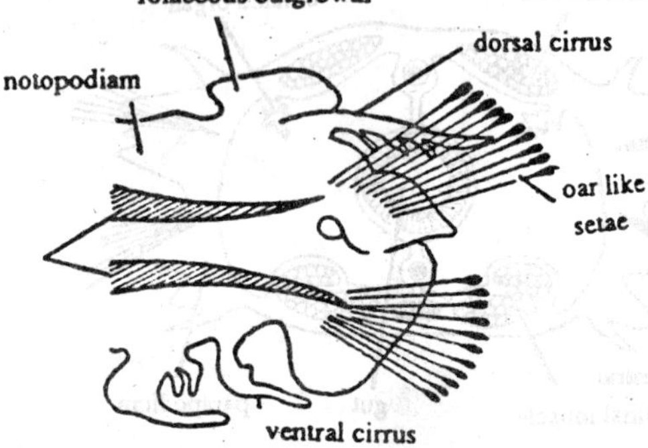

Fig. 4.11. Parapodium of Heteruereis.

becomes disseminated throughout the coelom. These cells undergo rapid divisions to form spermatids which undergo spermiogenesis to form spermatozoa. A mature spermatozoon consists of a minute rod-shaped *head* and a vibretile *tail*. Similarly the oogonia in female separate into the coelomic fluid and after divisions form the ova which are found floating in it.

The gametes are esceped through the nephridia or by the rupture of body wall. A pair of dorsal *ciliated organs* are found in each segment of the body. Each ciliated organ is a small, ciliated tract of folded funnel-shaped structure which opens into coelom by means of a wide aperture. It is believed that they are like coelomoducts and act as gonoducts. During breeding season, they open to the outside by temporary apertures.

Heteronereis Phase

N. dumerilli, undergo remarkable changes in external form and behaviour during breading season. At sexual maturity, most of the posterior segments are filled with gametes. The parapoida are also modified. The notopodia, neuropodia and cirri are enlarged and the formation of additional foliaceous outgrowths occur. The old setae are replaced with much longer, more numerous and our shaped setae which are arranged in a fan-like manner. The old parapodial muscles are also replaced by new muscles. This posterior region of body is sexual region or *epitoke* where as the anterior non-sexual region is referred to as *atoke.*

Following characteristics are noted in heteronereis phase:

1. Heteronereis does not crawl on sea bottom but it swims actively on water surface.
2. The eyes become enlarged and the animal properly precieves the light stimulus.
3. The peristomial cirri are comparatively larger than *Neanthes.*
4. The intestine becomes compressed and functionless due to much development of the gonads.
5. Special sensory papillae are formed on the anal segment.

Swarming

Neanthes swarms on moonless night during summer months after sunset. The males swarms rapidly about the females. Females peoduce some substance called *Ferlilium* (not the fertilizin produced by eggs) which attracts males. This fertilizin is responsible for stimulating and shadding of gametes. Swarming continues for about an hour or so. Females die after liberation of eggs and probably makes also.

TABLE 4.3

Defferences Between *Neanthes and Heteronereis*

Neanthes	*Heteronereis*
1. Neanthes is a non-sexual sluggish creature, creeping at the bottom of sea.	1. Heteronereis is an active swimmer on the surface of water.
2. Eyes normal and the worm is less sensitive to light. is very sensitive to light.	2. Eyes become enlarged and conspecuous so that the heteronereis
3. The parapodia are thin, flap-like and non-vascular.	3. The parapodia of the posterior or the sexual part are large, highly vascular and develop leaf-like outgrowths.

4. The setae on the parapodia are normal.	4. The normal setae are replaced by large, flattened and oar-shaped setae.
5. Setae are arranged in the form of bundles.	5. Setae are arranged in fan-like manner.
6. Sensory projections are well-developed.	6. Sensory projections become shruken and the anal segment develops special sensory papilae.
7. Dorsal cirri are less curved.	7. Dorsal cirri become altered.
8. The intestine is well formed.	8. The intestine is compressed and becomes functionless due to development of gonads.
9. Less developed for swimming.	9. Better adopted for swimming.

Fertilization

Fertilization is external and takes place in sea water except *Platynereis megalops* where internal fertilization takes place. Here the male inserts its and segment in the mouth of female. The sperms are passed into the coelom as the intentive is eroded. In the coelomic fluid fertilization takes place.

Development

The development starts after fertilization of eggs. Each egg is covered in a thick transparent gelatinous envelope. Inside this membrane there are two membranes :

(i) Outer membrane is very thin and delicate.
(ii) Inner membrane is thicker and distinctly striated radially, thererore, it is known as *zone radiata.* The cytoplasm contains numerous yolk spherules and oil drops.

Fertilization produces several changes in the egg. The yolk spherules move at the lower side or vegetal pole, leaving a granular protoplasmic area at the upper or animal pole.

Cleavage. The first two cleavages of the zygote are the equal and vertical and result in four cells or *blastomeres* lying in the same plane. The cleavage is *determinate,* i.e., the fate of

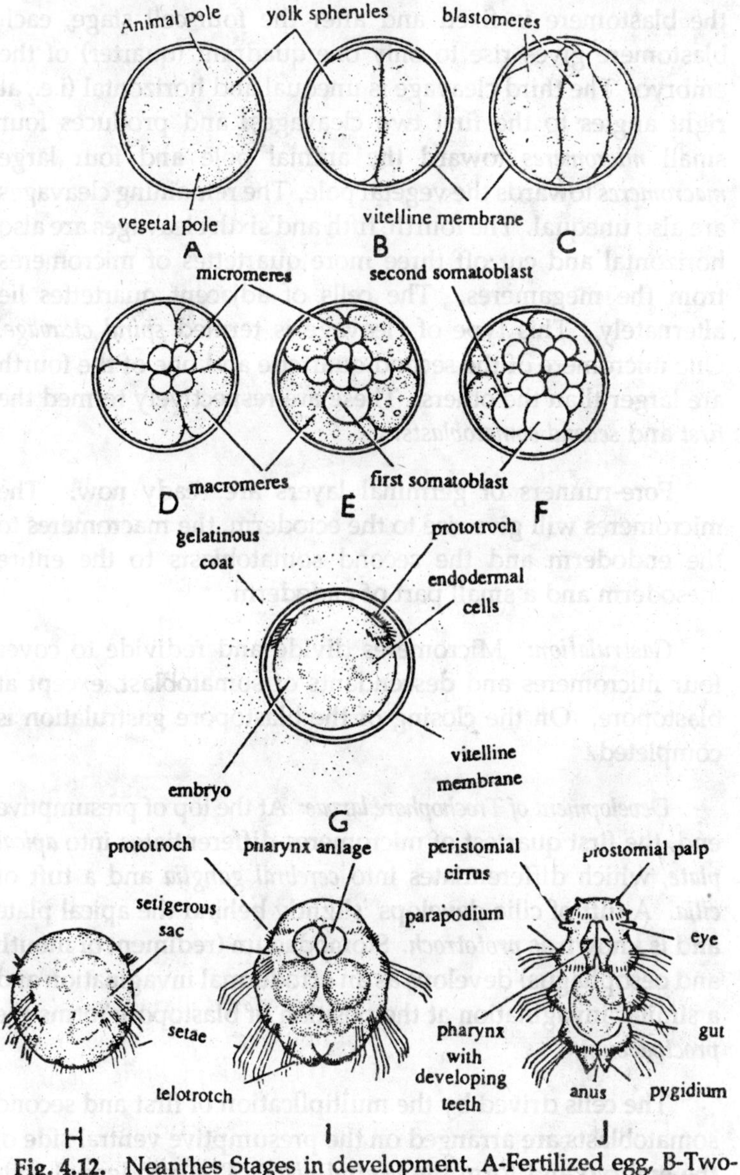

Fig. 4.12. Neanthes Stages in development. A-Fertilized egg, B-Two-cell stage C-Four-cell stage, D-Eight-cell stage, E-Ist somatoblast formed, F-2nd somatoblast formed, G-Young trochophore before hatching. H-Post trochophore larva with three setigerous segments, I-Older larva, J-Later larva, after three weeks.

the blastomere is fixed and after the four-cell stage, each blastomere gives rise to only one quadrant (quarter) of the embryo. The third cleavage is unequal and horizontal (i.e., at right angles to the first two cleavages) and produces four small *micromeres* toward the animal pole and four large *macromeres* towards the vegetal pole. The remaining cleavages are also unequal. The fourth, fifth and sixth cleavages are also horizontal and cut off three more quartettes of micromeres from the megameres. The cells of adjacent quartettes lie alternately. This type of cleavage is termed *spiral cleavage.* One micromere of the second quartette and one of the fourth are larger than the others. These are respectively termed the *first* and *second somatoblasts.*

Fore-runners of germinal layers are ready now. The micromeres will give rise to the ectoderm, the macromeres to the endoderm and the second somatoblasts to the entire mesoderm and a small part of endoderm.

Gastrulation: Micromeres divide and redivide to cover four micromeres and descendants of somatoblast, except at blastopore. On the closing of the blastopore gastrulation is completed.

Development of Trochophore larvae: At the top of presumptive end, the first quartest of micromeres differentiates into *apical plate,* which differentiates into *cerebral ganglia* and a tuft of *cilia.* A tuft of cilia develops slightly behind the apical plate and is known as *prototroch.* Stomodaeum (rediment of mouth and oesophagus) develops as an ectodermal invagination and a similar invagination at the position of blastopore forms the *proctodaeum.*

The cells drived by the multiplication of first and second somatoblasts are arranged on the presumptive ventral side of the embryo and form the *ventral plate.* The ectodermal cells present along the ventral side are thickened into *ventral nerve cord.* The deeply situated derivatives of somatoblast give rise to *mesodermal bands* which develop into muscles of body wall and of alimentary canal.

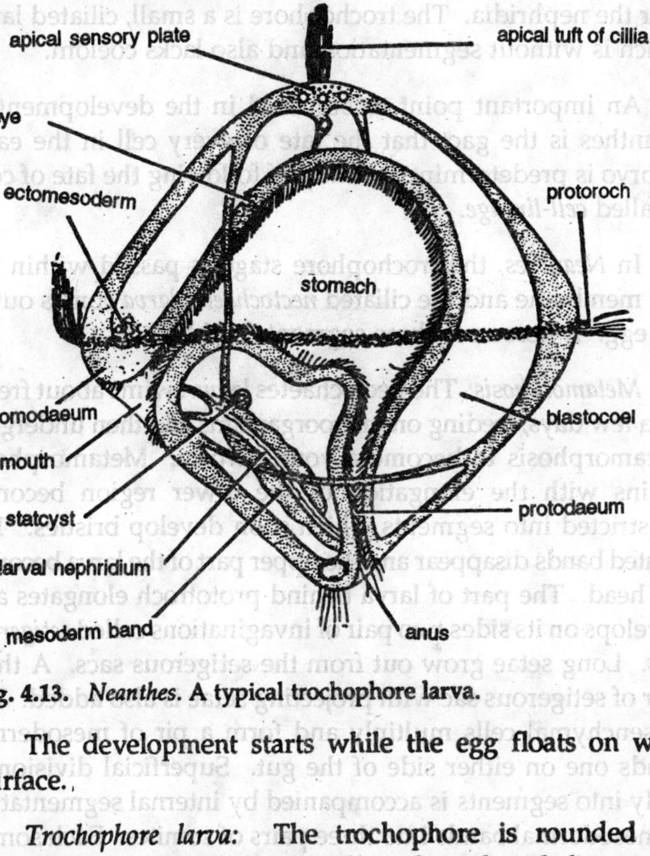

Fig. 4.13. *Neanthes.* **A typical trochophore larva.**

The development starts while the egg floats on water surface.

Trochophore larva: The trochophore is rounded and transparent, it has a thin external ectodermal epithelium which is thickened at the two ends and along an equatorial ring. There is a curved gut with a mouth, ectodermal *oesophagus* or stomodaeum, and endodermal *stomach,* and an ectodermal *hind gut* opening by an anus. On the thickened parts of the ectoderm is an anterior ciliated *apical organ* with an *apical ganglion* below which is an *eye spot,* at the posterior end and some large cilia and on the equatorial ring is a pre-oral ciliated band or *prototroch.* Between the ectoderm and gut is a large cavity, the *blastocoel* having *mesenchyme cells* larval *mesoderm* and a pair of larval *nephridia,* each made of two hollow cells, one of which contains a flame of cilia. There lies an otocyst near the nephridia. The trochophore is a small, ciliated larva which is without segmentation and also lacks coelom.

An important point to be noted in the development of Neanthes is the gact that the fate of every cell in the early embryo is predetermined. Study of following the fate of cells is called *cell-lineage.*

In *Neanthes,* the trochophore stage is passed within the egg membrane and the ciliated *nectochaetes larva* comes out of the egg. It possesses three segments and bristles.

Metamorphosis: The nectochaetes larva swims about freely for a few days, feeding on microorganisms and then undergoes metamorphosis to become a young worm. Metamorphosis begins with the elongation of the lower region becomes constricted into segments which soon develop bristles. The ciliated bands disappear and the upper part of the larva becomes the head. The part of larva behind prototroch elongates and develops on its sides two pair of invaginations called *setigerous sacs.* Long setae grow out from the setigerous sacs. A third pair of setigerous sac with projecting setae is also added. The mesenchymal cells multiply and form a pir of mesodermal bands one on either side of the gut. Superficial division of body into segments is accompanied by internal segmentation of mesodermal bands into three pairs of somites. Each somite develops a cavity, the *coelom.* Both the somites of each segment unite above and below the gut forming the dorsal and ventral mesenteries. Later on the ventral mesentery is disappeared. The coelom obliterates the blastocoel. The part anterior to prototroch modified into the prostomium. The part behind prototroch gives rise to the peristomium. The part bearing the anus is continually pushed backwards as new segments are formed in front of it. Tentacles, palps, parapodia with their cirri and permanent setae are formed. Larval setae are dropped. Larval nephridia are replaced by permanent nephridia. The young worm settles down on the bottom and changes into the burrowing adult worm.

Significance of Trochophore

The trochophore larvae has much significance since it is found widely in different groups such as Tubellaria, Mollusca, Bryozoa, Nemertines, annelida etc. It has much resemblence with adult rotifers. The early development of the animals of these groups is almost identical and their trochophore stages are also similar in many respects. The differences in their development arise after the trochophore stage. The affinities of annelid trochophore can be discussed as follows :

Reproduction and Development of Pheretima

The earthworms are *monoecious* or *haermaphrodite.* The reproductive organs are restricted to a few anterior segments. Though hermaphroditic, the self-fertilization not occurs in earthworm, but pair together (copulate) and later develop deposit eggs in cocoons. Moreover, the testes ripen earlier (protanday), make cross-fertilization necessary. Thus, haermaphroditism seems to be of no use to the earthworm.

Male Reproductive Organs: The male reproductive system in formed of testes, seminal vesicles, spermiducal funnel, vasa deerentia prostate glands and accessory glands.

Testes: There are two pairs of testes which produce sperms. There are minute, white and lobed structures. One pair is situated in the 10th segment and the other in the 11th segment on either side of nerve cord. Each testis is made up of a compact narrow base, from which arise 4-8 finger like processes containing spermatogonia.

Each pair of testes is enclosed in a thin-walled, fluid filled bag, the *testies-sac.* Thus there are two testis-sacs situated ventrolaterally, one behind the other, in segments 10th and 11th respectively. The posterior testies sac is larger than the anterior one.

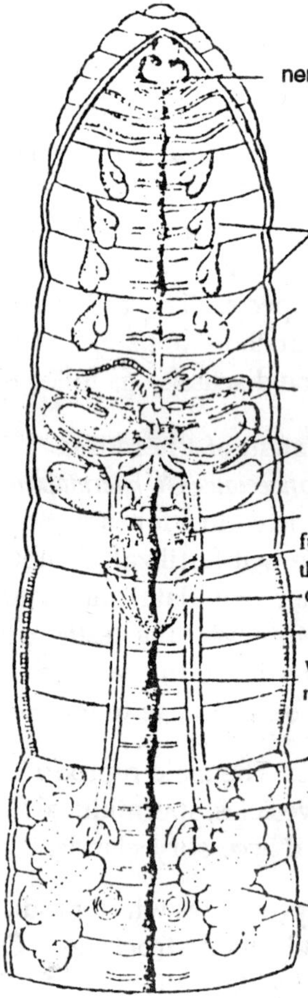

Fig. 4.14. *Pheretima.* Reproductive System.

There are two pair of spermatic funnel or spermiducal funnel in the two pairs of testies sacs. Each funnel is a ciliated structure and lies immediately behind its own testis and opens into a ciliated *sperm duct* or *vast deferens.*

Seminal vesicles: There are two pairs of large, while seminal vesicles one pair in each 11th and 12th segments respectively.

The testis sac of 10th segment communicates with 11th segment and the testis sac of 11th segment with the 12th segment. The seminal vesicle of 11th segment is not visible as they remain enclosed in the testis sac of that segment. However, the seminal vesicles of 12th segment lie free.

Jasadeferentia: Each spermiducal funnel leads postericaly into a fine ciliated duct, the vast deferens. The vasa deferential of the same side travel come together and meet a short prostatic duct of that side in the 18th segment.

Prostate glands: There is a pair of large, flat, white irregular organs situated one on either side of the intestine from 16th or 17th upto 20th or 21st segment, each prostate gland gives off a short duct or its inner side. The prostatic duct is joined by two vasa deferential, thus, forming three ducts are called *common prostatic* and *spermatic ducts* and open to the outside ventrally on the 18th segment by the male genital pore. The opening of the male genital pore is actually a group of three apertures, a large prostatic aperture and two small apertures of vasa deferential. The function of the vasa deferential is to convey the seminal fluid to the outside for fertilization after copulation.

Accessory glands: The accessory glands are two pairs of rounded masses lying in the 17th and 19th segments just internal to the prostate glands. They open to exterior by a number of ducts on the ginatal papillae situated on the 17th and 19th segments. Their secretion helps probably in uniting the two worms during copulation.

From the testes the spermatogonia are shed into the testies sacs, from where they pass into the seminal vesicles where they develop into spermatozoa by an elaborate process of spermatogenesis. Each sperm tozoon consists of a head and vibratile tail. The sperms come back into the testis sacs and find their way into the spermiducal funnel and travél along the vasa deferential and finally pass out, through the male genital prores during copulation, into the spermathecae of another worm.

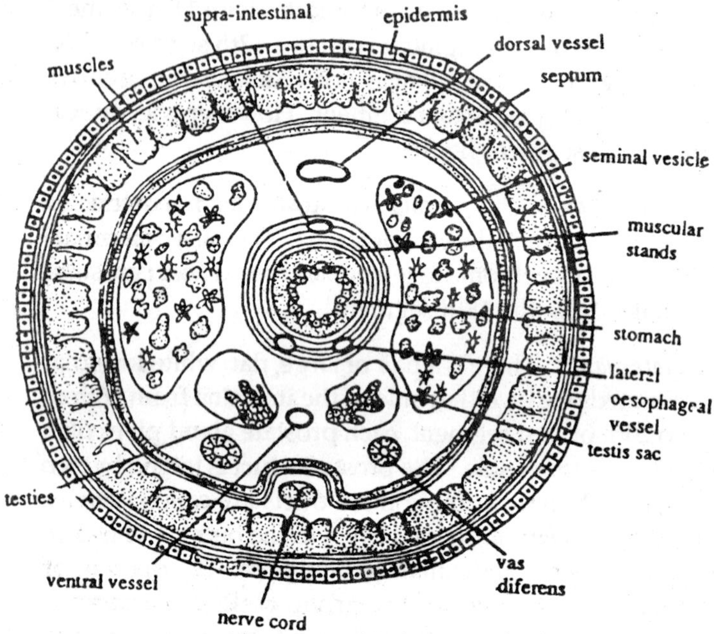

Fig. 4.15. *Pheretima.* T.S. through Seminal vesicle

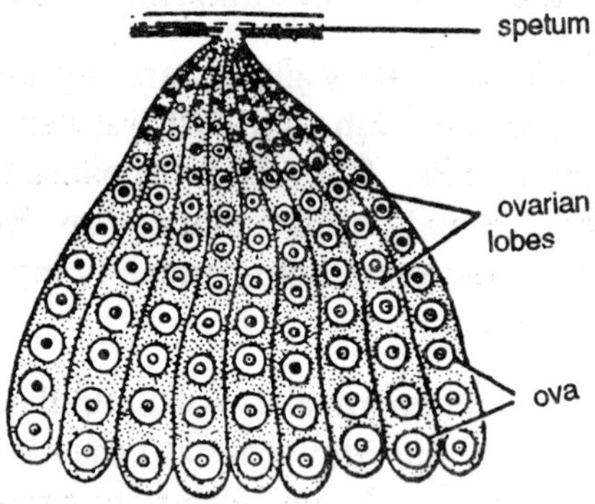

Fig. 4.16. *Pheretima* ovary.

Female Reproductive Organs: The essential organs of the female reproductive system consists of a pair of ovaries, oviduct and 4 pairs of spermathecae.

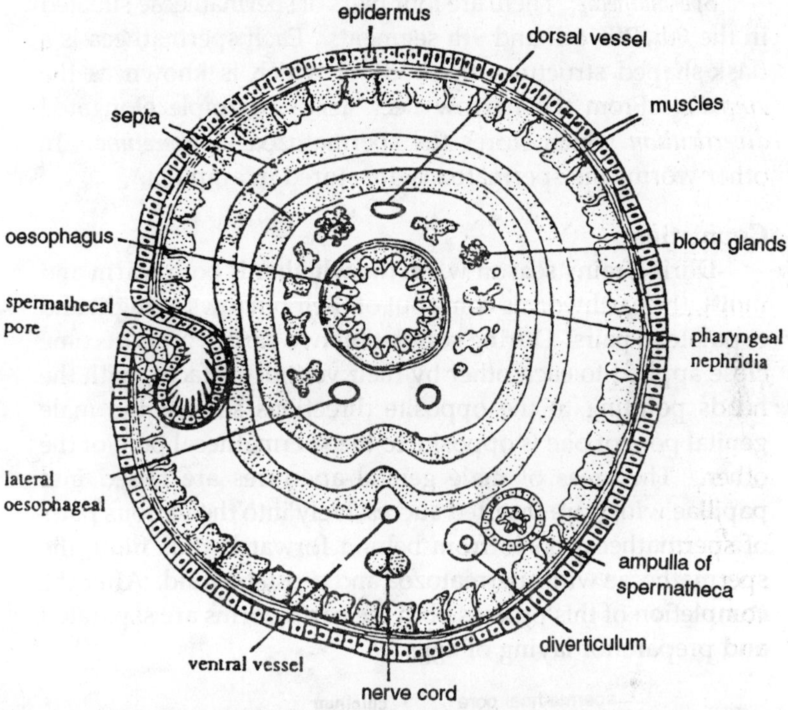

Fig. 4.17. *Pheretima.* **T.S. through spermechecal.**

Ovaries: The ovaries hang in the 13th segment, being attached to spetum 12/13, one either side of the nerve cord. Each ovary is a minute whitish mass in which the ova, in various stages of development, arranged in a linear series.

Oviducts: The oviducts are short conical tubes, each with a large oviducal funnel at its free end. Each oviducal funnel is more or less saucer-shaped structure with folded and ciliated margins, and lies immediately behind the ovary. The oviducts run behind, perforate the septum 13/14 and coverage meeting finally in the body wall beneath the nerve cord to form a short common duct that opens the *female genital opening* on the 14th segment. Internally the oviducts the oviduct are ciliated.

Spermathecae: There are four pairs of spermathecae situated in the 6th, 7th, 8th and 9th segments. Each spermatheca is a flask-shaped structure, the body is which is known as the *ampulla*. From the narrow neck arises a simple elongated *diverticulum* which stores the spermatozoa in *pheretima*. In other worms the spermatozoa are stored in ampulla.

Copulation

During rainy season when the whether is both warm and moist, the earthworms come out of their burrows at night and copulate in pairs. During copulation two earthworms become close applied to each other by their ventral surfaces, with the heads pointing to the opposite directions so that the male genital pore of one is opposite to the spermathecal pore of the other. The areas of male genital apertures are raised into papillae which are inserted successively into the various pairs of spermathecal pores from behind forwards thus filing the spermathecae with spermatozoa and prostatic fluid. After the completion of this process both the earthworms are separated and prepare for laying of eggs.

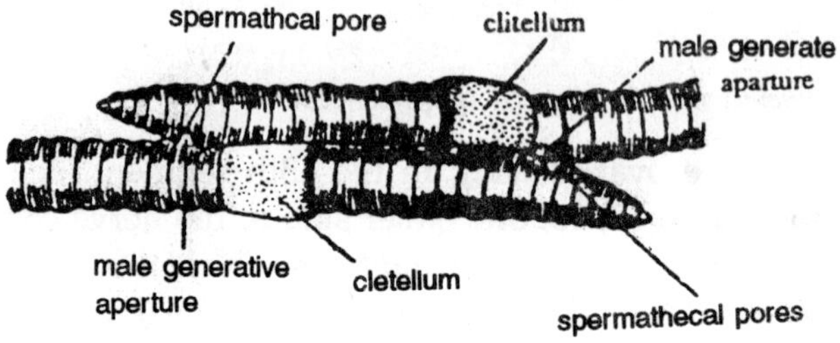

Fig. 4.18. Pheretima. Showing Copulation

Fertilization

Cocoon formation is not well-known is *Pheretima*, However, it has been studied in other worms like *Eisenia* and *Rhynchelmis* etc. The cocoons are formed usually during the months of August to October. If the conditions are favourable the cocoons are also laid in April, May and June.

The cocoon is a product of clitellum. The clitellum is a girdle around the segments 14-16. It has swollen epidermis containing three types glands in three distinct layers, they are unicellular mucous glands which secrete mucous for copulation, cocoon-secreting glands secrete the wall of cocoon the albumen glands produce albumen in which the eggs are deposited in cocoon. The cocoon-secreting golands produce a viscid substance which forms an elastic membrandeous tube around the clitellum on coming in contract with air. The earthworm now starts moving backward so that this tube is pushed forward. The eggs come out of female genital pore are collected in it. By gradual backward movement of the worm, this tube reach over the spermathecal pores and the sperms are released from spermathecae. Thus the fertilization takes place is external. As a soon as the earthworm withdraws itself from the tube, the openings due to elasticity closed at both the ends and, becomes a small rounded capsule, called the cocoon. The cocoon this formed is left in the soil, it is light yellowish in colour and about 2-2.4 mm in length and 1.5-2 mm in width.

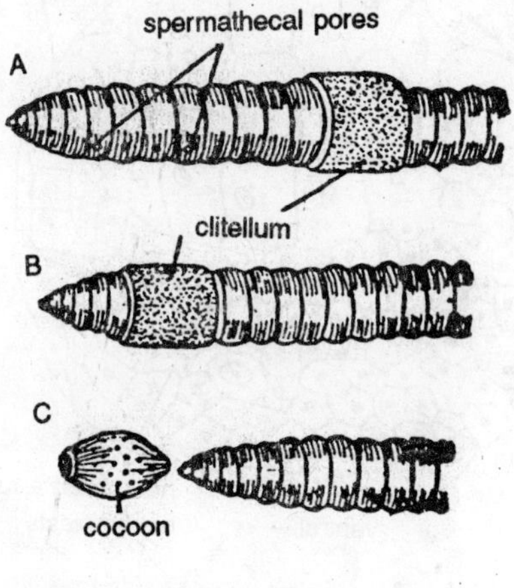

Fig. 4.19. *Pheretima.* Cocoon formation.

Development

The zygote contains very little amount of yolk. Cleavage

is holoblastic but unequal. Two types of cells formed, the small one are called *micromeres* and the larger ones are called *meagameres.* Continued segmentation results in the formation of a hollow spherical body, the *blastula.* The walls of baltula are composed of single-layer of cells, but the ones at the lower end are larger than those above. Two cells lying beside each other near the equator become distinguishable from the rest by the fact that they are quiescent, whereas the rest continue to divided. These are the *mesoblastic pole cells* or *pole-cell* of mesoderm. After sometimes these pole cells divide repeatedly to give rise to two rows of small cells, sivide repeatedly to give rise to two rows of small cells, which, together with the pole cells constitute the mesoblastic bands.

Now the blastula elongates and flatens forming a structure with the small cells above and large columnar cells below, while the mesoblast cells lie on the equatorial plane.

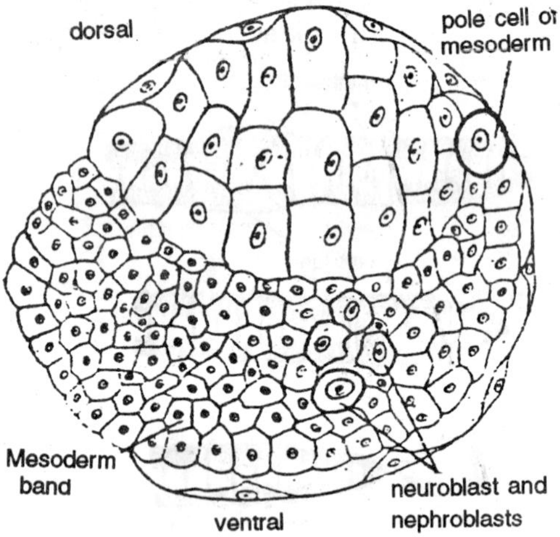

Fig. 4.20. Embryonic development in earthworm. Ventro-lateral view of young embryo.

Gastrulation beings at this stage by invagination. The lower surface made up megameres is pushed inwards. The micromeres multiply and keep on growing over the megameres.

Due to inrushing a cavity is formed, the *archenteron* and the *groove*, produced due to in vagination is a *blastopore* on lower surface. With the completion of process of gastrulation, the three germs layers are established. The outer embryonic cells constitute the *ecoderm*, the cells lining the archenteron, the *endoderm* and the mestoblastic bands, the *mesoderm*.

The blastoporal groove beings to close from behind forwards till only a small opening is left at the anterior end. The persists as mouth and leads into a small *stomodaeum* lined by ciliated ectoderm. The stomodeaum leads into the archenteron. The anus is formed in a very advanced embryo as an ectodermal in vagination fusing with the enter on in the last segment of the worm.

The mesoblastic bands are situated parrel to each other on either side of mid-ventral line. Each band segments into a row of blocks, the mesoblastic somites, which are solid at first, but develop each, a cavity in the interior secondily. The lateral somites with their cavities grow dorsally and ventral ultimately fusing with each other, with the result that a continuous space is formed in the mesoderm in contact with the body wall is the *somatoplure* and the inner layer is the *splanchnoplue.* The somatoplue forms the musculature of body wall and coelomic epithelium. The splanchnopluse forms the gut musculature and the coelomic epithelium surrounding it. The partitions between the successive mesoblastic somites persist as transverse septa.

In the meantime, the ectoderm cells of the ventral surface of the embryo become thickened and distinguished into four longitudinal rows of cells on each side of mid-ventral line, each row ending posteriorly in a large cells, called and *teloblast*. The two rows of cells immediately on either side on the mid-ventral line, give rise to the *ventral nerve cord,* hence the cells forming it are known as the neuroblast called nephroblast cells. All kinds of nephridia are produced by these cells but they differentiate later.

Blood-vessels are formed by the cells of the mesoderm, the ventral and sub-neural appearing in the mid-ventral line, and the dorsal at the top of the somites of each side. As somites of the two sides meet and fuse above, the paired dorsals fuse to give rise to the medium dorsal vessel.

Thus the ectoderm gives rise to the epidermis, nervous and excretory system as well as proctodaeum and stomiodeaum. Mesoderm gives rise to body musculature, coelomic epithelium and the blood vessels. The endoderm gives rise to the gut and its appendages, glands etc.

Further development of worm takes place by the formation of new mesoblastic somites from behind forwards till the embryonic development is completed.

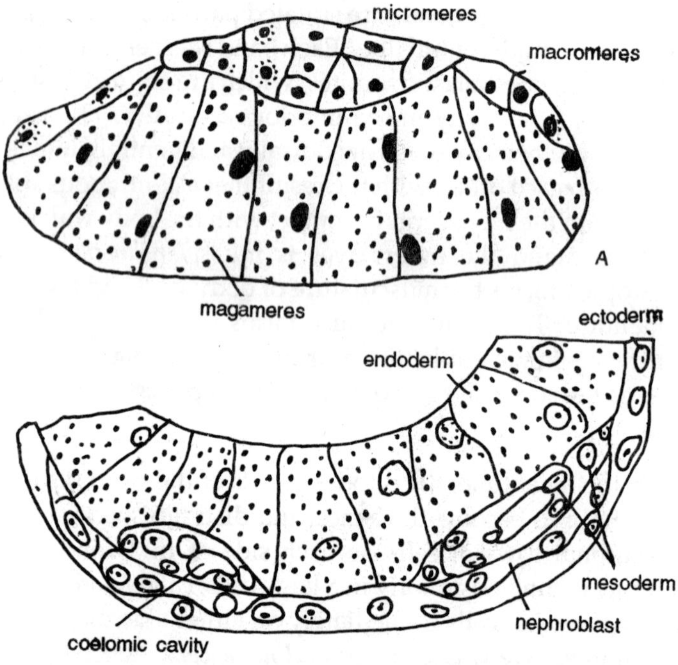

Fig. 4.21. Embryonic development. A-Transverse section of the oval embryo; and B-Ventral portion of a transverse section of an embryo showing the mesoblastic somites with coelomic cavities and the demarcation of nephroblast from ectoderm.

Regeneration

Earthworms posses the ability to regenerate lost part, but this ability is limited. A new head may be regenerated after the removal of upto 15 anterior segments, but the capacity to regenerate decreases rapidly behind the 9th segment. The posterior region are more easily regenerated. If the cut is made near the tail end, a new anterior end will not regenerate on the tail half, but instead another tail will develop from cut surface. When any portion of the tail region is cut off, the last parts readily regenerate. The number of segments regenerated is usually less then the number removed.

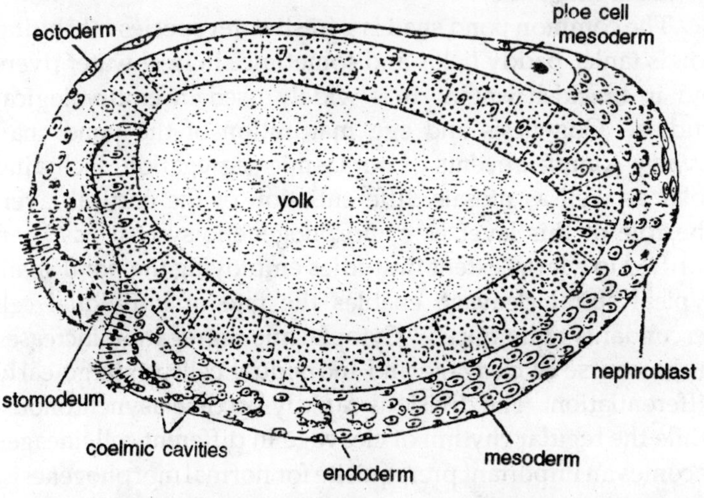

Fig. 4.22. Embryology development. A longitudinal section of the embryo showing the formation of the stomodaeum and the coelomic spaces.

5

Development in a Common Pond Snail

Lymnaea Stagnalis

The common pond snail is a freshwater species inhabiting ponds tanks, paddy fields and sometimes in freshwater rivers and in brackish water. It is widely used in embryological studies. Oogenesis and egg maturation of the pond snail involve complex events of ooplasmic segregation, including polar as well as mosaic differentiation of the cortical layer. The role of this layer as a morphogenetic system has been clarified to a large extent in studies employing the pond snail. Typical spiral cleavage enables the fate of individual cell descendants to be traced. Cytoplasmic segregation increases in the course of cleavage and individual cells undergo early differentiation. The divisions rapidly become asynchronous, while the regular rhythm of cleavage in different cell lineages becomes an important prerequisite for normal morphogenesis. These events can be conveniently studied using the pond snail because of the changing structure of the mitotic cycles. Early blastulation in the pond snail correlates with the early appearance of specialized cell junctions at the stage of two blastomeres; junctions play an important part in the physiology of the early embryo and during the critical periods of embryogenesis such as the emergence of the body plan and the dorsoventral organization. The structure of these junctions has been studied intensively at the submicroscopic level using various procedures, including freeze-fracturing techniques. Methods have been developed which enable a three-dimensional reconstruction of the whole embryo from its thin sections to be obtained; these techniques allow a specific determination of the types and number of contacts in individual blastomeres. The cell surface of early embryos has been studied using a scanning electron microscope.

The results of embryological and cytological studies conducted over the last 20 years suggest the existence of several different types of embryonic determination in the course of pond snail development. These studies demonstrate convincingly the central place of cellular interactions in the mollusc embryogenesis, which is generally considered a classic example of mosaic development. Mosaic features are manifested as the early differentiation of certain larval rudiments or as the early irreversible pinpointing of regions competent to respond to respond to induction stimuli. The most important processes of determination are associated with the contact interactions of blastomeres in the course of blastulation or of whole cellular sheets during gastrulation. It is noteworthy that early determination is associated with changes in embryo symmetry, followed by the appearance of regular differences in the rate and orientation of cell divisions. In contrast, biochemical specialization may precede the histological regionalization of the rudiments in the course of late development. Blastomere isolation in the common pond snail and other species of pulmorate molluscs, achieved by deletions or by separation using hair loops, enables certain regulatory capabilities of the early embryo to be revealed. It has been demonstrated, for example, that pulmonate molluscs belong to the small group of animals which regulate cleavage according to the whole egg pattern, and only in this case does subsequent embryogenesis proceed normally.

Detailed studies of the cleavage and cellular interactions responsible for the embryonic organizational pattern have served as the basis for the mathematical simulation of early pond snail development.

During cleavage the pond snail embryo is a suitable model for studying cytokinesis, mitosis, macro- and micropinocytosis, as well as cellular interactions. Removing the vitelline membrane (using proteases) enables the cell surface to be marked by carbon particles and its movements in the course of divisions to be traced. This method enabled the detection of the spiral deformation of the cleaving zygotes which have opposite directions in dextral and sinistral molluscs. A procedure for isolating the egg cortical layer has been developed; this allows an examination of the organization of contractile structures responsible for cytokinesis on large surface areas. By adding fragments of skeletal muscle myosin to cortices containing intact microfilaments, we have developed a contractile model simulating cytokinesis upon the addition of ATP. The possibility of testing various macromolecules makes this model a useful tool, not only when studying contractile mechanisms in the non-muscle systems, but also for elucidating mechanisms underlying dextral and sinistral cleavage. Zygotes of the pond snail can be used for studying the mechanisms of receptor-mediated endocytosis which may be easily evoked by various multivalent ligands. The factors responsible for the orientation of cleavage spindles in the pond snail may be studied by changing the shape and mutual orientation of the blastomeres; this can be achieved by the centrifugation of embryos and by changing the proportion of nuclear and cytoplasmic volumes in zygotes. Simple methods for isolating mitotic structures from the eggs and embryos of the pond snail allow the characteristics of multipolar mitoses to be examined, for example in the binucleate blastomeres which are produced by cytochalasin treatment and which have different initial distances between the nuclei. Cytokinesis and mitosis may also be studied using smaller zygotes with different cytoplasm composition.

The capsular and vitelline membrane can be permeated by a wide range of substances. This enables experimental studies of cell cycles to be performed and autoradiographic and radiometric experiments to be conducted. Studies of protein synthesis with the aid of electron microscopic autoradiography are also relevant because they throw new light on the function of yolk granules in development.

Pond snail embryos are being used to study genetics and the molecular biology of development. A large number of eggs present in one egg mass, which is undergoing almost synchronous development, enables nucleic acids at different stages of embryogenesis to be isolated. Changes in the spectra of synthesized proteins, activities of different enzymes, and the morphogenetic function of nuclei have been studied. Self-fertilization in the pond snail has been used to convert genes into a homozygous state; this enables their periods of active functioning in the course of embryogenesis to be detected. From the period much attention has been paid to the problem of genetic control underlying dextrality — sinistrality (dissymmetry) in pulmonate molluscs; this trait is inherited maternally and appears for the first time in the course of spiral cleavage cytokinesis. The latest studies performed with *Lymnaea peregra* enabled numerous genetic aspects of the problem to be defined. They also allowed us to demonstrate that dissymmetry of the dextral cleavage type is the result of a dominant gene product located in the egg cytoplasm.

Later stages of embryogenesis, cytodifferentiation in connection with polyploidization, postembryonic growth, and histogenesis in the pond snail are also the subject of numerous studies. Lastly, embryogenesis in the pond snail may be used as an inexpensive and readily available model for teratological experiments. These studies are aimed, for example, at testing reservoirs for industrial pollution.

The pond snail is quite common in natural freshwater habitats and may be easily cultured in the laboratory. The adult animals are used, for example, for studying the properties of enzymes, hormones, and calcium metabolism water metabolism, the function and composition of the hemolymph defense reactions and phagocytosis, neurophysiological processes, the physiology of nutrition, behaviour and the dynamics of natural populations. Considerable advance has been made in the study of the reproduction biology of the pond snail which shows endogenous rhythms under a complex hormonal control.

Obviously, the common pond snail is one of the popular laboratory animals and is widely used in experiment biology.

We now list some of the drawbacks of using the pond snail as a laboratory species. These include :

a. Difficulties in culturing embryos outside capsules, which is only possible up to gastrulation.

b. Few opportunities for performing microinjections and conducting electro-physiological studies because the plasma membrane of the egg or blastomeres is highly sensitive to mechanical lesions.

c. Rather high mortality rate among young snails, thereby making it difficult to maintain constant laboratory cultures.

d. High degree of infestation among natural populations by fluke larvae. This means that a considerable number of specimens collected in the field have to be discarded.

Some of these drawbacks can be overcome by using other species of the genus *Lymnaea,* also used in embryological studies: *L. acuminata, L. auricularia, L. exilis, L. elodes, L. ovata, L. palustris, L. peregra* and others. The embryos of *L. peregra,* for example, allow us to perform microinjections amounting to about 2-4% of the egg volume. This has been used in experiments to determine the temporal interval in which the genes responsible for coiling are active during development. Morrill has recently published a comprehensive survey, concerning the normal development and experimental embryology of *L. palustris* as well as other species of Lymnaeidae.

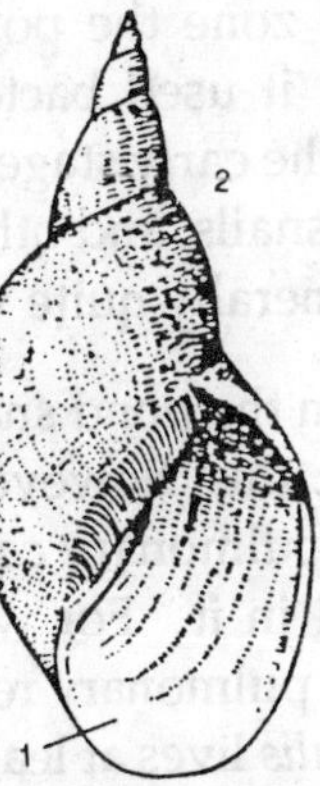

Fig. 5.1. Shell of *L. stagnalis* (natural size) 1) Aperture; 2) helix.

Taxonomy and Distribution

The genus *Lymnaea* (Gastropoda, Pulmonata, Basommatophora, Lymnaeidae) contains numerous species of the pond snail. There are several species, of *Lymnea*. They include *L. stagnalis L.*, which can be easily distinguished by its light-coloured shell, the helix of which has height no less than the diameter of the mouth, and whose last turn is greatly inflated. The largest shells attain 70 mm in height while the average height is 40-47 mm. The shape of shells may vary considerably. Normally the shell has right-handed coiling, exceptions to this being extremely rare. Further details on the anatomical and microscopic structure of different organs can be found in specialized studies. *L. stagnalis* is common in Europe, northern Asia, and North America. In the USSR it can be found throughout the entire European part as well as in the Trans-Caucasian region, central Asia, eastern and western Siberia, and in the Far East.

Ecology and Culture

The pond snail lives in the littoral zone of stagnant reservoirs or slow rivers (e.g., the Moskva). It prefers reservoirs where the pH of water is 6.0-7.0. Because of its capacity to withstand water pollution it is classed as a mesosaprobe.

In a moderate climatic zone the pond snail feeds from May to October. For food it used bacteria, algae, detritus, higher plans (frequently in the early stages of stagnation), and in summer it may eat ticks, snails, and other small animals, as well as insect eggs. It is generally quite voracious.

During the warm season the pond snail uses atmospheric air, breathing with its lung; this, however, does not exclude the possibility of filling the pulmonary cavity with water and using the oxygen dissolved in it. For wintering the animal digs into the ground and the pulmonary respiration is replaced by skin respiration. *L. stagnalis* lives at least a year on average, sometimes longer, from 2 to 5 years. It has been reported that the maximum age of the pond snail is 7 years.

In nature, pond snails often serve as the intermediate hosts of various flukes, including *Fasciola hepatica;* the larval forms of these flukes frequently infest the internal organs, such as the liver, of a considerable proportion (sometimes up to 40%) of molluscs in a given water reservoir. Since larval infestation affects fertility, the possibility of infestation should be checked immediately when animals collected in the field are destined for the laboratory. For this purpose the collected molluscs are placed individually into glass beakers containing 100-150 ml of clean tap water and kept without food. If they are infested with trematodes, the fluke larvae appear in the water as small white mobile points after 1-3 days. Such molluscs are then discarded. Warming the water by several degrees increases the reliability of the method. Although heavier body weight implies greater weight of gonads, moderately sized snails should be selected for the laboratory. This is because the largest specimens are usually infested by trematodes. If a constant laboratory population is needed, molluscs can be grown from eggs in the laboratory. This procedure, however, is rather cumbersome because the mortality rate of juvenile snails is high and correspondingly many batches should be cultured. However, pond snail progeny reared in the laboratory are always free from fluke larvae.

Sometimes reproduction is impeded because of infestation by semiparasitic oligochaetes such as *Chaetogaster lymnaei*, which resemble large rotifers. Conditions are keeping pond snails can vary depending on the purpose of the future experiments. Generally a constant laboratory population with minimal mortality and moderate reproduction activity is desirable. In the laboratory each animal should be provided with at least 0.5 liter of water. Increased population density not only adversely affects the feeding intensity and fertility, but may even result in cannibalism. The animals should preferably be kept in large aquariums with constant aeration and running water. Placing aquatic plants into aquariums is not a necessity, but may be desirable as a source of small animals for food. If the aquarium is stationary and there is constant aeration, the water need only be changed one every 2 weeks. Without aeration at 17-20° C the water should be changed weekly, and if plenty of food is supplied it should be done even more frequently. The following may be used to feed the molluscs: carrots, radish leaves, cucumbers, cabbages, various aquatic plants. Pond snails prefer lettuce leaves and *Hydrocharis*. Dead pond snails may be left in the aquarium because they are usually eaten by the remaining molluscs. Fraser recommends feeding with boiled wheat grains. According to some observations, food should contain mineral particles; in other words, some sand or river mud should be put into the aquariums. Plenty of food generally has a favourable effect on the growth and reproduction of the pond snail.

In winter, aquariums should be equipped with additional illumination facilities. An illumination period lasting 12 h is sufficient to maintain moderate intensity of metabolism; a lower temperature and a shorter illumination period may stimulate the snail's preparation for hibernation. Growth and gametogenesis proceed at temperatures of no less than 11-12° C. Genetic and embryological experiments may require self-fertilization in order to obtain homozygous embryos. For this purpose snails hatched from one egg mass are kept together until they reach 5 mm in length. Thereafter each animal is kept in an individual vessel.

Mass production of normal fertilized eggs for embryological studies requires the pond snails to be placed periodically under conditions which stimulate oviposition. Such conditions include a higher temperature, low content of metabolic products such as ammonia and urea in the water, no food, and the presence of green plants. Food, however, should be plentiful during the preceding period, and the illumination period should be sufficiently long; the latter generally maintains reproduction at a high rate. According to the author's experience, this can be achieved by keeping two or three pond snails together in special vessels with 2-3 liters of water but without artificial aeration. Each group of snails should be stimulated to lay eggs no more than twice a week. A small volume of water in the vessels leads to rapid water pollution, thereby preventing spontaneous egg-laying. Care should be taken to keep this contamination at an acceptably low level, avoiding water acidification and the appearance of bacterial growth on the surface of the water.

Reproduction and Stimulation of Oviposition

Pond snails are hermaphrodites but, as a rule, mutual insemination takes place during mating. Self-fertilization is less frequent. Furthermore, when the snails are cultured individually, egg-laying begins later than when several animals are placed together. The common pond snail reaches maturity approximately 1.5 months after hatching from the cocoon. When embryogenesis has been completed, the gonads pass through the juvenile stage, then the male maturity stage, and then achieve complete hermaphroditism. Ovulation first begins when the shell height attains about 25 mm. It can, however, be stimulated even in younger animals by injecting an extract prepared from the caudal-dorsal cells of the cerebral ganglia. The onset of reproduction may also be affected by the duration of the photoperiod. If the photoperiod is long (16 h), at 20° C the snails begin to lay eggs as early as 7 weeks after hatching. If its duration is 12 h, they start oviposition after 9 weeks.

In a moderate climatic zone adult pond snails begin to lay eggs late in May. Oviposition reaches its maximum in June and ends in August. The temperature at which pond snails lay their eggs is generally 15-18°C. In nature the reproduction of pond snails begins during their second summer as they only grow to 8-10 mm in size during the first summer. Under laboratory conditions (20-23°C) the molluscs lay eggs from March to November. Although the author managed to obtain eggs from pond snails in December and January under the conditions of a 12-h photoperiod, it is not usually recommended that embryological experiments be conducted during the winter months. As previously mentioned, reproduction is stimulated by a long photoperiod, abundant nutrition, optimal temperature, clean water, and moderate population density of pond snails. The function of gonads is quickly affected by starvation. According to Levina the pond snail may lay about 8700 eggs during its lifetime, although other data put the number at above 10,000.

Reproduction in the pond snail is controlled by several regions of the nervous system. Two groups of caudal-dorsal cells located in the cerebral ganglia secrete the ovulation hormone, usually no later than 2 h prior to egg laying. The dorsal bodies produce another hormone accelerating the differentiation of accessory organs of the female part of the gonad. The lateral lobes of the cerebral ganglia stimulate the secretion of both these hormones and inhibit the neurosecretory light-green cells which produce the growth hormone. A balance between growth and reproduction can be partly controlled by the duration of the photoperiod. Hormones of vertebrate animals have been used to control reproduction of the pond snail. Gametogenesis in the pond snail has been analyzed extensively. The duration of oogenesis from the completion of oogonial mitoses to oviposition is about a month. During the course of mating, sperms enter the spermatheca through the female genital pore and sometime thereafter small packets of spermatozoa are delivered into the vagina. They pass through the oviduct and fertilize the eggs in the straight part of the hermaphroditic duct.

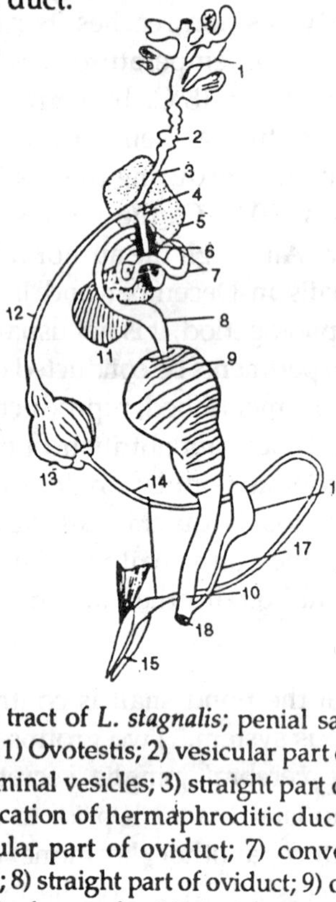

Fig. 5.2. Genital tract of *L. stagnalis*; penial sac and muscles are not shown. 1) Ovotestis; 2) vesicular part of hermaphroditic duct with seminal vesicles; 3) straight part of hermaphroditic duct; 4) bifurcation of hermaphroditic duct; 5) albumen gland; 6) ventricular part of oviduct; 7) convoluted part of oviduct (uterus); 8) straight part of oviduct; 9) oothecal part of oviduct; 10) vaginal part of oviduct; 11) accessory mucous gland; 12) pars afferens of seminal duct; 13) prostate; 14) vas deferens; 15) penis; 16) seminal receptacle; 17) seminal receptacle duct; 18) female genital pore.

Fertilized eggs from the hermaphroditic duct individually enter the ventricular part of the oviduct, which simultaneously receives a portion of the secretion produced by the albumen gland. Thereafter the egg, along with some of the liquid, enters the pouches of the convoluted part of the oviduct (the uterus), where a two-layered capsular membrane and the mucous capsule pellicle are being formed. In the straight region of the oviduct the capsules are glued together by the secretion of the accessory mucous gland and, finally, the envelope of the cocoon is formed in the nidamental or oothecal part. The egg masses are generally deposited onto green plants such as *Nymphaea alba*, *Nuphar luteum*, *Potamogeton natans*, and *Hydrocharis morsus ranae*. The envelope of the cocoon immediately after laying is opaque, but 1.5-2 h later its swells and becomes transparent. According to Raven the whole process at 17°C, from the entry of the first egg into the oviduct to laying, takes 2 h from the time the molluscs were stimulated.

When the cocoon is laid, it emerges blunt end first from the vagina. The eggs located at this end begin to develop earlier (the difference in the developmental stages of anterior eggs and those located in the posterior part of the cocoon may be 45-50 min. at 22°C). The first snails to hatch from the capsules, however, are those located in the external region of the central cocoon part. They are followed by those at the anterior tip and finally by the capsules of the posterior cocoon end. With regular feeding and aerated water the reproduction may be stimulated by several factors, such as lower water temperatures (by several degrees) or the replacement of rougher feed by lettuce leaves and a longer photoperiod (our observations). Other methods for stimulating reproduction are as follows :

a. Molluscs are kept in the aquarium without plants, and fed regularly; thereafter *Hydrocharis* leaves are placed into the aquarium and intense illumination is supplied.
b. Molluscs are kept at 17-18°C, fed regularly, and then the temperature is gradually raised to 25-26°C.

We use the following technique. Molluscs are kept in groups of 2-3 animals in aquariums or finger bowls without aquatic plants but with a small amount of river sand. The water is not changed for a week while the animals are regularly fed with lettuce. Two or three days prior to stimulation, cucumber pulp or cabbage leaves are added. This rapidly pollutes the water. The illumination is kept intensive at room temperature. On the day of the experiment the polluted water is changed for freshwater, an intense light is switched on, no food is given, and the water is gradually heated to 25-26°C. *Hydrocharis* leaves or transparent plastic dishes with a flat bottom are placed onto the water surface; the snails use them for oviposition, which takes place several hours after the water has been changed. The snail does not usually change its position for sometime prior to egg-laying.

THE GAMETES

Eggs

Pond snails lay eggs in elongated cocoons or syncapsules which at the beginning of the reproductive period contain several, and later up to 300, eggs each of which is located in an individual egg capsule.

The mean capsule length is 1.3 mm with a diameter of 0.8 mm. The presence of several eggs in a single capsule (polyvitelliny) is abnormal. Fertilized eggs are laid at metaphase I. The egg is somewhat irregular in shape immediately after oviposition. However, after about 5 min. it acquires the shape of a rotation ellipsoid; that is, it becomes flattened along the animal — vegetal axis. Its maximum diameter at this stage is 122 μm (the average is 120 μm). The egg volume varies from 1.1 to 1.6 nl. However, during the period between laying and the onset of cleavage, the egg volume slowly increases by about 45% owing to the uptake of water. A small part of the egg in the animal pole region is free of yolk and contains the first maturation division spindle. Usually the egg has yellowish-white pigmentation. In addition to a plasma membrane, the egg is covered by a thin (no thicker than 1 μm) vitelline or yolk membrane. The perivitelline space is virtually absent (0.5 μm or less). The vitelline membrane, which consists of loosely interwoven fibrils and small round bodies, is formed from the peripheral layer of the oocyte zona radiata. The egg may be regarded as telolecithal because of the distribution of yolk; the animal half is less rich in yolk and contains the nucleus.

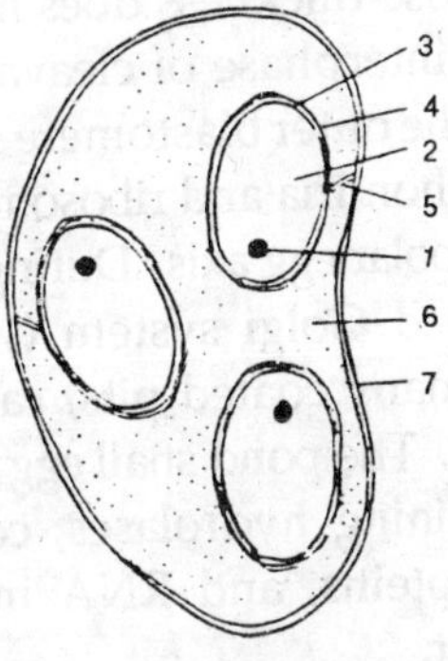

Fig. 5.3. Sketch illustrating structure of L. stagnalis cocoon. 1) Egg covered by vitelline membrane; 2) albumen fluid of egg capsule; 3) membrane of egg capsule; 4) outer membrane of egg capsule; 5) pedicle of egg capsule; 6) inner mucus of cocoon (tunica interna); 7) outer envelope of cocoon (tunica capsulae).

The plasma membrane forms microvilli, which contain bundles of microfilaments. Microvilli are first formed at the free oocyte pole corresponding to the future animal pole of the egg. The oocyte region attached to the gonad basal membrane corresponds to the vegetal pole. Under the plasma membrane of the mature egg, actin microfilaments form several successive layers, the microfilaments of adjacent layers being inclined to one another at twist angles of 30-60°C or 90°. Each layer is composed of many small domains containing 2-20 parallel microfilaments of identical polarity as revealed by decoration with myosin subfragment I. It cannot be excluded that such domain helicoid organization of the microfilament network is directly related to the spiral nature of cleavage cytokinesis (see below). The pond snail egg does not contain cortical granules. The growing oocyte vigorously takes up exogenous ferritin, along with other substances, and ferritin taken by endocytosis involving coated pits and vesicles forms paracrystalline inclusions in the yolk granules. According to our observations, these coated structures are also present in the mature egg cortex and may be intimately associated with microfilaments. Prior to cleavage microtubules are not numerous in the cortical layer, whose thickness does not exceed 0.1 μm; by contrast, during the interphase of cleavage divisions they may accumulate under the outer blastomere surface as parallel and oblique rows. Mitochondria and ribosomes are distributed non-uniformly along the polar egg axis. During the first cleavage division a well-developed Golgi system (with bulky trans-reticular elements and many coated pits) can be observed in the subcortical cytoplasm. The pond snail egg has yolk granules of different types containing hydrolases, complex proteins, phospholipids, basic proteins, and RNA in addition large-lipid droplets are present.

Yolk β-granules with a diameter of 1-3 μm and coated with a membrane are formed earlier during oogenesis than are the γ-granules. The diameter of the γ-granules exceeds μ4 m after oviposition. The γ-granules are surrounded by a vacuole which takes in water from outside. It is assumed that they play the role of lysosomes during development. β-Granules, which are consumed at a slower rate in the course of embryogenesis than the γ-granules, also appear to participate in the intracellular nutrition. There is evidence that yolk granules participate in the formation of cell membranes. Complex ooplasmic segregation may be traced cytochemically in the pond snail zygote. While the egg passes through the oviduct, a vegetal pole plasm containing abundant yolk β-granules is being formed. After oviposition this plasm moves to the animal pole under the membrane and is transformed into the subcortical plasm. Animal pole plasm rich in mitochondria appears soon after the extrusion of the second polar body. By the third cleavage division both plasms merge together. In the course of subsequent cleavage division the plasms undergo a non-uniform distribution between blastomeres in the form of ectoplasm surrounding the nuclei. At the sites where the oocytes contacts inner follicular cells, 6-9 subcortical accumulations (SCA) are formed under the equatorial streak. They are very rich in RNA, which does not, however, incorporate any labeled precursors after oviposition. The SCA partially fuse together in the course of synchronous cleavage divisions and can be located under the vegetal surface of the macromeres at the 8-blastomere stage. Later, concomitantly with the determination of the 3D macromere (see below) the SCA undergo compactization. They are transformed into ectosomes migrating to the center of the embryo along the inner contact zones of the blastomeres. It may well be that they participate in inductive interactions. Submicroscopically the SCA correspond to large accumulations of polysomes, endoplasmic reticulum, and small vesicles; the aggregation of these components appears to increase during the course of cleavage. Raven assumed that, prior to cleavage, the SCA system is related to the determination of the embryo symmetry plane, which always passes through the pair of blastomeres B and D.

The egg is suspended in the capsular fluid which is the main source of nutrition for the embryo. It contains galactogen (3-6% dry weight), proteins (6-8%), as well as salts and mucopolysaccharides. The content of individual components depends on the composition and quantity of food. At the 40-49 blastomer stage the capsular fluid undergoes pinocytosis in the animal zone, while at the 120-blastomere stage (corresponding to a late blastula) pinocytosis spreads all over the surface of the embryo. The extracellular digestion of the capsular fluid entering the gut lumen begins at the trochophore stage.

The pond snail egg is protected by the cocoon envelope, the jelly, the capsules, the capsular membrane, the protein-rich fluid present in the capsule, and finally the vitelline membrane. The content of the freshly laid egg is isotonic to a 0.093 M non-electrolyte solution. The salt content of the egg corresponds to 0.042 M NaCl or 0.05 M LiCl solution. The egg membrane as well as the vitelline membrane is permeable to numerous non-electrolytes, cytidine, thymidine, uridine, uracyl, amino acids, puromycin, colcemide, and inorganic ions such as alkali metal cations with the possible exception of lithium.

It has been shown that the intensity of sodium exchange between the embryo and the capsular liquid increases 50-100 times soon after the onset of cleavage, while water transport increases by a factor of 5. The cocoon envelope is permeable to inorganic ions, water, and various gases, but appears to be only slightly permeable to bile acid salts. The mucus present inside the cocoon appears to exert an absorptive action on alkali metal cations. The capsular membrane is permeable to water, inorganic ions, amino acids, cytochalasin, bile salts, sucrose, raffinose, ATP, EDTA, sodium azide, ethyl alcohol, and urea. The capsule membrane, however, is not permeable to polyethylene glycol (molecular weight 3000-3300). Actinomycin D has poor penetration and needs to be used in concentrations up to 200 μg/ml for a long period. Neyfakh has observed a rapid effect of actinomycin after puncturing the egg capsules. Puromycin penetrates the egg easily. The colloid osmotic pressure of the capsular fluid is 4-5 mOsm.

Spermatozoa

Characteristically, the pond snail spermatozoon possesses a very long intermediate region. The length ratio of the head, the intermediate region, and the tail is approximately 1 : 12 : 11. The intermediate region consists of an axonema and a strongly modified mitochondrial layer containing paracrystalline structures. The caudal part is rich in glycogen. The head and middle part of the sperm have been found to contain bundles of filaments with a diameter of 10-12 mm, which appear to be similar to the intermediate filaments.

Culture of Embryos

Observations of developments and certain other types of experiments may be performed using the cocoons. It is much more convenient, however, to work with individual capsules.

To obtain these, the egg mass is placed on a filter paper and cut into several parts with a surgical knife or a razor blade; then the capsules are pressed out with forceps onto filter paper. The capsules are rolled on paper to remove the jelly. They must be cleansed of the jelly may possess either protective or teratogenic properties after interaction with various agents.

The capsules may be cultured in ordinary tap water, which should be changed daily and, if possible, additionally purified of copper traces. The best results are obtained if the capsules are placed onto a 2% agar surface saturated with water vapor (cracks in the agar may be filled with water). In the latter case, the optimal temperature is 25°C. Under these conditions the spontaneous mortality of embryos does not exceed 2%. In order to facilitate permeability to different substances, the capsules may be punctured as described below. Although the capsular fluid undergoes gelatinization around the hole sometime after puncturing, the punctured capsule nevertheless becomes more prone to bacterial and fungal infection. Nevertheless, a certain proportion of the embryos in such capsules develop successfully up to hatching and almost all of them pass through the gastrula stage. Care should be taken to prevent too much of the fluid from leaking.

Finally, fixation and certain experimental manipulations require the embryos to be removed from the capsules. The best procedure for performing this is as follows : a batch of capsules is transferred by a pipette into a dry plastic dish and excess water is removed. The capsules are aligned by a glass rod along the wall under a dissecting microscope and then quickly punctured with sharp needles. During this procedure some of the fluid leaks out, thereby reducing the intracapsular pressure, and the majority of embryos remain inside. Some water is added and the fluid remaining in the capsules along with the eggs is carefully pressed through the hole in the torn capsular membrane. Decapsulated embryos are washed several times with water to remove the last traces of the viscous capsular fluid which may interfere with the fixatives. Such embryos, as well as the blastomere plasma membrane, are only covered by the vitelline membrane which is shed in the course of gastrulation. Up till now all attempts at a more or less long-term cultivation of embryos outside the capsules have been unsuccessful. This is probably because the embryo normally begins to absorb the capsular fluid at relatively early developmental stages, and this process continues almost up to hatching. It has not been possible to cultivate *Lymnaea* embryos in the capsular fluid although this has been possible in the case of *Limax*. According to Morrill, post-trochophore *Lymnaea palustris* embryos isolated from the capsules can develop normally in pond water.

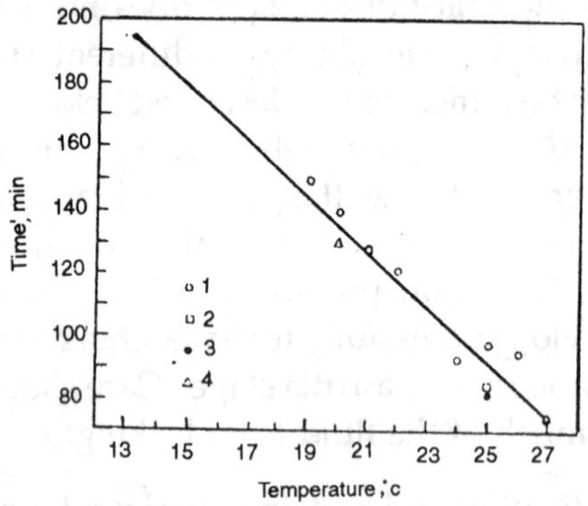

Fig. 5.4. Temperature dependence of the time between the appearance of the first and second cleavage division furrows or between the first and second division metaphases in the pond snail (mean values are given). 1) Our data; 2) Verdonk; 3) Labordus; 4) Jockusch.

We believe that the following approach is promising. After experimental manipulations with decapsulated embryos they are washed in a sterile solution. Each embryo is then transferred using a pipette with the tip of a 15-20 μm diameter into a punctured capsule where the egg was inactivated by a beam of UV irradiation. The compositions of the solutions used for the short-term culturing of early embryos are as follows :

Jockusch

	g/liter
NaCl	.060
KCl	0.42
$CaCl_2$	0.24
0.02 M tris-HCl	
(pH 6.7)	

Horstmann

	g/liter
Na_2HPO_4	1.246
$NaH_2PO_4 5H_2O$	0.468
NaCl	1.265
KCl	0.185
$CaCl_2$	0.115
$MgSO_4 7H_2O$	0.185

Taylor

	mmole/liter
$NaHCO_3$	2.00
$KHCO_3$	0.40
$CaCl_2$	1.72
$MgSO_4$	1.00
(pH 7.4)	

Jockusch has pointed out that eggs left inside the cocoon, eggs in capsules placed in water, and eggs decapsulated and incubated in the salt solution which she described develop almost synchronously during the early phase.

According to Raven early embryos develop normally in tap water or in 0.4 mM $CaCl_2$ solution. Our own observations confirm this. Sometimes development in water proceeds to late blastula, but more frequently to the stage of 24-29 blastomeres. From the 16 blastomere stage onward, a slight developmental delay as compared with the control may be observed. Thus, when studying early cleavage one can successfully culture embryos in tap water provided that the decapsulation was careful and accurate. Performing experiments and observations at 22-25°C is very suitable.

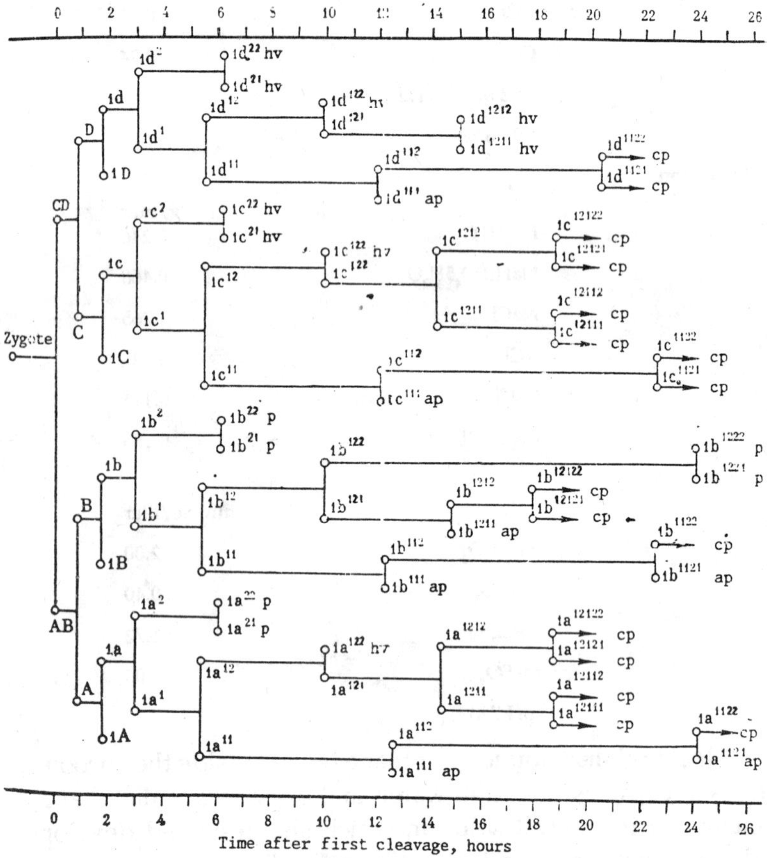

Fig. 5.5. Lineage of the first micromere quartet up to early gastrula in *L. stagnalis* at 25°C, ap, apical plate; cp, cephalic plate; hv, head vesicle; p, prototroch.

More refined experiments may require the eggs to be pooled at synchronous developmental stages. It has already been mentioned that eggs in a batch do not show quite synchronous development. More crude "synchronization" involves cutting the anterior and posterior fifth from the cocoon and using the middle part. According to another technique, the cocoon, immediately after laying but prior to the first cleavage division, is cut into 10-12 transverse parts. In each part 2-4 eggs are used as a control and the rest for experimental purposes. If the experiment begins after the first cleavage division, then groups of capsules are selected for which the appearance of the first cleavage furrow occurs within 1 min.

The first and last of such groups serve as controls. If the embryos are decapsulated, a similar method may be used even prior to cleavage. In this case, the extrusion of polar bodies may be used as a time marker, this can be observed using trans-illumination.

If it is not planned to follow up late morphogenetic events, which may be impaired after treating the early embryo at 37-38°C, one may successfully synchronize the embryos at the first cleavage division by heating and subsequent accumulation in the cold. For this purpose one requires a heating table with a temperature of about 35°C and an ice bath (preferably a plate of ice wrapped in a polyethylene film with a surface temperature of 4-6°C). The embryos with a furrow on the animal pole are immediately transferred to the cold (all manipulations are made in water droplets in plastic dishes), while the others are placed for 2-5 min. at 35°C. Thereafter the embryos with furrows are again selected and transferred to the cold. The procedure is repeated until all the embryos are in the cold. They can then be used for experiments. It should be pointed out that long-term contact with 0-6°C may prevent the appearance of the first or second polar bodies, and this whole process results in the formation of tetra- or tripolar mitoses during the first cleavage division. Therefore, temperature treatments for the sake of synchronization should be employed when the maturation divisions have been completed.

Effect of Temperature on Development

Pond snail development is possible at temperatures ranging from 10 to 30°C.

In nature oviposition usually takes place at 15-18°C and ceases at temperatures exceeding 27°C. However, in some pond snail species development may proceed at temperatures as low as 2-5°C. Richards has studied the duration of pond snail development as a function of temperature (from the outset of oviposition until 50% of embryos have been hatched); his results are presented below :

Temperature °C	*Percent of embryos hatched*	*Time of development (h)*
5	0	—
9	1	>2000
12	90	1262±41
15	94	866±18
18	84	688±14
20	93	588±20
22	96	489±15
25	97	448±17
28	96	416±27
32	0	—
35	0	—

If the time of development is T, then the relationship of 1n (1/T) to temperature is a straight line in the temperature interval 13-26°C. Thus, taking into account the percentage of successful hatchings, the optimal temperature range is 15-26°C. It should be remembered that the optimal temperatures is 25°C when capsules with embryos are cultured on agar. A special study dealing with the effect of different constant and varying temperatures on the duration of development has been published for *L. auricularia.* These authors point out that the embryos may be frozen for short periods without losing their capacity for subsequent development.

Since the first three cleavage divisions in the pond snail are synchronous, the interval between the appearance of the first and second cleavage division furrows (τ_0) may be used, as in other animal species, as a dimensionless unit to define the relative duration of development. Figure 5.5 presents available data on τo for the pond snail at different temperatures. Jockusch's data refer to the mean duration of the interval between the metaphases of the first and second cleavage divisions. In these experiments the eggs present in the cocoon, inside the isolated capsules, and decapsulated eggs in the salt solution underwent synchronous development. All other data are based on vital observation of cleavage, and they refer to the average interval of time between the initial appearance of first and second cleavage division furrows in the egg batch used. Temperatures quoted in the above account are accurate to only ± 0.5°C.

Normal Development

Cleavage in the pond snail is of the spiral type and is highly deterministic. In other words, the fate of individual blastomeres can be followed accurately until the early stages of organogenesis. Figures show the pattern of cleavage for the first quarter of micromeres, and for the whole embryo up to the stage of 49 blastomeres.

We shall now consider certain peculiarities in the cell cycle during early cleavage. According to Jockusch, the relative duration of the mitotic phases in cell cycles II and III at 20°C are as follows (figures in percentages):

	Interphase	*Prophase*	*Metaphase*	*Anaphase + telophase*
II	23.4	26.2	21.0	29.4
III	25.0	27.3	22.7	25.0

Van den Biggelaar has reported somewhat different figures for these cycles (in percentages):

	Interphase	*Prophase*	*Metaphase*	*Anaphase*	*Telophase*
II	46.5	12.8	11.6	5.8	23.3
III	44.2	16.2	11.6	4.7	23.3

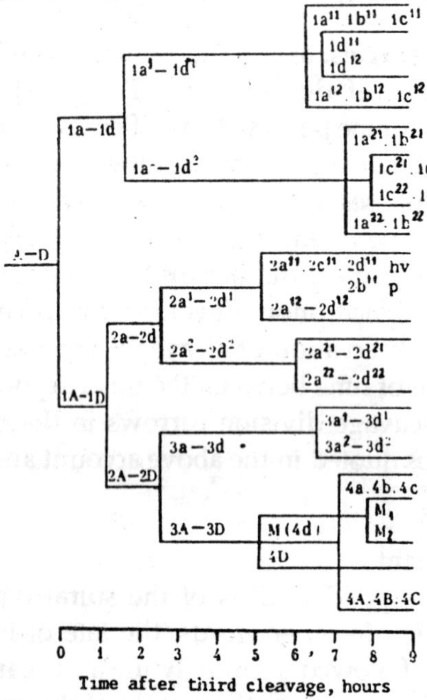

Fig. 5.6. Cell lineage in L. stagnalis embryos between the stages of 8 and 49 blastomeres at 25°C (19). Cells belonging to the prototroch (p) and the head vesicle (hv) do not divide any further.

The structure of the interphase has been studied in detail by van den Biggelaar using autoradiography and cytophotometry. In the course of the first cycles (whose duration in the course of synchronous divisions has been designated as T), the proportion between phases of the cell cycle is as follows: mitosis, 0.48 T; S-phase, 0.27 T; G_2 phase, 0.25 T. G_1 is absent during these cycles as well as during all subsequent ones. The duration of mitosis is identical for synchronous and asynchronous divisions. The karyomeric nuclei during cleavage cycles I-IV are irregular in shape and polymorphic even by the end of the G_2 phase. They only become rounded at the beginning of the prophase. The beginning of prophase in these cycles can be easily identified on the embryo sections by the concentration of chromatin on one side of the nucleus, close to the outer surface of the blastomere.

Until the 49-blastomere stage DNA synthesis begins at the telophase; at the 49-blastomere stage DNA is also synthesized in the cells which no longer divide. The duration of the S period begins to undergo significant changes only during cycles $1a^1$-$1d^1$ and $1a^2$-$1d^2$, that is, after the 16-blastomere stage. Any differences in the duration of cycles between individual animal blastomere prior to this stage are due to changes of G_2; for the vegetal blastomeres the same holds true until the stage of 49 cells. In the same $1a^1$-$1d^1$ and $1a^2$-$1d^2$ blastomeres the nucleoli appear for the firs time at the stage of 16 cells. At the stage of 24 cells they appear in other blastomeres, too. Although the first divisions of cells belonging to quartets 1, 2 and 3 are not synchronous. DNA synthesis procceds in them synchronously.

Three periods, each possessing certain morphological peculiarities, may be distinguished in the course of early cleavage by the criterion of cytokinesis synchrony. The first includes the synchronous divisions (from zygote to the 8-blastomere stage); the second involves synchronous divisions of cells with an identical index (e.g., $1a^1$-$1b^1$-$1c^1$-$1d^1$), that is, of cells located at an identical level relative to the poles of the embryo (8-24 blastomeres); the third is a period of general asynchrony involving differentiation of the quadrants A—D in the timing and direction of divisions (24-29 blastomeres). At first we will consider certain phenomena common to all three periods and associated with the contact interactions of newly appearing sister blastomeres. The vitelline membrane surrounding the whole embryo and providing anchorage for part of the egg-surface microvilli does not take part in forming cleavages furrows. The opposing membranes of new blastomeres begin to establish contact only when cytokinesis has been completed. The growth of contact zones at the place of earlier furrows ends at the G_2 period; it is achieved not by the adhesion of blastomere outer surfaces but through the intercalation of a new surface from the inside. Therefore there is a continuity between the surface of the zygote and the outer surface of the embryo. This continuity is partly broken only during the third period, when the new contact zones of the blastomeres emerge, not only in the course of cytokinesis but also as a consequence of interphase cell movements.

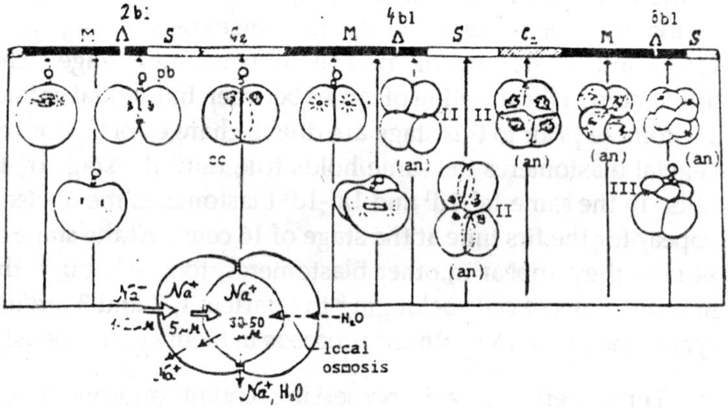

Fig. 5.7. Correspondence between the shape of the embryo and phases of the cell cycle during synchronous cleavage divisions in the pond snail. Open arrows on the cell-cycle scale indicate end points of cytokinesis. Overlapping of the S-phase and telophase is not shown for the sake of simplicity. Roman numerals refer to the number of the division where the corresponding contact zones are formed. pb, polar bodies (over the animal pole); cc, cleavage cavity (side view and animal view). Shown below is a sketch illustrating the active (open arrows) and passive (small solid arrows) flows of sodium ions and water accompanying the cleavage cavity formation. Concentration of exchangeable sodium is shown in capsular fluid, blastomere cytoplasm, and cleavage cavity, bl, blastomeres; an, animal view.

The so-called recurrent cleavage cavity appears at the 2-blastomere stage. It is formed as a result of the secretion of fluid from the vacuolized inner walls of the blastomeres. The cavity attains maximum size during the G_2 or prophase when its volume is 0.2-0.4 nl, while the volume of the embryo minus that of the cavity is about 1.1 nl. The cavity is emptied through apertures located on the poles of the embryo during metaphase, anaphase, or early cytokinesis. After each division, the recurrent cleavage cavity is rapidly restored along the previous contact zones, but forms slowly in the region of the growing contact zones. As the cavity continues to increase, the shape of the embryo becomes more regular. With experience the shape of the embryo, as well as the location and size of the sister nuclei during the first cycles, may provide information on the cycle phase. As early as 1850 it was assumed that the cavity plays a part in osmatic regulation, and this has recently been proved using radioactive isotopes. The fluxes of exchangeable sodium ions which serve to actively pump water into the cavity. It has also been assumed that the cavity serves to remove unnecessary metabolites and maintains a particular inner medium which is necessary for the functional interaction between blastomeres. A permanent central cavity forms during the third period of late cleavage, the local cavities, which are more stable in the animal hemisphere, coexisting alongside the central one.

The growth of the blastomere contact zones is accompanied by the appearance of specialized cell junctions as early as at the two-cell stage. Interlocking and unusually long microvilli have been seen in the contact zones. According to our observations, junctions of the unusual folded shape of the simple apposition type can often be found at the peripheral part of the contact zone at the four blastomere stage. Septate desmosomes and intermediate contacts of the zonula adhaerens type are the first of the more specialized contacts to be formed; it appears that they separate the growing cavity from the outer medium. The gap junctions appear about 60 min after the beginning of the second division. Contacts of this type are found ubiquitously at the 24-blastomere stage. We shall now describe more characteristic morphological features during each period of early cleavage.

Period I. Cleavage furrows are initially unilateral and displace the middle part of the anaphase spindle, which contains the stem bodies, so that the midbody and spindle remnants

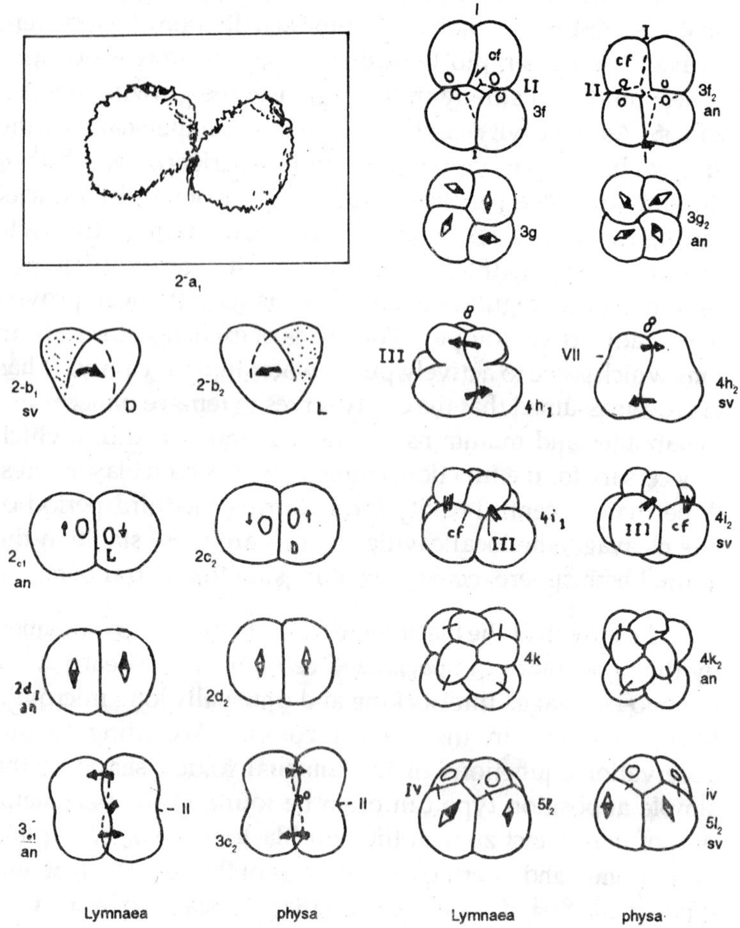

Fig. 5.8. Morphological manifestations of dissymmetry during synchronous cleavage divisions in the dextral species *L. stagnalis* and the sinistral species *Physa acuta*. 2a) Dexiotropic turn of the emerging blastomeres in the pond snail after the vitelline membrane has been removed; animal view. 2b) Sketch showing direction of the turn when viewing the embryo along the spindle axis; shaded halves of the cleavage spindles are closer to the observer. Arrows in 3c and 4th show direction of spiral deformations at the beginning of cytokinesis. I-IV, division number, cf, cross furrow; an, animal view; sv, side view; D, dexiotropic; L, laeotropic rotation around the spindle axis.

appear below the level of the sister nuclei. Such a "kink" in the spindle is no longer present during period III. The midbody may be seen until the prophase. This should be taken into account during experiments on blastomere isolation. Because of the telophase movements the sister nuclei are located close to the new contact zone of blastomeres and, therefore, they "mark" this zone. At the end of G_2 the nuclei are moved somewhat further away from the contact zone, while the distance between them and the outer surface of blastomeres is a few microns. With the exception of cycles I and II, when the location of the nuclei may depend on the rotations of blastomeres, the centers of the nuclei are located opposite each other.

In the course of normal development cleavage figures may sometimes be classified as left or right from the two-cell stages onward and always from the stage of 4-blastomeres. Furthermore, identical in dextral and sinistral species have opposite dissymmetry signs. This dissymmetry may be manifested as a helical deformation of the surface and contact zone as well as the position of sister nuclei the orientation of cleavage spindles and the location of sister pairs of blastomeres with respect to the polar axis. The sequence of figures with alternating signs of dissymmetry is determined during the first period. There is evidence that suggests that all manifestations of dissymmetry depend on the spiral field of stresses which appear during the metaphase of each cycle owing to the activation of cortical microfilaments. This field determines the specific inclined position of the second and third-division spindles as well as the helical deformations of the surface, which appear when blastomeres acquire an extended shape (anaphase of divisions I and II, metaphase of division III). Rotations of blastomeres in the course of cytokinesis at the first division can usually be visualized when the restricting vitelline membrane is removed. These dissymmetric movements, whose direction with respect to the spindle axis is identical in each division in a given mollusc species, are only possible during the first period, since the planes of cleavage furrows more or less coincide during this time. With the beginning of period II these rotations are naturally suppressed be cause the blastomeres are tightly packed together.

During metaphase and anaphase the cleavage spindles are located tangentially to the nearest outer surface of the zygote or blastomeres. They possess rather large asters. During unequal divisions the nonidentity of aster diameters appears no earlier than the anaphase.

Period II. Spiral surface deformations are no longer visible. Dexiotropic and laeotropic orientations of spindles and contact zones of sister blastomeres alternate during the consecutive divisions. The orientations of spindles with respect to the polar axis is completely determined by the shape of the external blastomere surfaces.

Despite the presence of general division synchrony for cells with identical in dices at the 16-blastomere stage, Raven found a slightly earlier onset of divisions in 2B, 2D and 2d, 2b as compared with the 2A, 2C and 2a, 2c blastomere pairs. On the one hand this correlated with the different number of adjacent blastomeres (six for 2B and 2D and five for 2A and 2C); on the other hand it correlates with the slightly greater volume for B and D as compared to A and C, as early as at the 4-blastomere stage. The quadrant A cannot be distinguished from C, nor B from D, during period II. Therefore, these designations should be regarded as arbitrary. The nucleoli appear for the first time during this period.

Period III. The cell cycles become markedly longer. This is accompanied by the most important event of the early embryogenesis of the pond snail; the determination of the dorsoventral axis and the quadrant D as a source of the mesoderm. At the 24-blastomere stage, the most vegetal macromeres which are in contact with each other along the cross-farrow begin to establish inner junctions with the blastomeres of the animal hemisphere. Eventually one of them

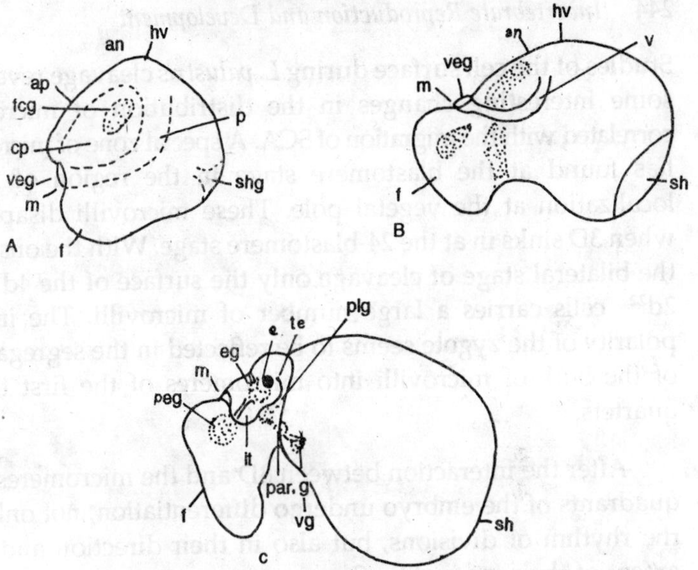

Fig. 5.9. Sketch illustrating embryogenesis in *L. stagnalis* (modified from [140]), side view, A) late trochopore (stage 21); B) mid-veliger (stage 23); c) late veliger (stage 24, "Hippo") an, initial animal pole; veg, initial vegetal pole; ap, apical plate; p, prototroch, v, velum; m, mouth; cp, cephalic plates; hv head vesicle, shg, shell gland; f, foot; fcg, field of cerebral ganglion; sh, shell; te, tentacles; lt, labial tentacles; cg, cerebral ganglia; plg, pleural ganglia; peg, pedal gantlia; par, g, parietal ganglia; vg, visceral ganglion; e, eye. Presumptive areas of the nervous system (a, B) or ganglia (C) are shaded.

disappears almost completely from the outer surface, filling the cleavage cavity and, in addition, increasing the number of its neighbors from 6 to 20. From this moment onward this macromere is determined as 3D. The RNA-containing ectosotnes (former SCA) move along the inner walls of all four macromeres in the course of their interaction with animal pole micromeres involving gap junctions. The ectosomes undergo progressive compactization in 3A-3C, but become dispersed in 3D. Each of the two vegetative macromeres may enter the cavity. This relates competition or "conflict" of the vegetal macromeres.

Studies of the cell surface during *L. palustris* cleavage revealed some interesting changes in the distribution of microvilli correlated with the migration of SCA. A special zone of microvilli has found at the blastomere stage in the region of SCA localization at the vegetal pole. These microvilli disappear when 3D sinks in at the 24-blastomere stage. With the onset of the bilateral stage of cleavage only the surface of the 4d and $2d^{22}$ cells carries a large number of microvilli. The initial polarity of the zygote seems to be reflected in the segregation of the bulk of microvilli into micromeres of the first three quartets.

After the interaction between 3D and the micromeres the quadrants of the embryo undergo differentiation, not only in the rhythm of divisions, but also in their direction and the extent of their inequality. Some accessory contacts remain in the cleavage cavity; they are realized through lobopodia or filopodia. The orientations of spindles perpendicular to the outer surface appear, latitudinal and meridional orientation with respect to the polar axis may be seen. Spindle asters become progressively smaller and eventually continue to appear only in divisions of the 4d mesentoblast descendants. As cytoplasmic segregation increases from the outer surface of the center of the embryo, the interphase nuclei are found deeper than during periods I and II; they are located roughly at the boundary between the ectoplasm and endoplasm. Thus, cleavage during period III acquires a bilateral nature and individual blastomeres begin to differentiate into larval structures.

Temporary larval head structures consist of the following elements; the apical plate — la^{111}-ld^{111}, la^{1121}, lb^{1211}; prototroch — lb^{1221}, lb^{1222}, la^{21}, la^{22}, lb^{21}, lb^{22}, $2b^{111}$, $2b^{112}$; head vesicle — lc^{21}, lc^{22}, ld^{1211}, ld^{1212}, ld^{122}, ld^{21}, ld^{22}, $2d^{11}$, $2a^{11}$, $2c^{11}$, la^{122}, lc^{122}. The right cephalic plate appears from the cells la^{12111}, la^{12112}, la^{12121}, la^{12122}, la^{1122}, lb^{12122}, ld^{1211}, ld^{1122}. The left cephalic plate originates from the cells $2c^{12111}$, lc^{12112}, lc^{12121}, lc^{12122}, lc^{1121}, lc^{1122}, lb^{12121}, lb^{1122}. The cephalic plates give rise to the tentacle epithelium, the dorsal part of the labial tentacle epithelium, eyes, and cerebral ganglia. The stomodeum or foregut is formed from the micromeres of predominantly the third and second quartets, and subsequently forms the oral cavity epithelium, pharynx, radular sac, tongue epithelium, salivary glands, esophagus, and buccal ganglia. The derivatives of the second and third quartet, 2d in particular, give rise to the foot, body wall ectoderm, and mantle. The foot epithelium produces statocysts, pedal glands, and pedal ganglia. The body wall ectoderm yields posterior head ganglia, copulatory organs with part of the outer region of the sperm duct, and a small region of the hindgut near the anus. The mantle forms the shell, the primary urethra, and the genital tract (with the exception of the vas efferens of the sperm duct). The development of the shell field and growth of the mantle have been studied in great detail by Timmermans and Kniprath. The larval muscles of the foot, head, and prototroch from the mesenchyme and probably have an ectodermal origin.

The following organs are formed from the 4d offspring corresponding to mesodermal embryonic streaks; heart, pericardium, vessels, definitive kidney, gonad, connective tissue integuments of the gastrointestinal tract, muscles of the radular sac and tongue, sublingual "cartilage," muscles of the copulative organ, the columellar muscle, the connective tissue layer of the skin (cutis), and, probably, in part, the body wall muscles. The so-called cells, which (like the protonephridium) are absent in the adult animal, also have a mesodermal origin. Finally, the endodermal derivatives 4A-4D, 4a, 4b, 4c, and enteroblasts (early derivatives of the 4D) result in the larval liver, hindgut, and stomach, whose walls later form the definitive liver, muscles covering the gastrointestinal tract also form from these derivatives. The lung, as well as the osphradium, appears independently on the mantle cavity from the body-wall ectoderm. Therefore the lung cavity of the pulmonates can not be considered a homolog of the mantle cavity of Prosobranchia. This outline of organogenesis is based on the studies of Fraser Raven, Regondaud, Verdonk, and Cumin.

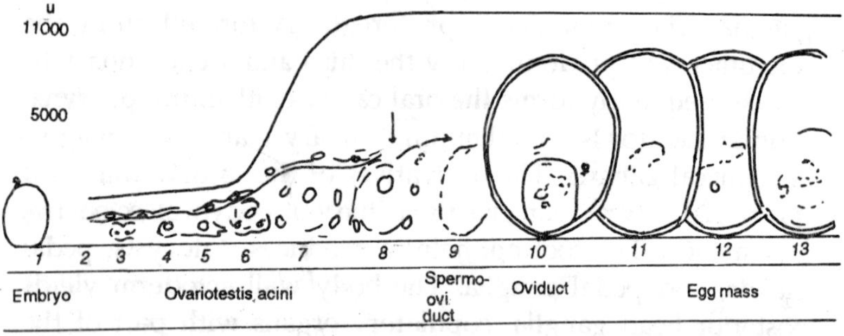

Fig. 5.10. Diagram of life cycle phase of *Lymnaea* egg (from [174] with modifications. 1) Embryonic development; 2) formation of primary germ cells; 3)multiplication of primary germ cells; 4) oogonia; 5) amoeboid phase (migration within the gonad); 6) formation of follicle; 7) growth phase; 8) rest phase; 9) ovulation and insemination (stage 0 and 1/1); 10)phase of prematuration (1/2); 11) first maturation division (1/3); 12) second maturation division (1/4) 13) first eleavage division and beginning of the embryonic development (2). Arrow shows time of complete formation of inner follicle.

To conclude this section, studies dealing with the molecular biology and developmental genetics of the pond snail are discussed briefly. Although RNA synthesis can be detected using biochemical techniques even in the zygote the first major increase is noted during the second cleavage period; this corresponds to the formation of nucleoli. According to Neyfakh, morphogenetic nuclear function in the course of pond-snail development begins at the 12-blastomere stage and supports embryogenesis when period II has been completed. A somewhat different conclusion has been reached by Verdonk, who studied the expression of pond snail lethal genes which are made homozygous by self-fertilization. Gastrulation is the first period depending on the embryonic genome, but a particularly high number of genes undergoes expression at the later stage of the intensive differentiation of rudiments in the early trochophore. The expression of a few genes essential to development may be observed after the early veliger stage. Morrill and associates on the basis of the results with *L. palustris,* believe that the bulk of proteins prior to the early trochophore stage is synthesized on long-living mRNA molecules; gastrulation involves molecular events which are critical to the differentiation of the definitive organs.

In the last few years some progress has been made toward understanding the genetic control of dextrality-sinistrality. Traditionally, such studies are performed genetic control of dextrality—sinistrality. Traditionally, such studies are performed using *L. peregra*, from which dextral and sinistral races have been described. It has been assumed for a long time that the dissymmetry sign for cleavage and adult animal organization depends on the dominant dextral allele and the recessive sinistral allele of one gene; the offspring phenotype is determined by the maternal genotype, that is, before maturation divisions. The problem was complicated by permanent formation of dextral snails in the sinistral "pure" strains. This has been explained by reverse mutations and the presence of the complex system of modifying genes. A thorough genetic analysis recently carried out by Freeman and Lundelius enabled a new mechanism to be proposed for the inheritance of dextrality—sinistrality and the formation of abnormal dextral specimens in the sinistral strains of *L. peregra*. The gene determining dextrality consists of two regions which have a dominant effect only when present in one chromosome in the "cis" position (S^{1+} S^{2+} or a tandem duplication of S^{1+} S^{1+} type). In sinistral snails this structure is formed spontaneously by crossing-over. This is possible in three variants: mitotic crossing-over at the oogonial stage; meiotic equal crossing-over during prophase I of meiosis; and meiotic unequal crossing-over at the same stage. Genetic analysis demonstrates the existence of all three types with a marked prevalence of the equal meiotic crossing-over.

Injections of the cytoplasm from phenotypically dextral eggs into sinistral ones (but not vice versa) result in the inversion of the dissymmetry sign of cleavage if these injections are performed prior to the extrusion of the second polar body.

It may be noted that the animal-vegetal polarity of the zygote appears to undergo irreversible determination at the same stage. Mocroinjection experiments have also enabled the relationship between the does of dextrality gene and its effect to be established, as well as dependence on the developmental stage of the cytoplasm donor. Methods are now available for purifying the products of chirality genes and for studying their mechanism of action, which appears to be due to the effect on the organization of the cortex action cytoskeleton.

Various Developmental Stages

A table of normal development is presented below for the first time. Stages 0-1/5 are based on Raven and Ubbels; this period is described by us in less detail. Here the description of stages is based on morphological peculiarities. Stages 2-18 are derived from Verdonk's study Stages 19-21 correspond to the description of embryogenesis taken from Raven but in far less detail. Finally, stages 22-29 are taken from the Normal Table of pond-snail development published by Cumin. They correspond to his stages E_4-E_{11}, recorded at daily intervals during development at 25°C. In Raven's and Verdonk's studies the stages have been identified by studying the structure of the embryo at certain standard intervals, with the exception of early cleavage where the stages are distinguished on the basis of the number of cells. During early developmental stages the time of onset of a given stage has been determined in our observations by the first appearance of its characteristic features. After early blastula the interval between observations was 3 h. In Table 5.2, we show the time up to a corresponding stage at 22 $\pm$ 0.5° C (our data and at 25° C in dimensionales units, and the size of the embryo (our data) and to stage 21 and further information according to a short diagnosis of the stage is attached. Drawing numbers correspond to those of stages; the drawings of stages according to Cumin are marked with the letter C (for example 22C). Our original drawings referring to stages presented according to Verdonk and Cumin are marked with the letter M.

It should be pointed out that our data are valid for the development of embryos in capsules incubated in tap water, while Verdonk and cumin have observed the development in capsules incubated in moist air upon agar surfaces. Therefore the data regarding the time of onset of a given stage at 22°C and 25° C should be compared cautiously. The time of onset of stages at 25° C, up to and including stage 24, is given according to Verdonk, and from stage 25 onward according to Cumin. At 22° C the value of to is 120 min, and 85 min at 25° C.

Stage 0: The Ovulated Egg

Ovulation proceeds with autolysis of the follicle and gonad epithelium. The oocytes pass into the gonad lumen, then into the hermaphrodite duct where they are accumulated in the straight part. The germinal vesicle is retained for some time after ovulation.

Stage 1: One-Cell Embryo

1/1. *Insemination* takes place in the straight region of the hermaphroditic duct. At the moment of insemination the germinal vesicle disintegrates. The inseminated eggs may remain in this part of the oviduct for up to 1.5 h.

1/2 *Metaphase of the first maturation division* : The deposition of eggs in the cocoon proceeds at this stage. Immediately after laying, the egg is somewhat irregular in shape but soon acquires that a rotation ellipsoid. A small part of the egg is free of yolk near the animal pole where the first maturation spindle is located. This region is semitransparent when seen with transmitted light.

1/3. *Extrusion of the first polar body:* At this stage the egg may display some amoeboid deformation. The first polar body, which has the appearance of a small transparent sphere, is extruded at the animal pole and is coated with its own membrane as well as by part of the vitelline membrane; it is located outside the perivitelline space over the animal pole.

1/4. Extrusion of the second polar body : The second polar body is extruded directly under the first; it is somewhat smaller and becomes located between the vitelline membrane and the egg plasma membrane. Some amoeboid activity of the egg can be observed during extrusion.

1/5. Fusion of pronuclei : This stage is preceded by the fusion of the female karyomeres into the pronucleus. True fusion of the male and female pronuclei does not take place. On the sagittal section it can be seen that the pronuclei contacting each other are located almost directly under the animal pole.

The axes of the two pronuclei may have different orientation. Sometimes the pronuclei, as well as the spindle of the first cleavage division, may be seen in the course of vital observation using strong oblique illumination or when a dark-field condenser is being used.

TABLE 5.1

Intervals Between Developmental Stages of the One-Celled Embryo 22°C—our data; 25°C—data of Labordus [102] or Van Den Briggelaar [18]; in min)

Stage	22oC	25oC
1/2-1/3	Time varies; usually it is	
45-60	53	
1/3-1/4	60	38
1/4-1/5 (Up to the beginning of the prophase of the 1st cleavage division	—	53
1/4-2	102	76, 90

We shall now present data on the duration of individual periods of this stage (Table 5.1). A detailed cytological description of the one-cell stage can be found in Raven.

Stage 2: Two Blastomeres

A: first the furrow appears at the animal pole of the egg which has become elongated in the equatorial plane (stage 2-). Later the furrow divides the whole egg (stage 2).

Stage 3. Four Blastomeres

The furrows of the second laeotropic cleavage division are formed synchronously in both blastomeres at a small angle to one another and to the animal-vegetal axis. Consequently, blastomeres A and C contact each other at the animal pole along the so-called polar or cross-furrow, which originates at the bending of the first furrow; at the vegetal pole the blastomeres B and D are in contact, and can not be distinguished from each other.

All blastomeres usually appear to be identical in size. However, it has been established by Bezem et al. that the volume of A or C(24.34% of the whole) is significantly lower than that of B or D (25.66%).

Stage 4: Eight Blastomeres

The cleavage division is dexiotropic. Four micromeres of the first quartet la-ld and macromeres 1A-1D are formed synchronously.

This is the last division with helical deformation in the course of which cross-furrows are formed on the lateral cell surface. It is also the last synchronous division. The ratio between the volume of blastomeres 1q (q = quartet), 1A (or 1C), and 1B (or 1D) is 5.58:18.46:20.39 [17].

Stage 5: Twelve Blastomeres

The lA-lD macromeres undergo laeotropic divisions forming the second quartet of micromeres 2a-2d, which are slightly larger than la-ld.

Stage 6: Sixteen Blastomeres

The micromeres of the first quartet undergo laeotropic division which results in more animal la^1-ld^1 cells and the four primary trochoblasts la^2-ld^2. Nucleoli appear for the first time in these eight blastomeres, their number not exceeding three per nucleus. RNA synthesis may be detected by autoradiography for the first time at this stage. The ratio between the volumes of blastomeres lq^1, lq^2, 2q, 2as (2C), 2B (2D) is 3.34:2.15:7.4011.40:12.83 [17].

Stage 7: Twenty-four Blastomeres

The macromeres 2A-2D and the micromeres of the second quartet, that is, all blastomeres of the vegetal hemisphere, undergo synchronous and dexiotropic divisions. From the surface the size of 3a-3d is greater than that of 3A-3D. The 24-blastomeres stage is remarkable because it involves a cessation of cleavage for about 3h. Nucleoli appear in all blastomeres. The 3D blastomere differs from 3A-3C in its rounded shape. It fills almost the entire cleavage cavity, where it contacts blastomeres la^1-ld^1, la^2-ld^2, 2a^1-2d^1, , 3a and 3b. At 25° C these contacts become fully established 80 min after the beginning of this stage. After 110-170 min the top of 3D, located inside, is oriented toward lb^1 and 2b^1; the material of the ectosomes in the 3D, located inside, is oriented toward 1b^1 2b^1; the material-of the ectosomes in the 3D is concentrated along the contacts with these micromeres. Two hundred minutes later 3D enters mitosis;l the spindle is formed in the centre of the blastomere, which is then displaced toward the vegetal pole, providing an inequality of the 3D division. Ectosomes from 3A-3C are then found in micromeres 4a-4c. The volume of 3D at this stage, according to Bezem et al. exceeds that of 3B by about 15%. The ratio between the volumes of lq^1, lq^2, 2q^2, 3q. 3A (3C), 3B, 3D is 3.07::2.10:3.47:3.68:6.75:5.01:6.40:7.38.

Stage 8: Twenty-five Blastomeres

The 3D macromere is divided by a plane forming a sharp angle with the polar axis, yielding the 4D macromere and 4d mesentoblast (M) which serves as the future source of most of the mesoderm. 4d is much larger than 4D.

Stage 9: Thirty-three Blastomeres

Contacts between 4d and 3a, 3b and 2a^1-2d^1 are partially lost while 3A-3C cells are displaced closer to the centre of the embryo. 2a^1-2d^1 cells are the first to divide, forming the so-called tip cells of the future cross, 2a^{11}-2d^{11}; this is followed by a laeotropic division of $2a^2$-$2d^2$ cells, $2q^{21}$ blastomeres being larger than $2q^{22}$.

Stage 10: Forty-nine Blastomeres. Early Blastula

This stage is achieved by several sequential divisions closely following each other. Initially $1a^1$-$1c^1$ micromeres undergo dexiotropic divisions (36 blastomeres); this is followed by dexiotropic divisions of $1d^1$ and by almost radial divisions of 3a-3d (41 blastomeres); the size of $3a^2$ is greater than $3a^1$, $3b^2$ is greater than $3b^1$, but $3c^2$ is smaller than $3c^1$ and $3d^2$ is smaller than $3d^1$. The primary trochoblasts $1a^2$ and $1b^2$, taking part in forming the prototroch, divide almost in the horizontal plane; 3A-3C macromeres divide laeotropically (46 blastomeres); eventually the 4d primary mesoblast divides, yielding two identical mesoreloblasts M_1 and M_2 and the remaining primary trochoblasts $1c^2$ and $1d^2$. It is noteworthy that the delayed divisions of $1d^1$, $1c^2$, and $1d^2$ correlate with their longer contact with 4d.

Verdonk [174] distinguished the 48 blastomere stage with eight trochoblasts located pairwise one on top of the other and the-as-yet-undivided 4d. The trochoblasts do not divide for a long time after this. This stage completes the early cleavage period.

The subsequent stages 11-15 are characterized by the existence of the so-called molluscan cross, formed by the cells of the first quartet in addition to trochoblasts and the tip cells $2a^{11}$-$2d^{11}$. Beginning with these stages, the details of the cell localization may be distinguished predominantly on fixed embryos.

Stage 11 : Resting Stage of the Early Cross

After the primary trochoblasts divide (preceding stage) cleavage is interrupted for about 3h. During this period $1a^{11}$-$1d^{11}$, the larger part of the animal cells move away from the surface, while the basal $1a^{12}$-$1d^{12}$ cells are displaced toward the middle of the animal hemisphere and become wider. Their

Fig. 5.11 to 5.19. Stage of *Lymnaea stagnalis* L. normal development. Numbers correspond to stage numbers. Letters accompanying numbers designate; an, animal view; veg, vegetal view; m, view from the mouth; bl, view from the blastopore; sv, side view; al, anterolateral view; lef, view from the left; r, view from the right; d, dorsal view; df, dorsofrontal view; sag, sagittal section.

Fig. 5.11 to 5.13. Sketches 2-24 are from preparations impregnated with silver [175].

Fig. 5.14 & 5.15. Sketches 1/3 M1/5 M and 15 M sand 15 M-24/2 M are outlines of the pond snail embryo at consecutive developmental stages, an, animal pole; veg, vegetal pole; 1(2) pb, first (second) polar body; vitmem, vitelline membrane; pr, pronuclei; bl, blastopore; p, prototroch; ap, apical plate; cp, cephalic plates; m, mouth; f, foot; shg, shell gland; sh, shell; ma, mantel; vel, velum; e, eyes; te, tentacles; lte, labial tentacles; h, heart; pcs, post-trochal ciliate strand; pco, pulmonary cavity orifice; hv, head vesicle. Some of the abbreviations here also appear.

Fig. 5.16 to 5.19. Normal development stages of the pond snail at daily intervals during 2-11 days of embryogenesis (25°C) [49]. 1) Stomodeum (mouth); 2) gastral cavity; 3) intestinal plate consisting of small cells; 4) hind gut; 5) albumen sac; 6) left albumen sac; 7) right albumen sac; 8) stomach; 9) gastric epithelium; 10) gastric wall; 11) dorsal gastric aperture; 12) ventral gastric aperture; 13) large gastric aperture; 14) small gastric aperture; 15) large liver ruliment; 16)small liver rudiment; 17) radula; 18) exophagus; 19) crop; 20) anus; 21) prospective foot region; 22) transverse ventral groove; 23) foot; 24) cerebral ganglia; 25) cerebral commissure; 26) pedal ganglia; 27) visceral ganglion; 28) buccal ganglia; 29) parietal ganglia; 30) pleural ganglia; 31) mesoderm; 32) prototroch (velum); 33) nuchal cells; 34) protonephridium, 35) remnants of protonephridium; 36) kidney an age; 37) kidney; 38) renopericardial channel; 39) heart; 40) shell; 46) pulmonary cavity; 47) respiratory orifice; 48) mantle cavity; 49) osphradium; 50) statocyst; 51) eyes; 52) tentacle; 53) oral lobe (labial tentacle). The digestive tract is shaded.

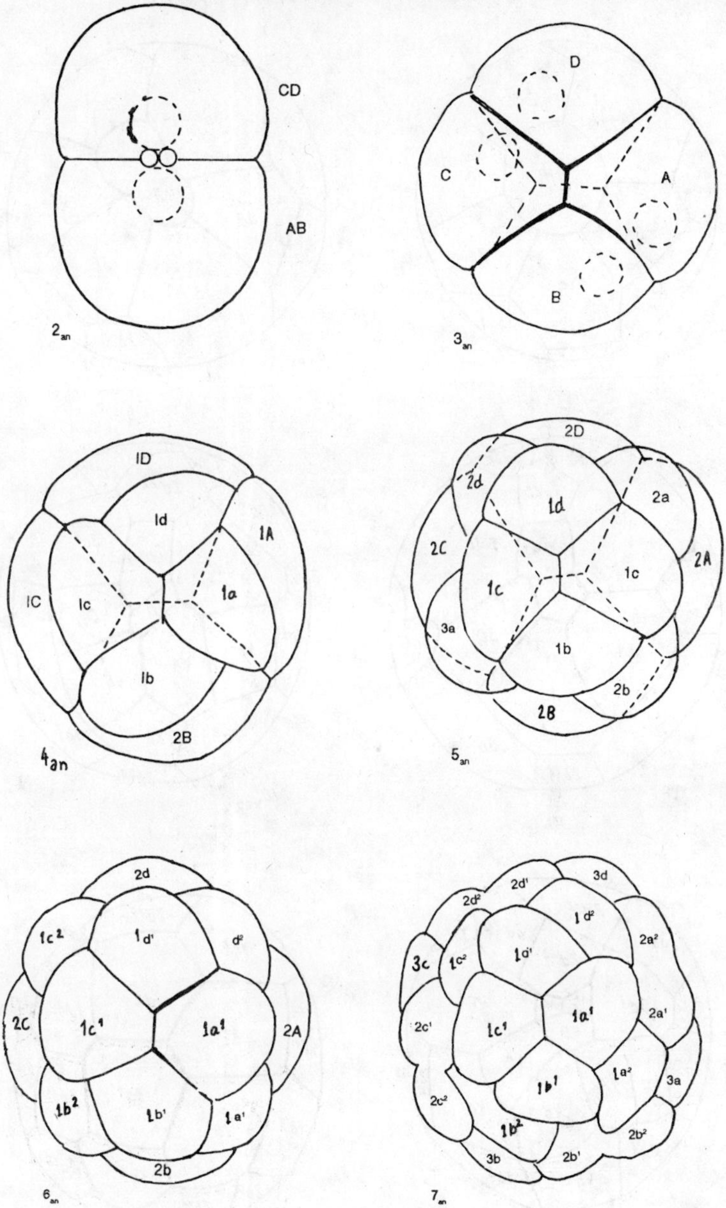
CD
AB
2_{an}
D
C
A
B
3_{an}
ID
Id
1A
IC
Ic
1a
Ib
2B
4an
2D
2d
1d
2a
2C
2A
1c
1c
3a
1b
2b
2B
5_{an}
2d
1c²
1d'
d²
2C
1c¹
1a¹
2A
1b²
1b'
1a'
2b
6_{an}
2d'
3d
2d²
1d²
2a²
3c
1c²
1d'
2a'
2c'
1c¹
1a¹
1a²
3a
2c²
1b¹
1b²
2b²
3b
2b'
7_{an}

Fig. 5.11

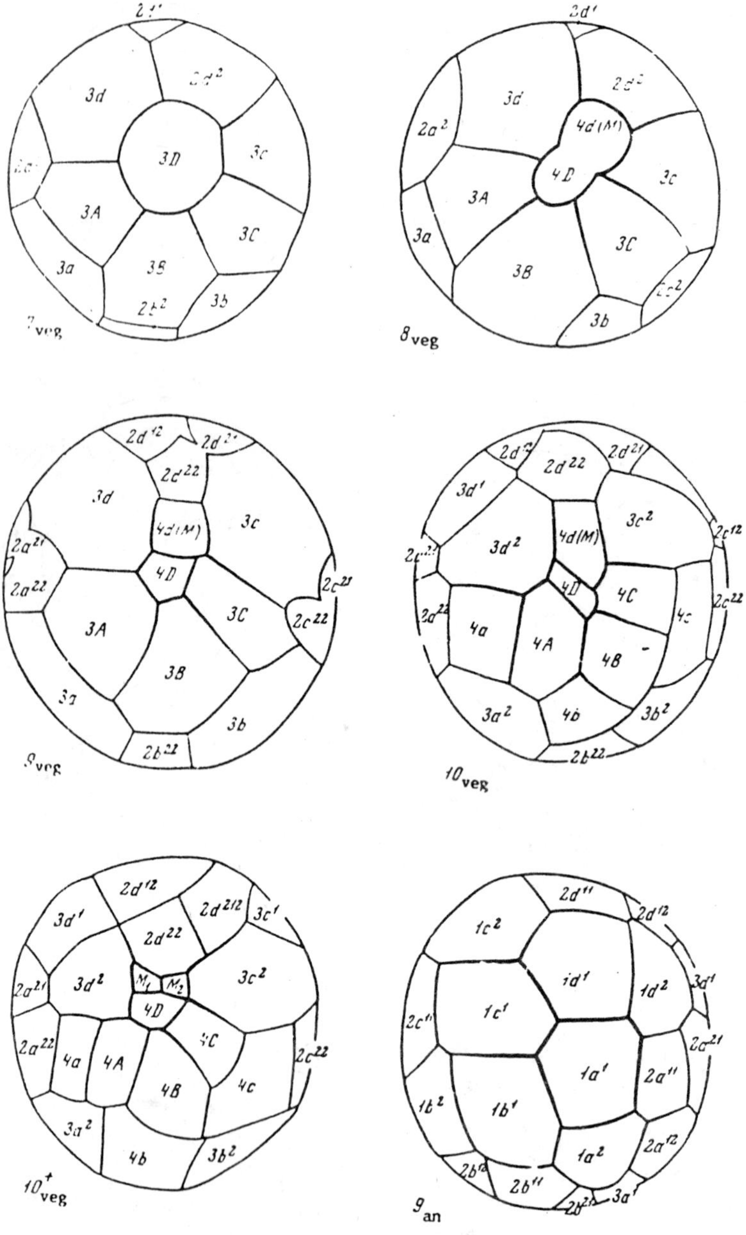

Fig. 5.11 (continued)

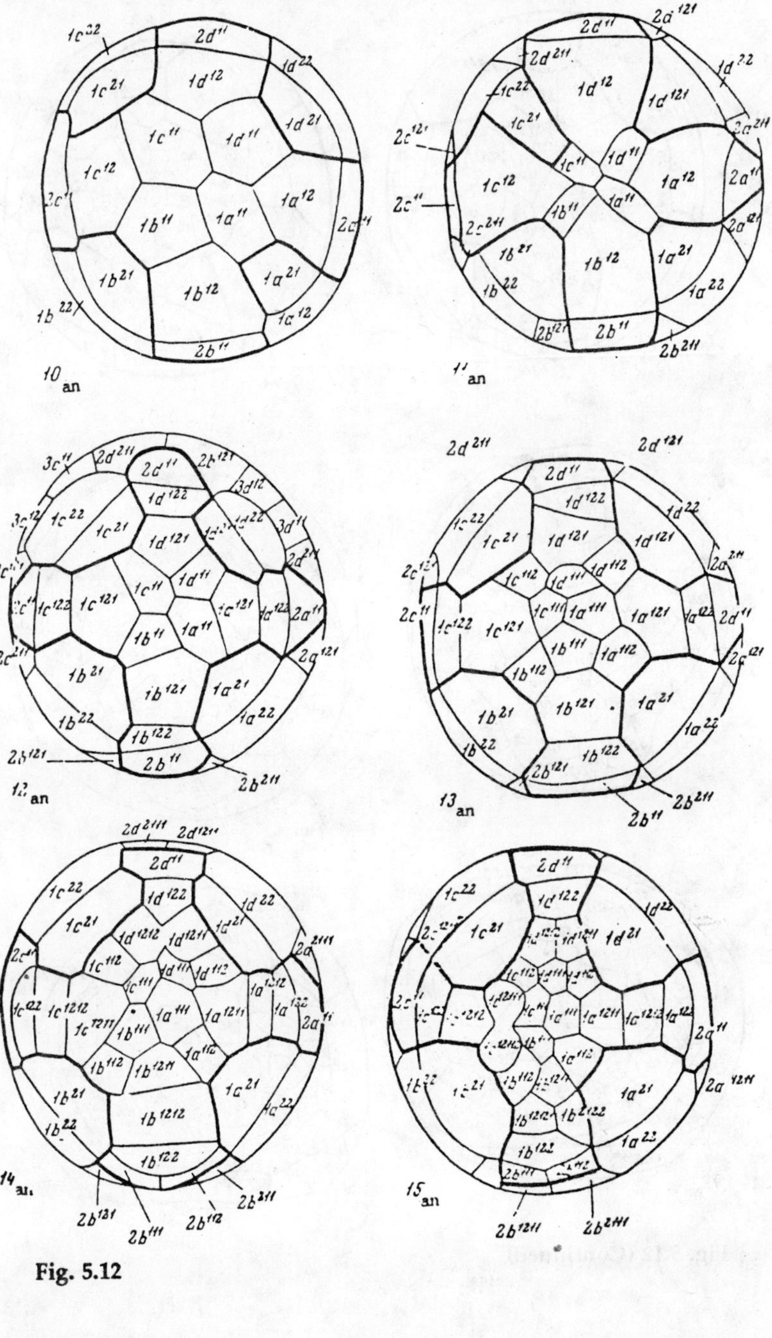

Fig. 5.12

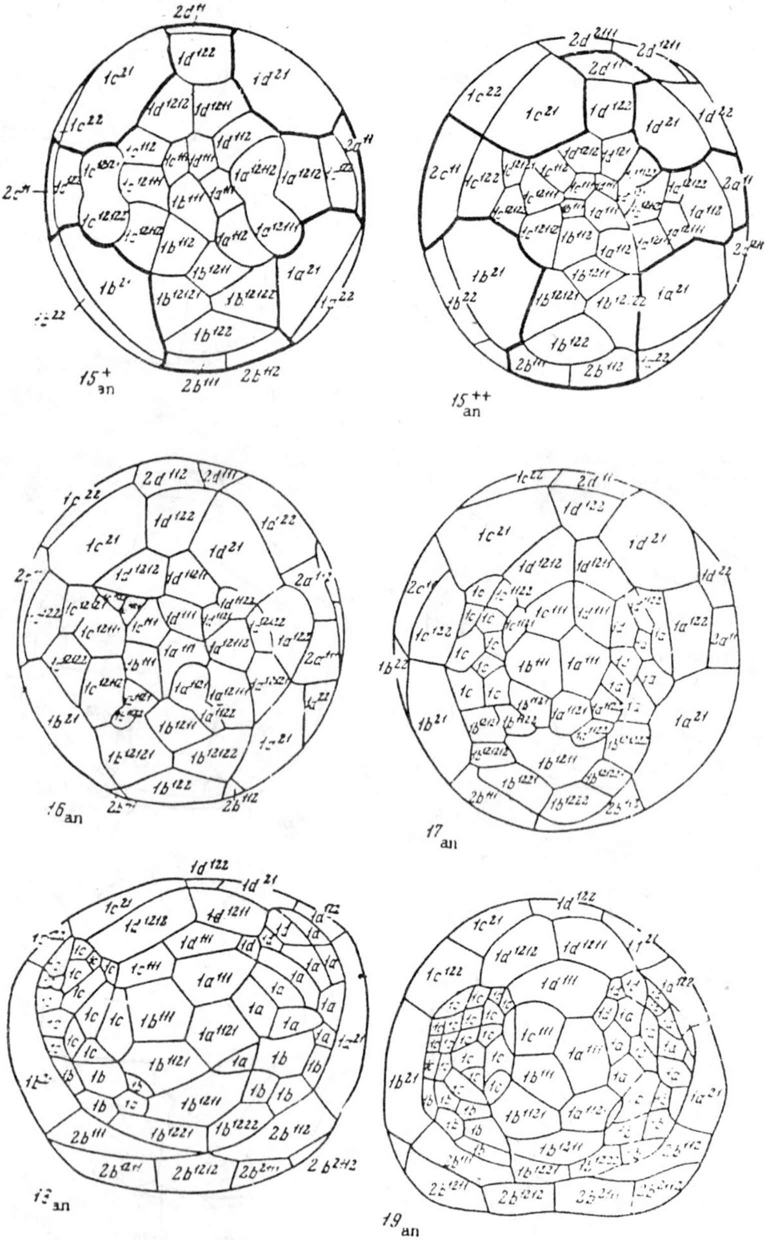

Fig. 5.12 (Continued)

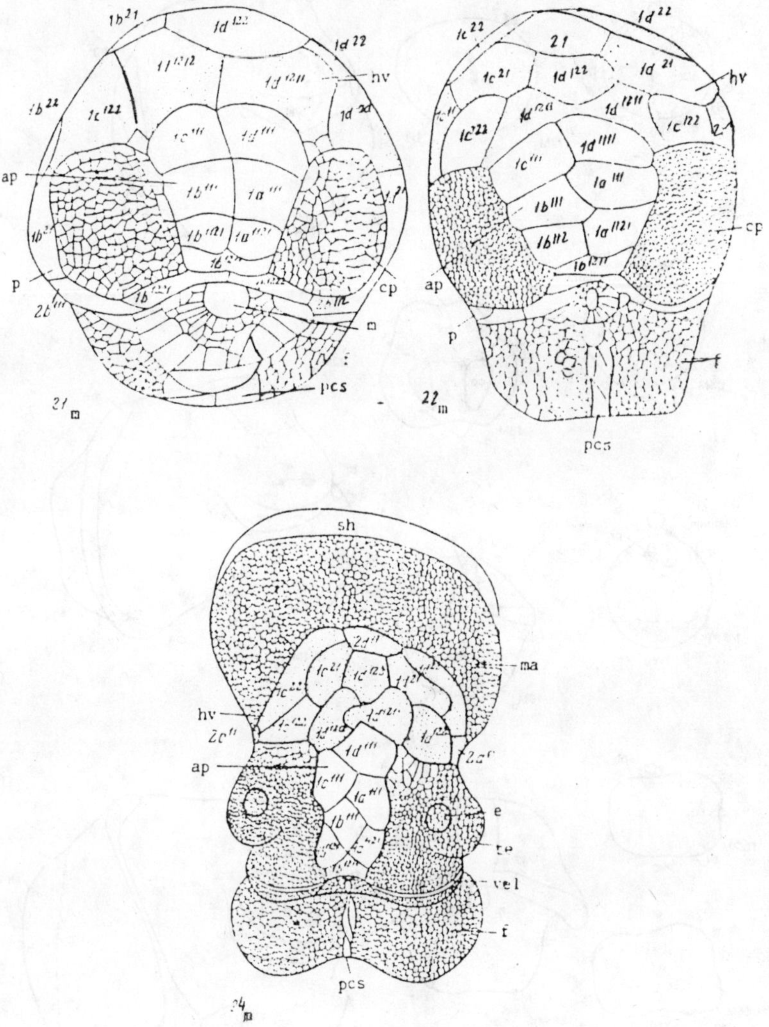

Fig. 5.13

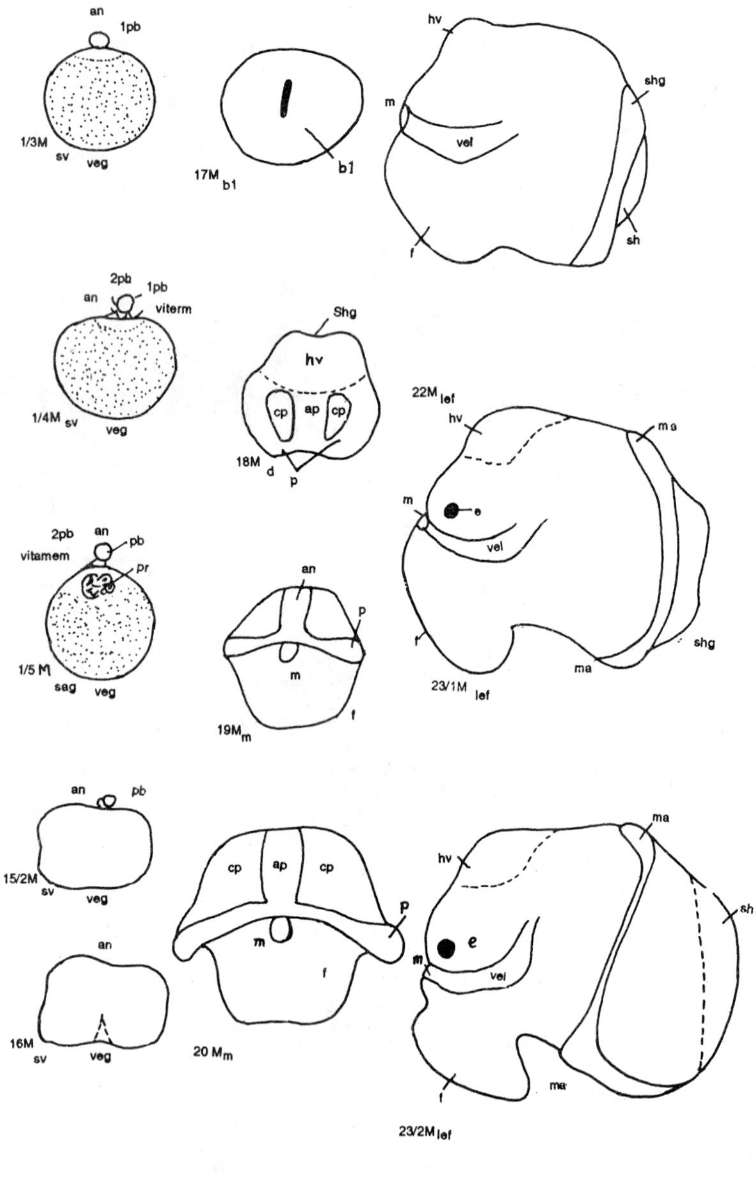
an
1pb
1/3M
sv
veg
17M b1
b1
hv
shg
m
vel
f
sh
2pb
1pb
an
viterm
1/4M sv
veg
Shg
hv
cp
ap
cp
18M
d
p
22M lef
hv
ma
m
e
vel
f
shg
ma
23/1M lef
2pb
an
pb
vitamem
pr
1/5 M
sag
veg
an
p
m
f
19Mm
an
pb
15/2M
sv
veg
an
16M
sv
veg
cp
ap
cp
p
m
f
20 Mm
ma
hv
sh
e
m
vel
f
ma
23/2M lef

Fig. 5.14

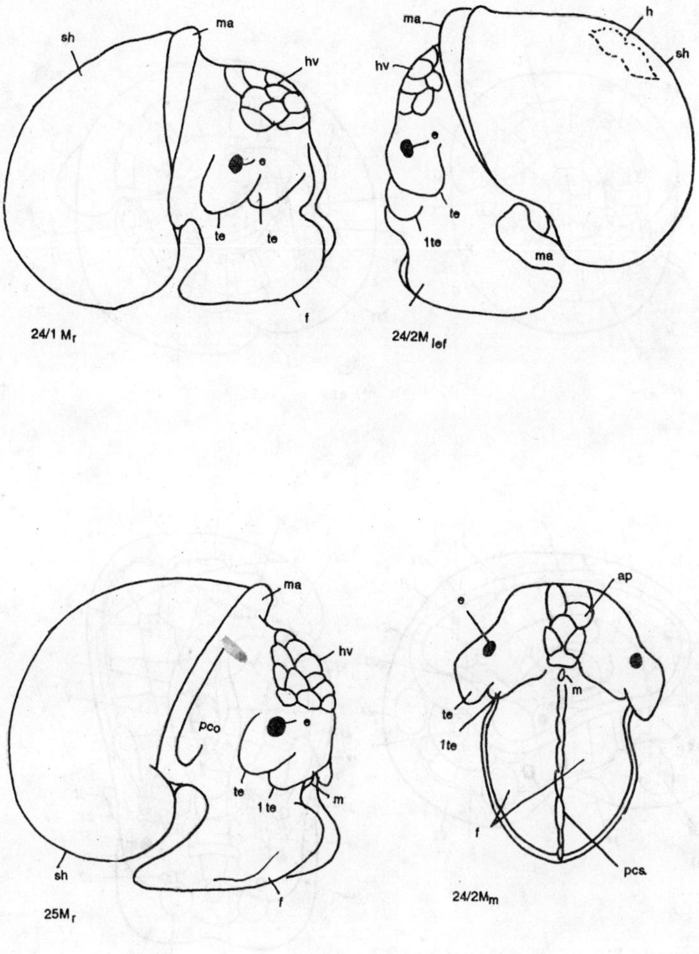
sh
ma
hv
e
te
te
f
24/1 Mr
ma
h
sh
hv
e
te
1te
ma
24/2M lef
ma
hv
pco
e
te
1 te
m
sh
f
25Mr
ap
e
m
te
1te
f
pcs.
24/2Mm

ig. 5.15

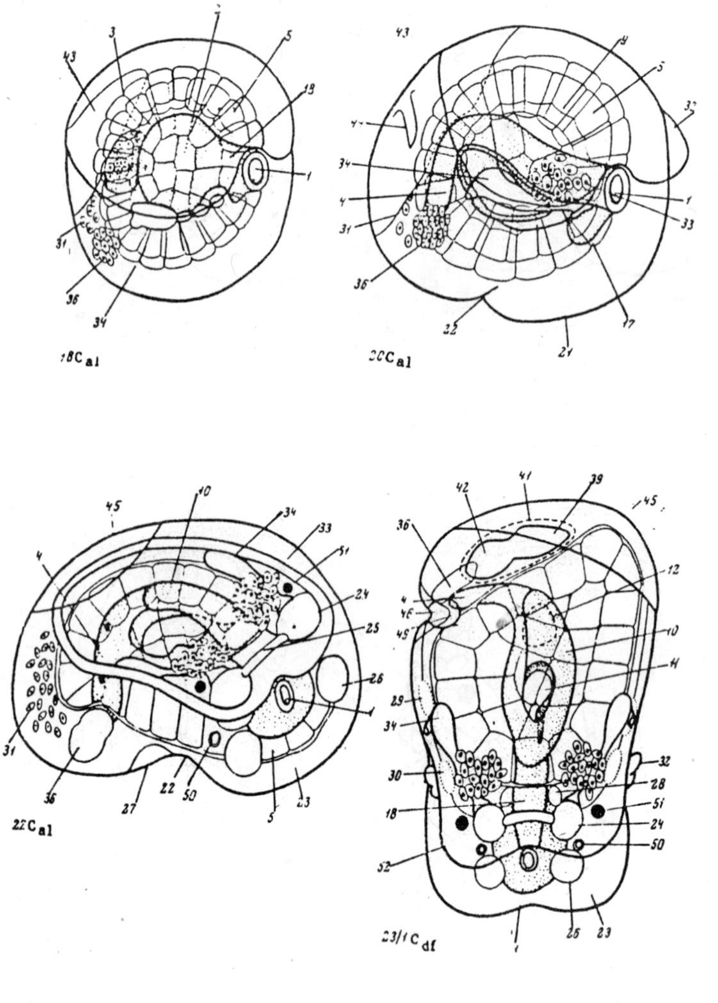

Fig. 5.16

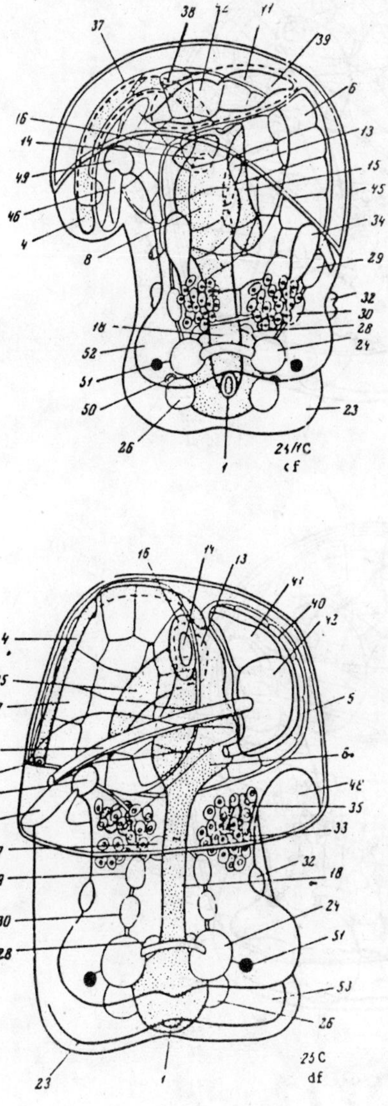
38
11
37
39
6
16
13
14
15
49
45
46
34
4
8
29
32
30
18
28
52
24
51
50
23
26
24/1C
cf
1
16
14
13
41
40
42
4
15
5
7
37
6
20
48
49
35
46
33
27
32
29
18
30
24
51
28
53
26
23
1
25C
df

Fig. 5.17

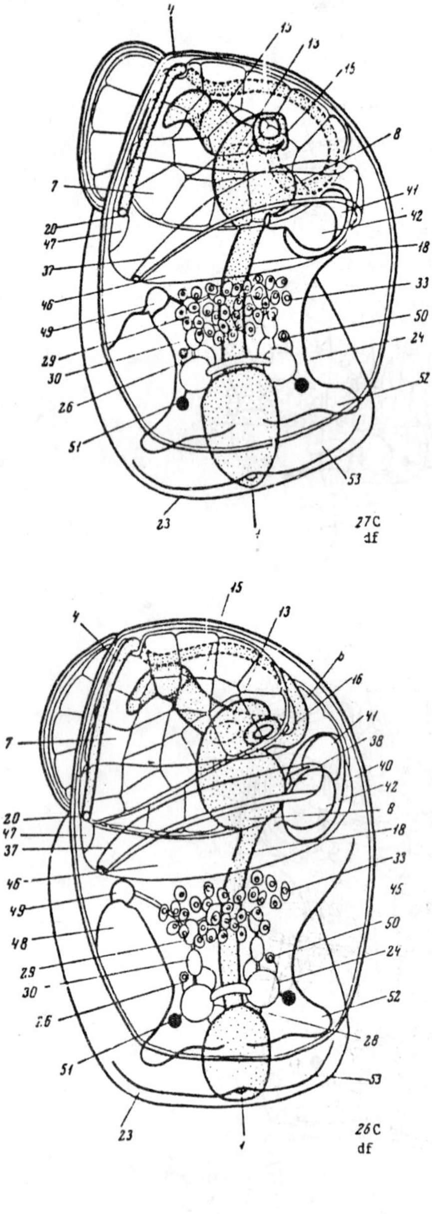

Fig. 5.18

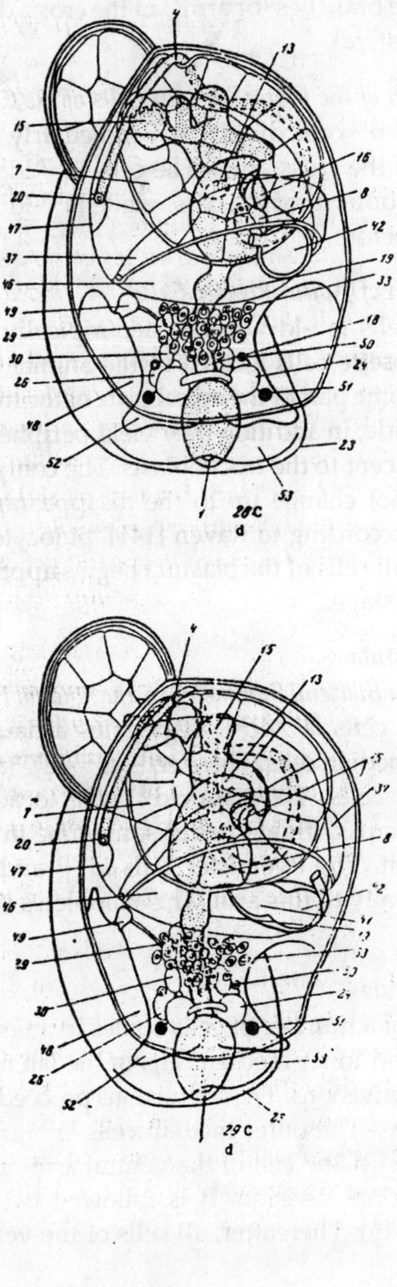
4
13
15
16
7
20
8
47
42
37
19
46
33
49
18
29
50
30
24
26
51
48
23
52
53
1
28C
d
4
15
13
7
8
20
47
42
46
49
29
30
24
48
51
25
33
52
29C
d

Fig. 5.19

position marks the branches (or arms) of the cross: d, dorsal; b, ventral; a and c, lateral.

Stage 12: Formation of the Outer Median Cells of the Cross

The basal $1a^{12}$-d^{12} cells divide perpendicularly in relation to the direction of the cross arms. The cells la^{122}-ld^{122} are now referred to as the outer median cells, while la^{121}-ld^{121} cells are termed basal, as before.

Stage 13: Formation of Apical Rosette Cells

The animal cells la^{11}-ld^{11} divide laeotropically and form la^{111}-ld^{111} apical rosette cells located at the animal pole; later these cells will become part of the apical plate of the trochophore and no longer divide; in addition they yield peripheral rosette cells la^{112}-ld^{112} adjacent to the trochoblasts. The composition of the rosette does not change up to the disappearance of the cross (stage 16). According to Raven [141], pinocytosis of the capsular fluid by all cells of the blastula begins approximately at this or the next stage.

Stage 14 : Mid-Blastula

Appearance of a Bilateral Symmetry in the Animal Hemisphere Each of the basal cells la^{121}-lc^{121} segregates a daughter cell toward the outer median cells. These cells la^{1212}-lc^{1212} are termed the inner median cells. The basal ld^{121} cell, located in the dorsal arm of the cross, divided after some time in the plane parallel to the arm. The two sister cells ld^{1211} and ld^{1212} are located symmetrically to this arm. This completes the lineage of the dorsal arm.

Stage 15: Late Blastula

The division of a number of cells comes to an end by this stage. The basal and inner median cells in the lateral arms of the cross undergo divisions. These divisions proceed in planes parallel to the arms. The outer median cells la^{122} and lc^{122} no longer divide. The tip $2b^{11}$ cell in the ventral arm of the cross undergoes meridional division. It is followed by the inner median cell lb^{1212} [15]. Thereafter, all cells of the ventral arm, with the exception of the basal lb^{1211} cell, begin to expand in the transversal direction. Descendants of the lb^{1212} cell gradually reach the lateral arms. The cross is still distinctly discernible at the stage 15/1 (15^{++}). Prior to gastrulation (stage 15/2) the embryo, now consisting of about 200 cells, becomes markedly flattened at both poles.

Stage 16 : Early Gastrula

The cross is no longer discernible. Cells located at the vegetal pole elongate in the animal direction and gradually begin to migrate inside. Mesoblasts and their descendants are located in the blastocoel by this time. At the beginning of gastrulation the blastocoel may be almost completely full of cells. The blastopore is located at the vegetal pole and has a rounded shape as seen from the surface. At the animal pole one frequently sees a flat depression formed by the apical rosette at the base of the dorsal arm of the cross. The ventral arm of the former cross is quite wide. The la^{112}-ld^{112} peripheral rosette cells begin to proliferate while the apical rosette cells widen considerably.

Stage 17: Mid-gastrula

The blastopore moves to the ventral arm of the cross and acquires a fissure-like shape. Its borders appear to consist of 2d derivatives. According to Verdonk, the embryo at this stage emerges from the vitelline membrane but remains inside the capsule. The apical and cephalic plates are already discernible in the animal hemisphere. In other words, bilateral symmetry of the embryo is already quite distinct.

Stage 18: Late Gastrula

The blastopore is at the definitive position opposite the shell gland rudiment; it looks like a small depression (closure of the blastopore takes place in the posteroanterior direction). The rudiment of the shell gland is composed of a wide depression of the thickened ectoderm. An open prototroch is located immediately above the blastopore; the prototroch does not as yet rise over the surface of the embryo. Cilia are active in the prototroch, foot, and several other regions of the body. The embryo rotates inside the capsule at a speed of 1 rev/105 sec. The head vesicle —a translucent structure consisting of 12 flattened cells—is located under the animal pole. The apical plate, extending from the head vesicle to the prototroch, divides two regions of moderately sized ectodermal cells which are known as cephalic plates. Digestion of the protein-containing fluid begins in the large "albumen" endodermal cells. Two protonephridia consist of four large cells each. The midgut plate consisting of small cells is located caudal to the albumen cells, which are covered by the common envelope, giving rise to the albumen sac and delimiting the gastral cavity. The midgut plate later yields the hindgut, stomach, and part of the definitive liver. The definitive kidney exists as a loose complex of mesodermal cells. A more detailed description of the development between stage 19 and stage 25 can be found in Raven.

Stage 19 : Early Trochophore

The lateral parts of the prototroch begin to rise over the body surface. The first cilia appear on the surface of the apical plate cells. The rudiment of the shell gland has a somewhat depressed middle part. The stomodeum, which originates from the ectoderm, comprises the outer part of the blastopore and begins to differentiate into the oropharyngeal part and the esophagus. The alkbumen cells discharge secretion into the gut lumen. Two protonephridia develop. The embryo is only barely transparent at this stage.

The ultrastructure of the trochophore has been studied in detail by Arni.

Stage 20: Mid-trochophore

Lateral parts of the prototroch rise above the surface of the body as two bands. A row of larger cells corresponding to the post-trochal ciliate strand is obvious in the middle of the future foot region. The embryo rotates, each rotation taking about 5 sec. It is more transparent than in the preceding stage.

The shell gland is deeply invaginated toward the gut rudiment, and there is a thin translucent shell over it. Both the gland and the shell are slightly displaced to the left from the sagittal body plane. (In our description of pond-snail development we will, as usual define as right and left the sides of the embryo which are located respectively to the right and left of the cranial-caudal axis of the animal when viewed from the dorsal side of the body.) Functional protonephridia are formed. Two groups of 15 nuchal cells are located over the protonephridia. Cell proliferation begins in the cephalic plate ectoderm and in the part of the foot which is adjacent to the mouth. The radular sac segregates from the stomodeum wall. At this stage the sac has a lumer; radular muscles form around the sac. Two epithelial strands connected with the esophagus in the frontal region detach from the gut plate on its right and left sides. Later they form the stomach walls and muscles. The newly formed stomach is open at the sides and is in contact with the cells of the albumen sac, which are no longer active in secretion. The hindgut rudiment has appeared.

Stage 21: Late Trochophore

The shell gland depression has almost reached the outer surface of the posterior part of the body. The middle part of the gland is covered by a thin transparent shell. The foot region is well formed. The formation of cerebral ganglia begins in the cephalic plates, and eye rudiments appear as not-very-prominent fossae. The anlage of statocysts appear in the foot region. The stomodeum begins to differentiate anteriorly in the oral and pharyngeal cavities. The head vesicle occupies about half of the pretrochal region and laterally is in contact with the dorsal tips of the prototroch. The ventral part of the prototroch on each side consists of the two cells la^{21} and la^{22} on the left and lb^{21} on the right. Generally, the embryo is still rounded.

Stage 22: Early Veliger

The embryo makes one complete turn in 10 sec. The foot rudiment is distinctly separated from the visceral sac. The protoiroch is about 0.1 mm in length; laterally it forms strongly protruding thickenings which dorsally join up with the head vesicle. The mouth is located in the middle of the prototroch. From this stage onward the prototroch begins to reduce gradually. The shell gland almost flat; thickenings or mantle folds have formed near its borders and are in contact with the shell. The body's cranial-caudal axis is longer than the dorsal-ventral axis. The nuchal cells proliferate. Cerebral ganglia are connected by a commissure. Rudiments of the pedal ganglia are present, and the formation of pleural, parietal, and visceral ganglia is under way. The pulmonary cavity is present as thickening of the ectoderm near the rudiment of the definitive kidney. The radular sac is not connected to the oral cavity. A cavity or lumen appears in the albumen sac on both the right and left sides of the stomach; these cavities accumulate the proteinaceous fluid which comes from the stomach. This fluid is pumped to the stomach from the mouth with the help of a cilliate thickening of the esophagus.

The process of torsion begins at this stage. It is characteristic for the development of all gastropods and involves a turn of the whole visceral complex with respect to the head region around the cranial-caudal body axis. In the pond snail this turn proceeds clockwise if the embryo is viewed in the direction from the head to the shell. This is particularly well seen in the movements of different regions of the gastrointestinal tract. Thus, parts of the albumen sac turn by 180° between stages 24 and 29; the small stomach orifice turns by 270°, and the large one by 90° — The heart and genital duct rudiment move together along the arc in the frontal plane and are displaced from their position and shifted somewhat to the right side of the sagittal plane into the left dorsal position. The gut rudiment makes a loop and opens to form the anus near the right edge of the dorsal mantle fold.

State 23: Mid-veliger

This stage can be divided into two substages :

23/1. The embryo makes a complete turn in 15 sec. The foot shows spontaneous movements for the first time; it is separated from the visceral complex by a deep furrow. The middle region of the shell gland is still flat. The mantle fold envelops about one-third of the visceral sac. The heart, consisting of two chambers and pulsating at a rate of 44-48 beats/min, is located somewhat to the right of the sagittal plane in the caudal direction. The ventricular and atrial walls are composed of one layer. The pericardium has not yet formed. On the right side a small open pulmonary cavity is located, and the osphradium has not yet formed. On the right side a small open pulmonary cavity is located, and the osphradium is situated near it. The length of the prototroch is about 0.1 mm. Protonephridia are functional. The kidney does not yet show a lumen. The nuchal cells have stopped dividing. The pedal ganglia are connected by a commissure, while the buccal ganglia are being laid down. Other ganglia have not yet segregated from the ectoderm. Red-brown eyes are located at the base of the barely protruding head tentacles. A dental plate has appeared in the radular sac, which is still closed; the jaw has appeared between the oral cavity and the pharynx. Owing to torsion, the stomach apertures are now located dorsally and ventrally rather than laterally. The hindgut is connected to the stomach cavity.

23/2. The mantle and shell surround two-thirds of the visceral complex. The foot is divided longitudinally into two halves by a grove containing a stand of large ciliated cells. The heart is located in the sagittal plane approximately under the middle part of the shell.

Stage 24: Late Veliger or "Hippo"

24/1. The rotation of the embryo is irregular. The foot is the posterior part has not yet reached the shell; it may extend in the direction parallel to the cranial-caudal body axis. The prototroch is still present, but has been displaced more caudally. The shell covers the whole posterior half of the body. The first evidence of its helical coiling is present. The eyes have a reddish colour. A lens begins to form from the secretion supplied by the eye vesicle cells. The contracting heart, which is located in the center dorsal region, can be seen through the wide aperture of the pulmonary cavity. The heart ventricle has a multilayered wall. The pericardium has not yet completely formed. The kidney and pericardium are connected by the renalpericardial channel which is formed at this stage. The protopnephridia still function; their flame cells are no longer in contact with the cerebral ganglia. The buccal ganglia are connected by a commissure. The pleural ganglia are connected to the parietal and cerebral ganglia through connectives, which are longer on the right side of the body than on the left. The osphradial ganglion has formed. Visceral—parietal and osphradial parietal (right) connectives are present.

According in Kruglyanskaya and Sakharov biogenic amines can now be detected in the carebral and pedal ganglia for the first time.

Radular teeth begin to form. The stomach has turned even further to the right along the longitudinal axis; the large aperture is now located laterally on the left side. The epithelium has started to grow at the borders of these apertures and large slowly). These lobes divided the albumen sac into two parts, the left being greater than the right.

24/2. The foce reaches the shell at the posterior end. The eyes are dark.

Stage 25: Veuconcha

The embryo sometimes turns slowly in the proteinaceous fluid which is present in the capsule, and sometimes uses its foot to move along the inner capsule wall. It is still unable to withdraw into the shell. The shell now covers the whole visceral complex. A small mantle cavity has been formed by ectodermal cell proliferation and ectoderm wrapping under the shell edges. The tentacles as well as the mouth lobes are well developed. The prototroch is in the final stage of reduction. Resorption of the promnephridia and head vesicles has begun. At the beginning of the eyes are dark, and the lens is formed and yet pigmented. The heart has been displaced to the left body side and beats 56-60 times aq minute. The ventricle is larger than the atrism, and the pericardium is completely formed. The distances between the ganglia have decreased on the right side of the body owing to torsion, and the whole ganglionar ring possesses a plane of symmetry. The radular cavity is connected to the orthcavity. The jaw has formed completely. Salivary glands have been laid down as diverticular of the oral cavity walls.

From this stage onward torsion affects all organs of the visceral complex. The larger stomach apermre, together with the large liver lobe, has been displaced to the ventral side and the small one to the dorsal side. The larger albumen sac is located under the heart and stomach, while the smaller one is in the upper right region of the body. The hindgut perforates the pulmonary cavity and opens outside as the anal orifice. The oviduct has been laid down.

Stage 26: Transition to Foot Movement

Owing to the presence of a larger mantle cavity, the embryo can now withdraw into its shell and use only its foot for movement. The shell is coiled. There is intensive pigmentation of the occiput, tentacles, and mouth lobes. The prototroch has been completely resorted. The degrading protonephridia are still visible. The kidney is located on the roof of the pulmonary cavity and reaches the pericardium on the left side. Its position does not change any further before hatching. The nuchal cells are located as a single group and are active in holocrine secretion. The heart has been further displaced to the left, its main axis located horizontally. When the embryo moves, the contraction rate of the heart is 60 beats/min, while at rest it is 39 beats/min. The osphradium cavity is still open outside.

The pulmonary cavity extends from the right to the left body wall, its orifice bent toward the left. The caudal tip of the large liver lobe borders the shell epithelium. The left albumen sac protrudes into the depression formed by the first turn of the helix. The hermapiroditic gland appears to the right of the stomach in the ventral body region. The gonad begins forming after only two primary germ cells have arisen.

Stage 27:

The first helical turn of the shell is twice as high as at the previous stage. The kidney is bent over the pericardium to the ventral side of the embryo. The main axis of the heart is no longer horizontal, the ventricle now being located above the atrium. The heartbeat rate is 82 beats/min. The osphradium sac is closed and begins to divide into two parts. Diverticula form in the large lobe of the liver. The left albumen sac fills the entire caudal-ventral region of the visceral complex and is located perpendicularly to the main body axis. Some albumen cells divide.

Stage 28:

The first helical turn of the shell is twice as high as at stage 26. The heartbeat rate is 88 beats/min. The pulmonary cavity undergoes a marked extension in the caudal direction, its anterior wall bordering the shell. The small orifice of the stomach is on the left side owing to torsion, which generally reaches completion at this stage. Three diverticula are present in the large liver lobe. A dilatation appears in the caudal part of the esophagus. The right albumen sac is displaced to the left of the body. A penis appears at the caudal border of the right entacle.

The hermaphroditic gland opens into a short hermaphrodite duct.

Stage 29: Hatching

The embryo occupies the entire capsule. The capsular membrane breaks owing to movements of the young mollusc's radula. The nuchal cells are present in roughly a third of their initial number (which was maximal at stage 24). The degeneration of nuchal cells does not reach completion until the third week of post-embryonic development. The main axis is the heart is vertical and the heart is positioned well away from the stomach. The dorsal thickening with cilia, previously present, disappears from the esophagus. The large liver lobe consists of four diverticula. After hatching, the cells of the albumen sacs transform into epithelial cells, becoming part of the definitive liver. The left sac participates in forming the large liver lobe, and the right contributes toward the smaller lobe. The shell has one and a half turns and its retractor is well developed. The hermaphrodite duct is connected to the rudiment of the oviduct. Generally the genial system at the time of hatching is at an early stage of development. According to Cumin [49], the pulmonary cavity is closed at this stage; the respiratory orifice first opens into the atmosphere when the mollusc creeps up to the surface of the water. On the other hand, we have observed a wide open respiratory orifice in young molluscs well under water.

When embryos are cultured in capsules on agar, all molluscs hatch within 12 h. In our experiments capsules were incubated in water and the interval between the hatching of the first and last mollusc was also 12 h.

After hatching, young snails feed on slime and the envelope of the cocoon, but even during the first day they can eat green plants. According to van der Steen, the mortality rate of young snails during the first 11 weeks after hatching is 25%, whereas during the subsequent 3 weeks it does not exceed 10% at 20°C.

6

The Development of Drosophila Melanogaster

Drosophila is the common fruit fly and has been extensively studied since the time of T.H. Morgan. Because of the well-known genetic and the accessibility of the embryo and larva to manipulation, the fruit fly is extensively used in developmental studies. *Drosophila* embryo complete development about 22 hours after fertilization. The resulting larva hatches and enters a four-day growth phase. The larva is then transformed into a pupa. The pupa undergoes a metamorphosis during the next four days and finally emerges as an adult fly. The life cycle is relatively short compared to other embryonic systems; from fertilized egg to adult takes about nine days. During pupal metamorphosis, the larval tissues are for the most part destroyed. Most parts of the surface of the adult fly are formed from small nests of larval cells known as imaginal discs. There is a disc for each leg, eye, the genitalia, wing, and so forth. These cells are set aside during larval development and never function during larval life. They grow in size as the larva grows, but remain undifferentiated. Developmental studies on *Drosophila* have centered on the early development of the egg into the larva and on the development and determination of the imaginal discs.

GAMETO-GENESIS

Oogenesis

Nurse and follicle cells: Oocytes of insects must be provided with a yolk reserve, as well as a reserve of informational molecules involved in protein synthesis. Many insects are provided with highly polyploid nurse cells, which synthesize the required compounds, and a system of canals through which these materials pass into the oocyte. The oocyte is organized with the yolky material occupying the central portion of the oocyte cytoplasm (centrolecithal). Each egg chamber in a *Drosophila* ovary consists of 15 nurse cells surrounding the oocyte and a single layer of follicle cells surrounding the nurse cell. The population of 16 cells destined to become oocyte and nurse cells is so arranged that only 2 cells of the 16 form 4 cytoplasmic bridges with neighbouring cells. Only these two cells are potentially oocytes; normally one of these dominates and the other becomes the fifteenth nurse cell. In situation where the cellular arrangement provides no cells with four bridges, only nurse cells will develop and the female will be sterile.

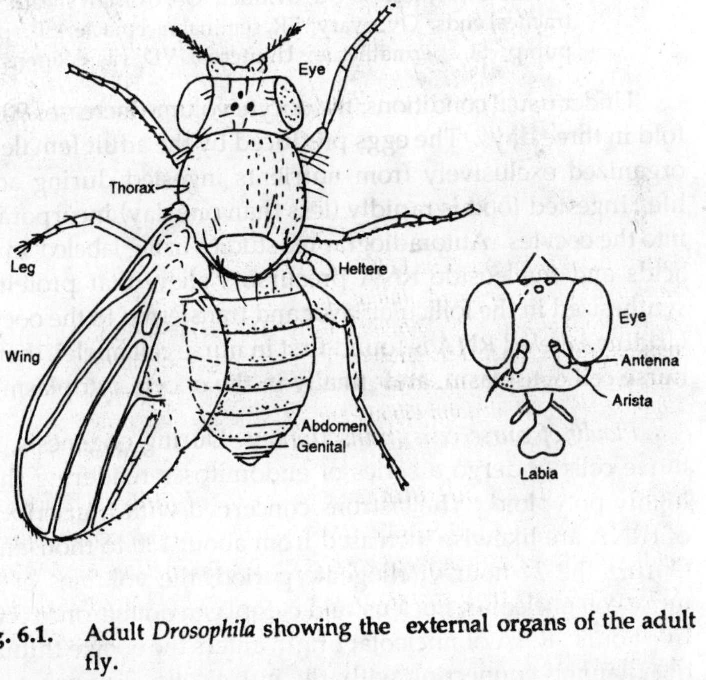

Fig. 6.1. Adult *Drosophila* showing the external organs of the adult fly.

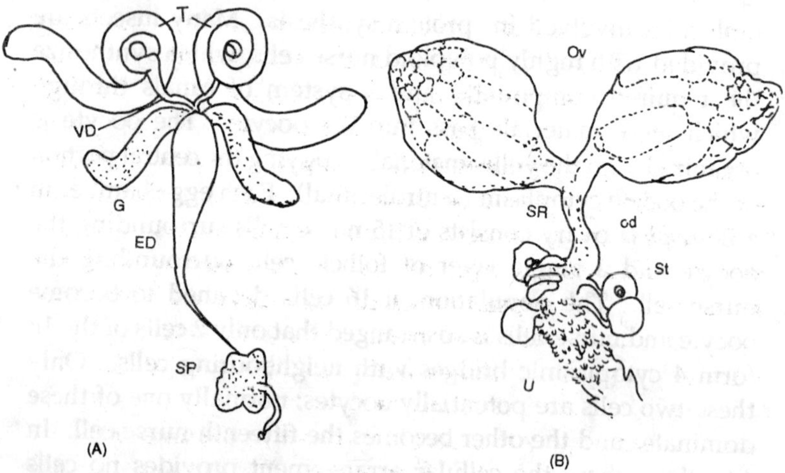

Fig. 6.2. (opposite) Genital systems of *Drosophila melanogaster*. *(A)* Male genitalia. (B) Female genitalia. ED, ejaculatory duct; G, Glands; T, testes: Od, oviduct; OS, ovarian sheath with tracheal ends; Ov, ovary; SR, seminal receptacle; SP, sperm pump; St, spermathecae; U, uterus; VD, vas deferens.

Under usual conditions, the oocyte volume increases 90,000 fold in three days. The eggs produced by the adult female are organized exclusively from nutrients ingested during adult life. Ingested food is rapidly (less than one day) incorporated into the oocytes. Autoradiographic studies using labeled amino acids and nucleoside RNA precursors show that protein is synthesized in the follicular layer and transferred to the oocyte. Uridine-labeled RNA is found first in nurse cell nuclei, then in nurse cell cytoplasm, and finally in the oocyte cytoplasm.

Ploidy of nurse cells; transcription: During oogenesis, the nurse cells undergo a series of endomitoses rendering them highly polyploid. The cistrons concerned with transcription of rRNA are likewise increased from about 130 to thousands. During the 24-hour vitellogenic period, the volumes of the nurse cell nucleolus, nucleus, and cytoplasm double once every five hours. RNA of nucleolar origin enters the oocyte through the channels connecting with the nurse cells. The oocyte

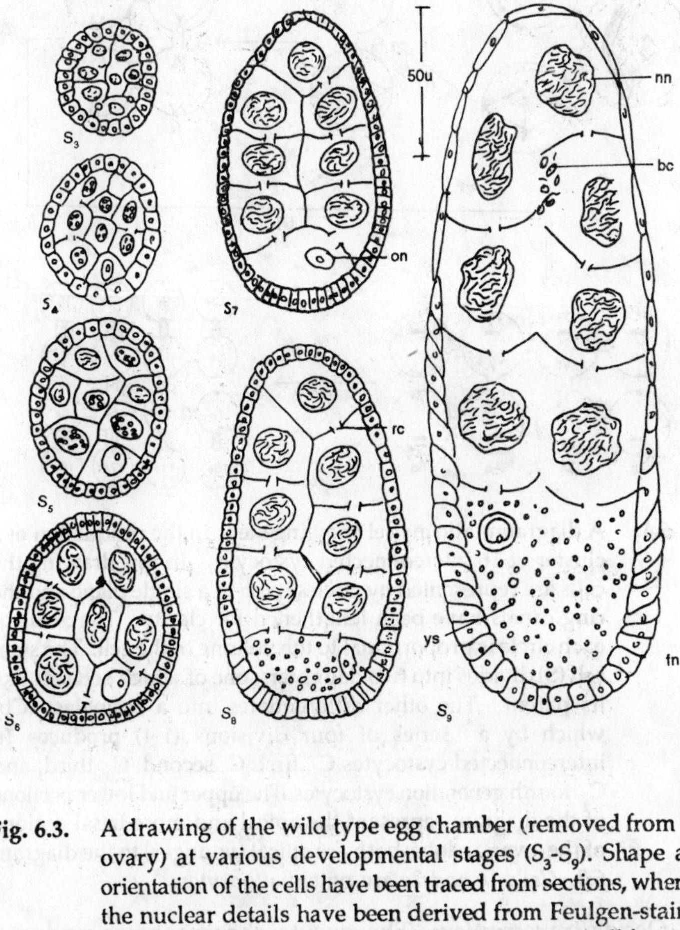

Fig. 6.3. A drawing of the wild type egg chamber (removed from the ovary) at various developmental stages (S_3-S_9). Shape and orientation of the cells have been traced from sections, whereas the nuclear details have been derived from Feulgen-stained whole mounts, be, border cells (these migratory cell form the micropylar apparatus); on, oocyte nucleus; ys, yolk sphere; nn, nurse nucleus; fn, follicle cell nucclllеus; rc, ring canal.

doubles its volume every two hours during the 24-hour vitellogenic period. The ribosomes of the oocyte, as well as mRNA and tRNA, are derived from the nurse cells. The cytoplasm of the mature oocyte contains approximately 2 x 10^{10} ribosomes.

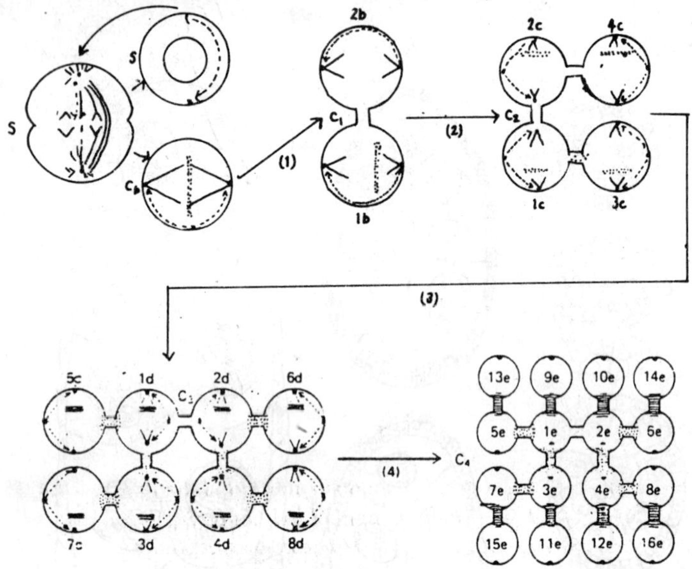

Fig. 6.4. A diagrammatic model showing steps in the production of a cluster of 16 interconnected cystocytes. In this drawing the cells are represented by circles lying in a single plane and the ring canals have been lengthened for clarity. The areas of each circle is proportional to the volume of the cell. The stem cell (S) divides into two daughters, one of which behaves like its parent. The other differentiates into a cystoblast (Cb) which by a series of four divisions (1-4) produces 16 interconnected cystocytes C_1, first: C_2 second: C_3, third; and C_4. fourth generation cystocytes. The upper and lower portions of the diagram represent the apical and more basal regions of the ovarian lobe; both are apical (younger) to the diagram 6.3. Cells 1e and 2e are potential oocytes.

Oocyte transcription: The oocyte chromosomes undergo synapsis and form synaptonemal complexes. Each complex is composed of dense lateral elements parallel to either side of a medial complex. The medial elements are made up of interdigitating rods extending from the lateral structure towards the midline of the complex. The lateral elements possess lateral loops extending out from the complex. The complexes may be involved in transcription of various RNA molecules.

As pointed out above, the oocyte rRNA is derived almost entirely from the nurse cell nucleoli.

Oocyte structure: The oocyte within the "uterus" of the female is bilaterally symmetrical. The anteroposterior, dorsoventral, and left-right axes of the egg and future embryo are coincident with those of the parent female. The micropyle, a point provided for sperm entry, is located anteriorly on the dorsal side. The egg measures about 420 micra long by 150 micra wide. The oocyte is surrounded by a tough extracellular membrane, the chorion. Anterior-dorsal extensions of this membrane form paired chorionic filaments, seemingly designed to aid in keeping the egg from sinking into a soft substratum after oviposition. Inside the chorion the vitelline membrane covers the oocyte plasma membrane. Internally, the oocyte contains yolk, concentrated in the center of the cytoplasm. The oocyte nucleus (meiosis 1) is eccentrically placed toward the anterior—dorsal side. Strands of material are positioned between the nucleus and the clear cortical cytoplasm. The posterior cortex is distinguished by a number of RNA-containing granules, marking the area of the germ plasm (pole plasm). The pole plasm area gives rise to the primordial germ cells : The ooplasm also contains, in addition to proteinaceous yolk platelets, numerous mitochondria and lipid droplets. The lipid originates from the nurse cells, as do some of the mitochondria. The mitochondrial DNA is replicated during the late phases of oogenesis, indicative of mitochondrial replication. The endoplasmic reticulum is studded with ribosomes and organized into long tubes. Free ribosomes are also present in large numbers in the ooplasm.

Centrifugation studies separate the cytoplasm into three layers : a centripetal layer of fat : a region midway between the centrifugal poles, containing mitochondria and endoplasmic reticulum; and a centrifugal zone of yolk platelets. During centrifugation, the oocyte membrane remains attached to the overlying vitelline membrane.

Spermatogenesis

The testis: The primary spermatogonia are located in cysts within the testis of the larva. Meiosis is not observed until pupal and adult life. Primary spermatogonia divide mitotically to form secondary spermatogonia. These cells form, through four mitotic divisions and two meiotic divisions, 64 spermatids, which are housed in an individual cyst of the testis. The development within a cyst is synchronous. The number of cells undergoing meiosis reaches a maximum during midpupal life; thereafter, the testes contain smaller numbers of dividing spermatocytes. The apical cyst retains spermatogonial cells that proliferate during the reproductive life of the male. These cells replenish the supply of spermatocytes as matings subtract mature sperm from the basal end of the testis, which opens into the vas deferens.

Sperm Maturation

The transformation of spermatid into mature sperm (spermiogenesis) follows the general outline provided in the introductory chapter. Included are the development of the axial filament (axoneme) from the distal centriole, the development of a "midpiece" (nebenkern), a body which extends over the entire length of the tail from the mitochondria; the formation of an acrosomal body by the Golgi apparatus; and the formation of cytoplasmic bridges between spermatids occupying the same cyst. The mature spermatozoan is tremendously elongate, the axial filament with mitochondrial bodies measuring 1.76 mm in length by 1.0 micron in diameter. The head measures 13 micra by 5 micra. The length of the sperm is indeed tremendous when it is recalled that the mature egg is but 0.42 mm in length, and that the entire sperm including the tail penetrates the cytoplasm of the egg at fertilization. The syncytial bridges seem to be involved in the loss of spermatid cytoplasm and cytoplasmic organelles. This process begins in the head of the spermatids of a given cyst and proceeds to the axoneme. The bridges are lost as the cytoplasm and enclosed organelles are discarded in a large common sac of degenerating material.

Role of the Y Chromosome

In males lacking a Y chromosome the XO condition prevails. Files of XO genotype are phenotypically male, but contain no viable sperm. Spermatocytes of males carrying a Y chromosome contain chromosomal structures called Y loops. These loops are considered to be active in transcription. A component of RNA complementary to Y chromosome DNA has been found in XY males; this component is not evident in XO males. Actinomycin prevents the formation of Y loops in XY spermatocytes. The implication is that stable transcripts of the Y chromosome are stored and parcelled out during meiosis to all spermatids, where they are active in protein synthesis during spermiogenesis. The situation is analogous to that found in oocytes containing stable, stored mRNA for use in early development. Transcription of Y-DNA in the spermatid is ruled out since only half the spermatids contain a Y chromosome during the process of spermiogenesis and since protein synthesis occurs without RNA synthesis in spermatids.

It is of interest that male *Drosophila* form no synaptonemal complexes. An obvious correlation between this lack of complexes during pachytene and the lack of cross-overs in the male *Drosophila* comes to mind. The significance of the lack of these structures is conjectural at this point.

EMBRYONIC DEVELOPMENT

Fertilization

Maturation divisions: Fertilization occurs in the female genital tract and is usually monospermic. The sperm are stored following mating in a spermatheca. The nuclear membrane of the primary oocyte nucleus has dissolved by the time the sperm (including the tail) has entered the oocyte. The first meiotic division metaphase is located on the anterior—dorsal side of the egg; telophase is reached as the sperm approaches the division spindle. No centrioles are visible during egg maturation. The second division closely follows on the first and produces three pronuclei within the egg. The haploid nucleus furthest from the egg surface forms the female pronucleus, the remaining three nuclei remain in an anterodorsal island of cytoplasm and eventually degenerate. Centrioles for succeeding mitotic (cleavage) divisions are provided by the sperm. Completion of meiosis is independent of sperm penetration and occurs in uninseminated eggs during laying.

Cleavage: DNA replication occurs before pronuclear fusion in both male and female pronuclei. A mitotic apparatus is formed around each pronucleus as both nuclear envelopes dissolve. Two daughter nuclei are formed following the first cleavage division. The zygote nucleus undergoes nine rapid divisions, without cytoplasmic division, resulting in a syncytial stage. Two hours after fertilization these cleavage nuclei migrate to the egg cortex and become separated by cell membranes after three more divisions. The nuclei begin RNA synthesis at this time and nucleoli appear. Evidence presented below suggests that these cleavage nuclei are equivalent and become determined upon migration to the cortical region. The basic pattern for embryogenesis is established in the cortex prior to the arrival of the cleavage nuclei and is imposed on them when they arrive to form the cells of the blastoderm.

Blastula and Gastrula

The pole cells and the blastoderm: Three hours after fertilization, the egg is covered by the blastoderm layer. The yolk is centrally located, and the pole cells are located posteriorly and dorsally. The pole cells are not dividing and are pressed between the flattened blastoderm and the perivitelline space.

Gastrulation: The initial indication of gastrulation (mesoderm formation) is the appearance of a longitudinal ventral furrow. This furrow extends from a point near the anterior end, all the way to the posterior end. The furrow remains open for only a short time. It is deeper in the midventral portion, forming the germ band which differentiates anteriorly and posteriorly from this region. Following closure of the furrow, the inner tubular structure thus formed flattens and the lumen within it is obliterated. The stomodeum forms as an anterior ventral invagination; the hindgut and posterior midgut form as an invagination in the dorsal surface midway between the anterior and posterior ends. The pole cells are carried into the embryo via the hindgut invagination and lie within the posterior midgut region. During the sixth hour of development, the pole cells move to their final position, four segments from the caudal end of the development larva. From this position, they are included in the developing gonad as the primordial germ cells.

The proctodeal opening is moved posteriorly from its middorsal position. The posterior midgut portion encircles the yolk mass and extends forward to join with the anterior midgut rudiment. This rudiment is continuous with the stomodeal invagination. The larva becomes segmented at about this time.

The nervous system develops from the ventral ectoderm as a longitudinal nerve cord with segmental ganglia. The salivary glands, famed for their polytene chromosomes, appear at the eighth hour of development. They first appear as ventrolateral ectodermal plates immediately adjacent to the ventral nerve cord. The glands form as deepened pits that lose their connections to the outside and join with the pharynx through a common duct. The tracheal system of air tubes and the imaginal discs are also developed from surface ectodermal cells.

Larval Organogenesis

Summary of germ layer derivatives: The imaginal discs, skin, salivary glands, and tracheae are derived from the lateral ectoderm; the nervous system, foregut, and hindgut are from the medial ventral ectoderm. The dorsal ectoderm is involved with extraembryonic membranes, which in *Drosophila* are transient structures. The mesoderm formed from the ventral furrow gives rise to an outer layer, forming the body musculature, and an inner layer, forming the gut musculature.

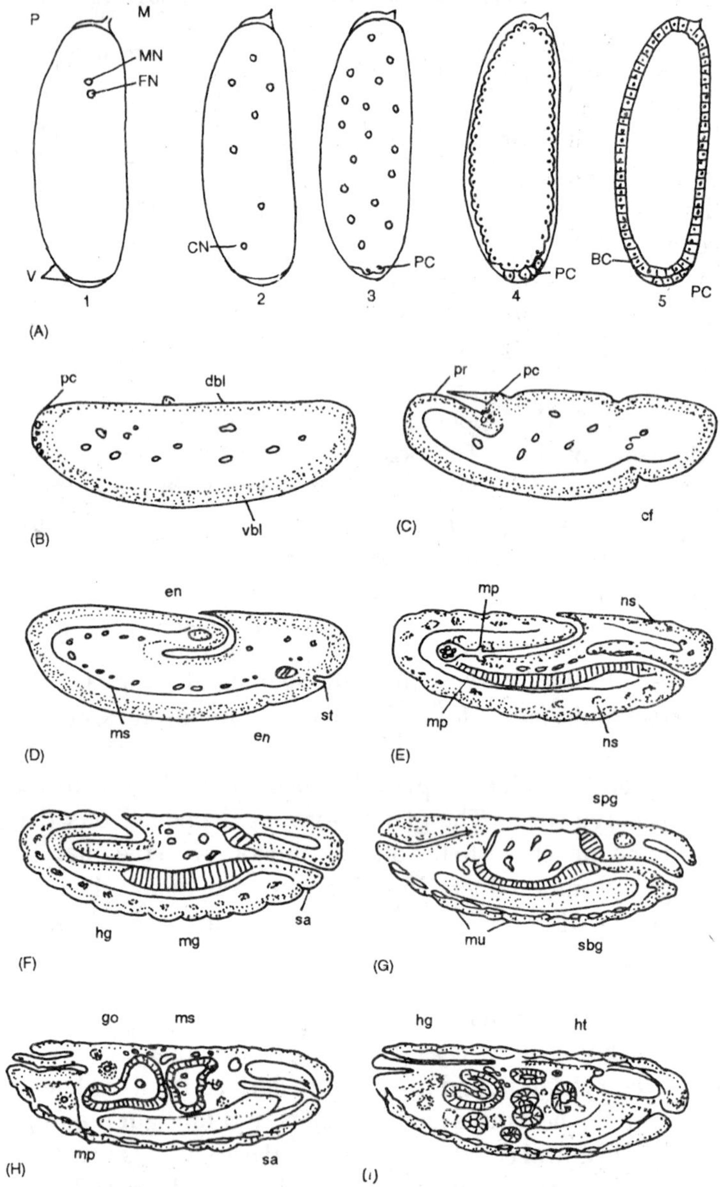
P
M
MN
FN
CN
PC
PC
BC
PC
V
1
2
3
4
5
(A)
pc
dbl
vbl
(B)
pr
pc
cf
(C)
en
ms
en
st
(D)
mp
ns
mp
ns
(E)
hg
mg
sa
(F)
spg
mu
sbg
(G)
go
ms
mp
sa
(H)
hg
ht
(I)

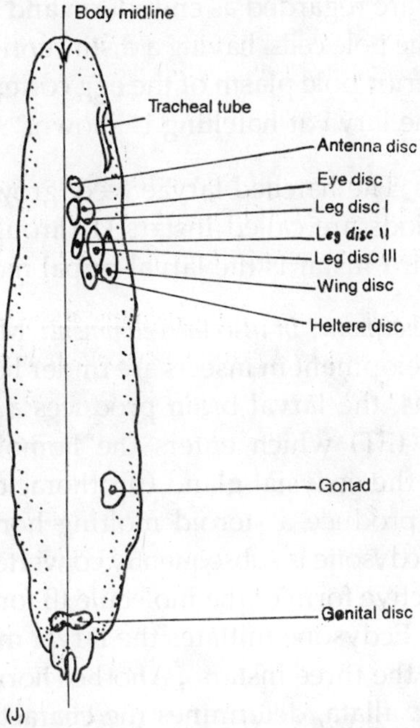

Fig. 6.5. (A) Development to larva series. Early embryogenesis of Drosophila : (1) Fertilized egg (2) "cleavage," (3) pole cell formation. (4) syncytial preblastoderm, and (5) blastoderm. M. micropyle: P. plasma membrane: V. vitelline membrane: MN. male pronucleus: FN. female pronucleus: CN. "cleavage" nucleus: PC, pole cells: BC, blastoderm cell. 0-2 hours (B) Blastoderm (2nd hour). (C) Gastrulation (3rd hour). (D) Expanded germ band (4th - 6th hour). (E) Later germ band showing nervous system (8th hour). (F) Contraction of germ band (11th hour). (G) Embryo following contraction (12th hour). (H) differentiation in embryo (16th hour). (I) Larva before hatching (22 hour). Abbreviations: cf, cephalic furrow : dbl, dorsal blastoderm: en: endoderm rudiments of midgut: go. gonad: hg, hindgut: ht. heart on dorsal blood vessel: mg. midgut: mp. malpighian tubule: ms, mesoderm of germ band; mu, muscle; ns, nervous system; pc, pole cells; pr, proctodeum; sa, salivary gland; sbg, subesophageal ganglion; st, stomodeum; vbl, ventral blastoderm. Sagittal sections.

The invaginating tips of the proctodeal and stomodeal invaginations are regarded as endoderm and form the midgut epithelium. The pole cells, having a distinct origin in association with the posterior pole plasm of the egg cortex, give rise to the germ cells. The larva at hatching is shown.

Larval life: The hatched larvae feed, grow and molt. The intermolt periods are called instars (1 through 3). The final molt of the third instar is the larval-pupal molt.

The role of hormones in insect development: Molting, pupation, and adult development in insects are under hormonal control. In most insects, the larval brain produces a hormone called ecdysiotropin (ET) which enters the homolymph and acts ultimately on the ecdysial gland (prothoracic gland) causing this gland to produce a steroid molting hormone called α–ecdysone. α-Ecdysone is subsequently converted to β-ecdysone, which is the active form of the molecule (β form = 20-hydroxy α–ecdysone). Ecdysone initiates the larval molts as the larva grows during the three instars. Another hormone, produced by the corpora allata, determines the character of each molt. This hormone is called the juvenile hormone (JH). It is high in concentration during the larval molts, causing preservation of larval structures. A decrease in JH in the third instar brings about the larval-pupal molt under the influence of ecdysone. The pupal-adult molt is initiated by ecdysone alone in the absence of JH. In *Drosophila*, the endocrine glands mentioned above (excepting the brain) are fused into a single structure termed the ring gland (corpora allata and prothoracic or ecdysial glands).

Ecdysone causes development of imaginal discs, the puffing of polytene chromosomes (RNA synthesis), protein synthesis, and finally metamorphosis into the pupa. When sufficient JH is present, ecdysone induces larval molting, but not metamorphosis.

Larval imaginal discs: These are basically folded sacs that originate as invaginations of the embryonic epidermis. During

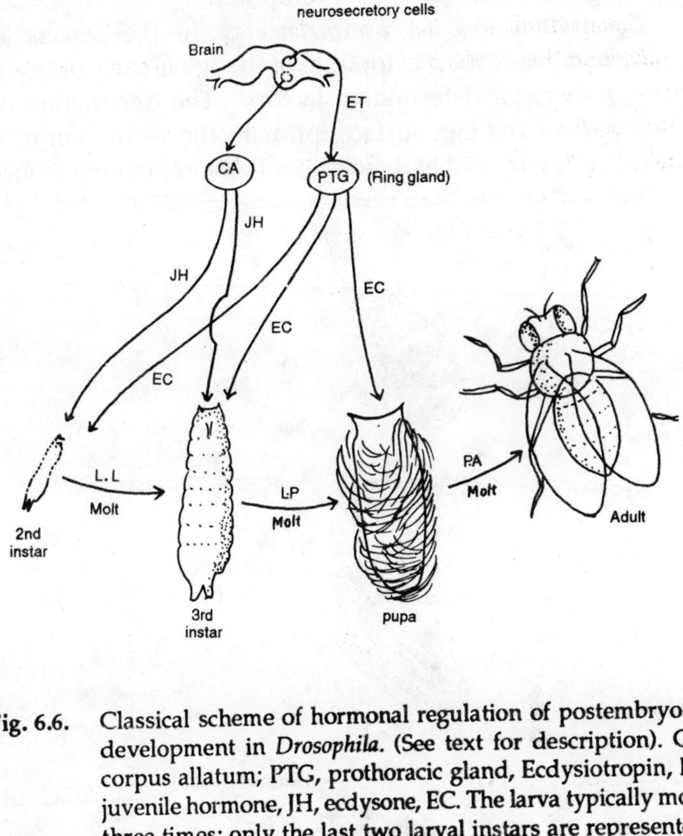

Fig. 6.6. Classical scheme of hormonal regulation of postembryonic development in *Drosophila.* (See text for description). CA, corpus allatum; PTG, prothoracic gland, Ecdysiotropin, ET; juvenile hormone, JH, ecdysone, EC. The larva typically molts three times; only the last two larval instars are represented.

pupation, they undergo eversion to form adult structures (wings, antennae, legs, etc.), a process that involves change from columnar to squamous cell shape. The nuclei of the imaginal discs are the cleavage nuclei, which move randomly into the specific cortical regions following the period of mitosis. The discs do not develop beyond sac-like structures of the larva, but continue growth (cell division) through the three larval instars until the final larval-pupal molt. During differentiation, cell proliferation ceases and dramatic morphogenetic events the place.

EXPERIMENTAL EMBRYOLOGY

Regulative Versus Mosaic Development

Constriction and defect experiments: In the *Diptera* and *Lepidoptera*, the cortical cytoplasm of the egg is early organized into a pattern of determining factors. The destruction of a small part of the egg surface, prior to the formation of the blastoderm, will lead to deficiencies in the remaining embryo.

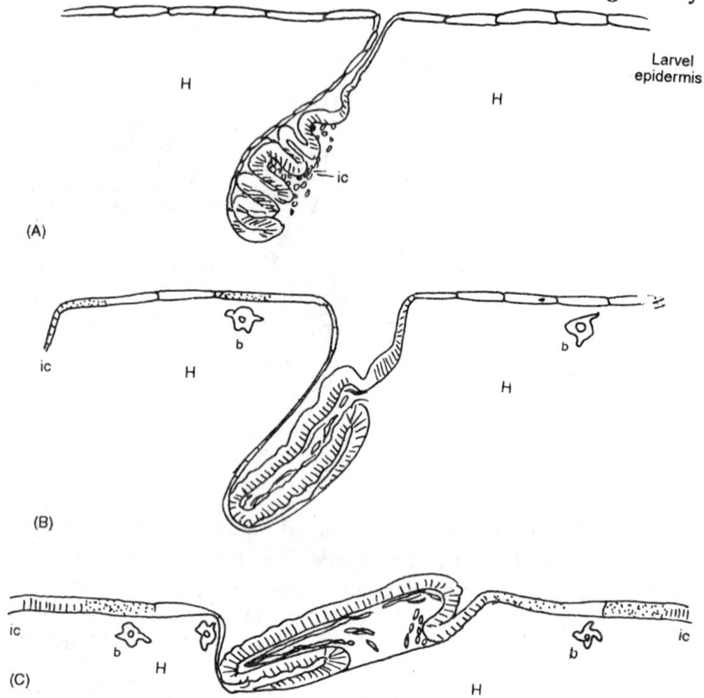

Fig. 6.7. Diagram to illustrate the process of eversion in a leg disk. (A) the disk epithelium in the mature larva is a single layer of epidermal cells, continuous with the larval epidermis through the disk stalk. (B) 3-5 hours after initiation of metamorphosis, the disk is partially unfolded and partially uncovered. (C) 5-6 hours after initiation of metamorphosis, the developing leg is almost completely unfolded and uncovered. Imaginal disk cells (ic) sequentially replace the larval epidermal cells. H, Hemolymph; b, blood cell.

Other insects, such as the honey bee and dragon fly, exhibit more regulative development in that whole embryos can be obtained from ligated portions of the egg, under the proper conditions. Longitudinal separation of egg halves leads to twinning in these forms.

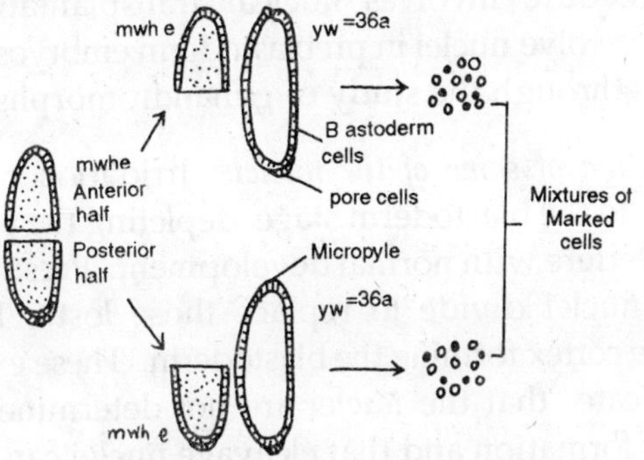

Fig. 6.8. Experimental procedure for analyzing the state of determination of blastoderm cells. Genetically marked blastoderm embryos are cut in half and separately intermixed with whole embryo carrying different genetic markers. Subsequently the embryos are dissociated into single cells, reaggregated and cultured in vivo.

Early determination: Dissociated embryonic cells of *Drosophila* can be reaggregated by centrifugation and cultured *in vivo.* The reaggregated cells will form various anterior and posterior imaginal structures following transplantation into a larval host, and following the metamorphosis of the host. It is possible to extend this technique by using an anterior or posterior half dissociated along with a whole embryo of a recognizably different genotype. It was found that cells of anterior halves mixed with cells of whole embryos gave rise only to anterior structures; posterior halves treated in the same manner gave rise only to posterior structures. These results indicate that the *Drosophila* blastoderm cells are determined, at least with respect to their ability to form anterior or posterior structures. The germ cells are also determined early in development. "The cortex and early determination").

Nuclear Equivalence

The problem of nuclear equivalence and nuclear totipotence has been studied using three kinds of experiments. The first involves irradiation of cleavage nuclei in such a way as to inactivate at random some but not all of the nuclei. The second procedure involves nuclear transplantation. Both procedures involve nuclei in preblastoderm embryos. The third approach is through the study of gynandromorphs.

Elimination of some of the nuclei: Irridation of cleavage nuclei prior to the blastoderm stage, depleting the population, does not interfere with normal development. Presumably, the remaining nuclei divide to replace those lost. The nuclei populate the cortex forming the blastoderm. These experiments clearly indicate that the nuclei are not determined prior to blastoderm formation and that cleavage nuclei can substitute for each other in embryogenesis and are, thus, equivalent.

Nuclear transplants: Transplantation experiments involve either the transplantation of cleavage nuclei to unfertilized eggs, or the transplantation of nuclei from one cleaving egg to another. The interpretation is somewhat complicated in both instances since the resident nuclei of the recipitent eggs are not removed and since normally more than nucleus is injected. These difficulties are largely overcome by using host and donor nuclei of known genotype , so that tissues arising from the donor nuclei will be marked in the larva or adult. Control experiments show that puncture or injection of cortical cytoplasm alone does not stimulate parthenogenesis of the unfertilized eggs that serve as hosts.

Transplantation of cleavage nuclei to unfertilized eggs initiates development and results (in about 1 per cent of the cases in larvae of the donor genotype. These larvae never metamorphose into adults. However, the imaginal discs of these larvae will give rise to all adult structures when transplanted into unoperated larvae and allowed to metamorphose with the host. Moreover, even eggs that are arrested at cleavage produce larval and adult tissues *in vivo* culture in pre-metamorphic (first and second instar) larvae. Nuclei from the posterior pole plasma region and from other regions of the preblastoderm embryo give similar results. These experiments indicate that at least some of the nuclei are totipotent. The failure of metamorphosis is not understood at present.

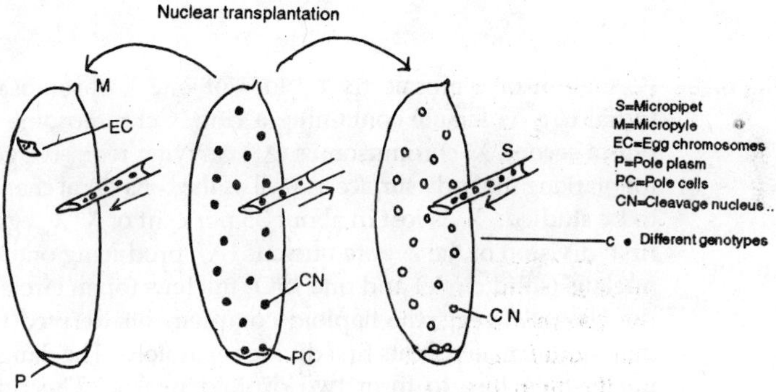

Fig. 6.9. Diagram of nuclear transplants in *Drosophila*. Cleavage nuclei are transplanted to either unfertilized eggs or cleavage stages from genetically marked donor cleavage stages (dark nuclei). S, micropipet: M, micropyle; EC, egg chromosomes: p, pole plasm: PC, pole cells: CN, cleavage nuclei. Open circles, recipient nuclei; filled circles, donor (marked) nuclei.

When nuclei of cleavage stages are transplanted into fertilized eggs, genetic mosaic flies result, derived in part from the zygote genome of the host egg and in part from the transplanted cleavage nuclei. Two of the mosaics in one study were male and produced offspring carrying the marker genes of the donor. This indicates that the transplanted nuclei found their way to the pole plasm and participated in gametogenesis. The transplanted nuclei were not derived from the pole plasm region of the donor and could not have been influenced by this portion of the cortex. Again, the totipotence of cleavage nuclei is implied.

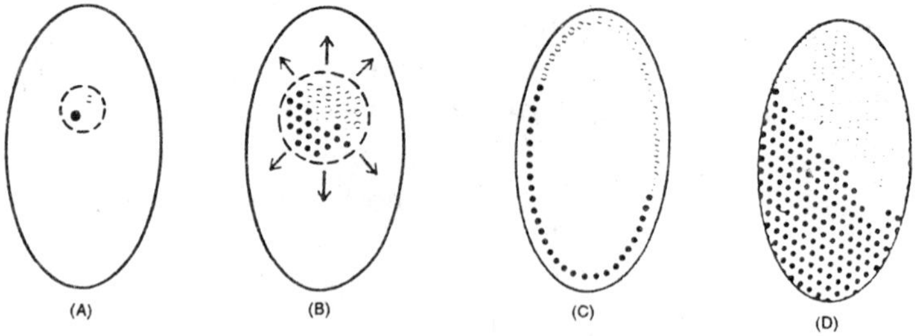

Fig. 6.10. Formation of a mosaic fly by loss of one X chromosome. Initial egg is female containing a ring X chromosome (X_R) plus a second X° chromosomes (X°) carrying recessive genes for marking the body surface as well as the behavioral character to be studied. X_R is lost in about 35 per cent of X° X_R eggs at first division of the zygote nucleus (A), producing one X°X_R nucleus (solid circle) and one X°O nucleus (open circle). In the *Drosophila* egg, the haploid complements derived from male and female parents first divide separately. The daughter nuclei then fuse to form two diploid nuclei. The ring X chromosome used here tends to be lost from one of these, probably the non-disjunction in the next division so that half the cells of the embryo become X°O. For discussion of the mechanism see the papers by Hinton (1955, 1957, 1959). Loss of X_R may occasionally occur also at later divisions. Nuclei divide in cluster (B) then migrate to egg cortex, where cell membranes form to produce a blastoderm one cell thick (C). D shows the surface as seen from outside; part of the blastoderm area is, covered by each cell type, the mosaic boundary separates the two areas.

Evidence for equivalence from gynandromorphs: Gynandromorphs are mosaic flies composed of both male and female cells. The mosaic XX/XO is produced by loss of one X chromosome during the first cleavage. It appropriate recessive markers are used on the X chromosome retained by the male cells, male and female tissues can be easily distinguished. The recessives will be expressed in those cells containing the single X chromosome. In adult gynandromorphs of this type, about half the epidermal cells are male, the other half are exclusively female. The dividing line between male and female epidermis is randomly oriented. Considering a number of gynandromorphs, this randomness indicates that the migration of the cleavage nuclei to the periphery of the egg is random, save that male and female nuclei tend to stay together. The interpretation has been verified by direct cytological examination. Since the parts of the embryo are always derived from specific parts of the egg cytoplasm and since the migration to the periphery is random, the nuclei must have originally been equivalent. The nuclei later acquire positional information from the cortex of the egg cytoplasm in which they reside following migration.

In the direct tests for potency of cleavage nuclei, the possibility that the tested nuclei have undergone some differentiation has *not* been ruled out. The experiments show only that any differentiation that might have occurred in the test nuclei must be reversible upon transplantation.

The Cortex and Early Determination

Cortical defects: Since there is evidence that the early blastoderm is determined as to anterior and posterior parts and since the cleavage nuclei are equivalent, it is reasonable to assume that the cortex has an early determining influence on the fate of the nuclei. Observations of the cortical pole plasm, which acts to determine the primordial germ cell nuclei, would support this concept. Irradiation of the pole plasm with ultraviolet light produces sterile adults. The idea that regions of the cortex other than the pole plasm might act in similar fashion has been tested by analyzing the developmental consequences of defects made in the cortex of the young embryos prior to the migration of cleavage nuclei to the cortex.

In one series of experiments, regions of the egg cortex were irradiated with an ultraviolet microbeam during the early cleavages. The nuclei were unaffected by the irradiation, since intervening cytoplasm would absorb the ultraviolet light. Adults with various defects were produced by this procedure, but there was no correlation between the location of the embryonic defect and the resulting abnormality in the adult.

A second series of investigations made on later cleavage stages during the preblastoderm stage, gave somewhat contrasting results. In this case, mechanical punctures made in the early embryo were correlated with defects in adult structures known to arise from the treated area. A third technique, microcautery, in which a small portion of the cortex is burned with a fine, hot wire, was used to show that some regulation took place at the nuclear multiplication stage (one hour of development). No regulation took place at the blastoderm stage ($2\ ^3/_4$ hours of development).

Assuming the methods are comparable, then the cortical organization must become fixed sometime in the two-hour period between the early cleavages and the syncytial blastoderm stage.

In the germ cell cortex, determination has been demonstrated, in recent reported work, to occur in the pole plasm in prefertilization ovarian eggs. Transplants of pole plasm from prefertilization oocyte to anterior regions of cleavage stage eggs induces germ cell formation in this abnormal position in the recipient. These germ cells are cycled through the germ line of genetically different host embryos and are found capable of forming functional gametes, as deduced from the phenotypes of the offspring of recipient flies. These results show that the posterior pole plasm of the unfertilized egg has *already* acquired its germ cell determinative ability.

The pole plasm and germ cell determination: The pole cells are the first cells of the blastoderm to be formed. From two to eleven nuclei penetrate the polar region and are budded off as pole cells. The polar cytoplasm is characterized by the presence of polar granules that are incorporated into the pole cells. The pole cells divide until there about 50 per cent at the posterior pole as the blastoderm is completed. About half of these cells migrate to the egg interior, passing between the nuclei of the syncytial blastoderm on their way to the developing gonad. Ultraviolet irradiation prevents pole cell formation and results in sterile adults. It has been found more recently that the injection of posterior pole plasm from non-irradiated embryos will repair the induced defect in the pole cells. Only injections from the posterior pole are effective; injections using cytoplasm from the anterior region of the egg are not effective. It has been subsequently noted that the effect of the pole plasm is not species specific because pole plasm from other species of *Drosophila* will also work. It has been assumed that the polar granules are the sensitive elements of the pole plasm, primarily because it was assumed that their content of RNA would render them sensitive to ultraviolet irradiation. Centrifugation studies seem to indicate that both polar granules and polar cytoplasm are essential for normal pole cell formation. These experiments are difficult to control, in that other parts of the system may also be altered by the centrifugal field.

Genetic analysis of the pole plasm: As interesting mutant in *Drosophila subobscura* is called "grandchildless." Females with this autosomal mutation in homozygous condition lay eggs which do not form pole cells and that thus give rise to sterile adults. Homozygous males are apparently normal. The mutation affects oogenesis and brings about a specific defect in polar cytoplasm. This effect can be obtained without the use of microsurgery or irradiation, and for this reason, the mutation is most valuable. Polar granules are present in the mutant as well as in the wild type. Whether or not these granules are *biochemically* normal in the mutant is unknown. Superficially, it would seem that the polar granules may not be involved in the determination of the polar cytoplasm. In the development of the pole plasm area, the mutant embryos exhibit a vacuolated cytoplasm in the polar area, and the cleavage nuclei do not enter the polar plasm. The pole plasm is sealed off from the rest of the developing embryo, and since the pole plasm is not cellular, there is nothing to migrate through the blastoderm into the developing gonad. It is not known whether injection of polar cytoplasm from wild type eggs will correct the mutant gene defect.

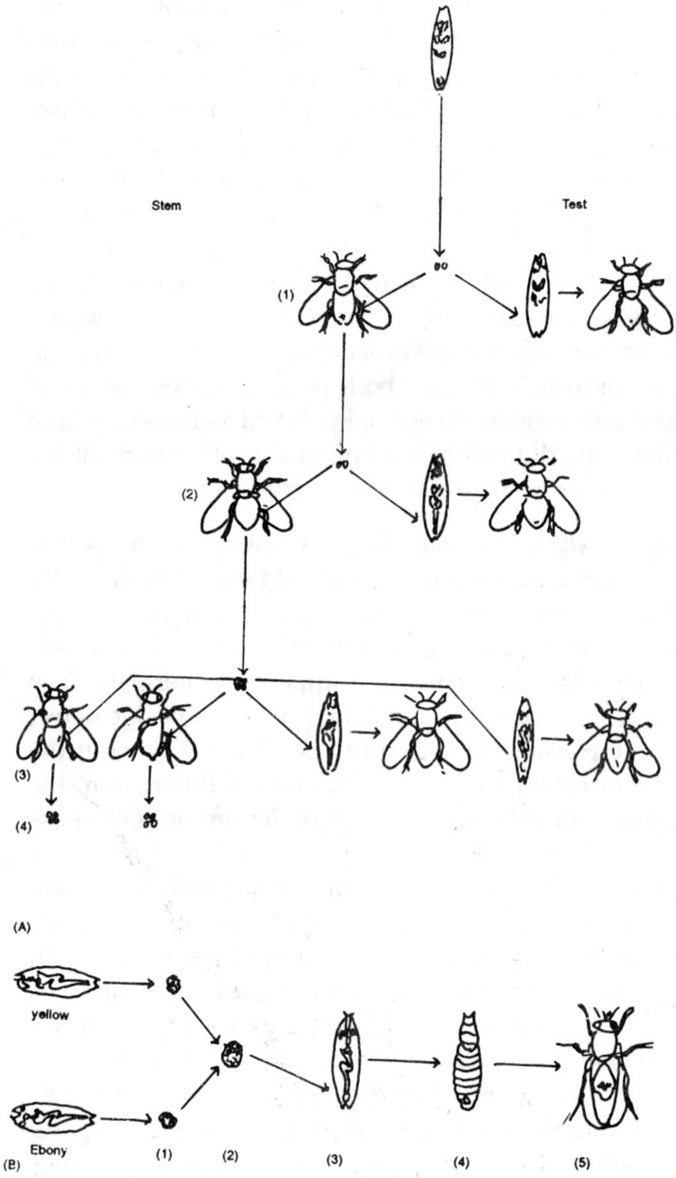
Stem
Test
(1)
(2)
(3)
(4)
(A)
yellow
Ebony
(B)
(1)
(2)
(3)
(4)
(5)

Fig. 6.11. (A) In vivo culture of imaginal disc cells. Successive transfers of cells grown from a single bit of imaginal disk is diagrammed over four generations. A bit of genital disk from a larva is divided between an adult host (1, left) and a larva (right), which is allowed to mature so that the condition of the transferred cells can be checked. The process is repeated (2) by retransplanting the disk cells of the "stem" host. By the next transplant (3) the stem disk cells have increased sufficiently to provide two transplants for testing and two transplants into stem hosts. In the next generation (4) the number quadruples. Disk cells have survived and multiplied for more than 150 transfer generations. (From E. Hadorn, "Transdetermination in Cells, "*Scientific American* 219 : ©110. 1968 by Scientific American, Inc. Reprinted with permission). (B) Mixture of imaginal leg cells from the larvae of fruit flies that belong to separate colour strains (yellow and ebony) is formed by removing part of a leg imaginal disk from each larva. (1, left). The bits of disk are then separated into their individual cells, mixed together (2), and implanted in the body cavity of a host larva (3). The host pupates (4) and the implanted cells mature. When removed from the adult fly (5), the leg organs of the complete leg are partly yellow and partly ebony in colour.

Studies on the Imaginal Discs

Culture procedures: The test system for imaginal discs involves the removal of a specific disc from the larva (first, second, or third instar) and its placement into a suitable host for either growth or metamorphosis. If growth is desired, the disc tissue is injected into a first or second instar larva, or more usually, into an adultfly. In such situations, the disc will increase in size, and portions of the disc may be removed and subcultured. When expression of the developmental potential of the disc is desired, the tissue is either left in the larval host or transplanted from the adult larval host into a third instar larva about to pupate. As the larval host metamorphoses, so does the implant of imaginal tissue. The different responses of a transplanted imaginal disc to the different environments is dependent upon the host's hormones. In the young larva, JH will prevent metamorphosis of the imaginal disc during larval molting, permitting growth but not differentiation. Transplantation from either adult or larval host to a third instar larva about to undergo the larval-pupal molt will result in differentiation of both host and implanted imaginal discs because of low JH and the presence of ecdysone. An imaginal transplant to the body cavity of the adult permits growths, but differentiation does not occur due to the absence of ecdysone. Imaginal transplants growing in adult flys can be sub-divided and subcultured in other new hosts or tested by transplant to a third instar larva where they are allowed to undergo pupation. Tests of growing imaginal subcultures reveal that they remain true to their origin over many generations of subculture in adult flies.

Determination in imaginal discs: Extirpation of any of the imaginal discs of the larva leads to a loss of the structure in the emerging adult. Other regions of the larva cannot compensate for the loss of any imaginal tissues. Whole imaginal discs implanted into larval hosts invariably develop according to their prospective fate. The disc thus appears to be a rigidly determined stable primordium. Exceptions to this statement are discussed below. Discs derived from third instar larvae can be subdivided and implanted as disc fragments into other third instar larvae. This procedure results in production of partial adult structures. Removal of small pieces from a disc under similar experimental conditions results in deficient adult structures derived from the larger portion of the disc, which was left in place, and missing the small extirpated section. The conclusion is that discs in third instar larvae are aggregates of cells, or cell groups, each determined to develop into specific portions of the adult structure.

In these later discs, single cells seem determined: Given that imaginal discs generally remain true to their fate during serial culture in adults, the question arises as to the ability of mixed, dissociated discs to differentiate. Using two genetically marked discs (e.g., yellow versus wild type), it has been shown that after culture in vivo reaggregated discs will form mosaic

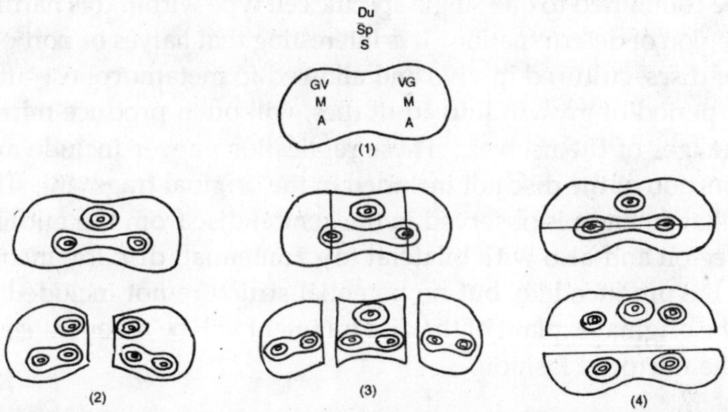

Fig. 6.12. Regeneration potential of portions of the male genital disc after *in vivo* culture and induced metamorphosis. Note patterns, particularly in 2.3 and 4. Discussion in text.

structures faithful to the original disc in homologous combinations (leg + leg; wing + wing). The cells of the two homologous discs may have had to modify their fate to some degree from the normal situation, since co-operation between cells of two imaginal discs is needed to produce a single structure. In heterologous mixtures of marked cells, the different disc cells sort out, so that wing cells give rise to wing, separate from the leg cells which form leg. The cells of the two primordia sort out following mixing and before differentiation. Occasionally, marked wing cells will be trapped in a body of differentiating leg cells, and they can be distinguished as wing disc cells by the genetic marker. These islands of trapped cells may consist of a single cell or a very few cells. The trapped cells differentiate autonomously according to the disc of their origin. The possibility that individual cells are responsible for the determination, rather than the primordium as a whole, has emerged from these observations. The exact state of determination of individual cells of the late third instar larval discs is difficult to ascertain using the methods described. Some findings seem to indicate that though the cells may be committed to a narrow developmental fate, they may not yet be committed to one single specific cell type within this narrow region of determination. It is interesting that halves of portions of discs, cultured in vivo and allowed to metamorphose after a period of growth into adult flies, will often produce mirror images of themselves. These replications never include any portion of the disc not included in the original fragment. The phenomenon is observed in the genital disc from the midline region and also with bilateral (e.g., antennal) disc fragments. Like produce like, but no potential structure not included in the original explant of the disc fragment will be regenerated in the mirrored fashion.

Determination of the discs proceeds in a stepwise manner: Given the final rather specific determination of the imaginal cells in the late third instar larva, examination of clonal markers, produced by x-ray-induced somatic crossing over in the four-strand stage of mitosis with irradiations carried out at both early and later developmental stages indicates a gradual specification of disc areas. Early (24 hours after oviposition), in the wing, clones of homozygous marker cells occupy both the upper and lower portions of the wing blade. Later (48 hours, after oviposition), the cells are confined to either the upper or the lower wing surface. This determination occurs by the end of the first or beginning of the second instar. By the end of the third instar, determination between the margin of the wing (bristle cells) and the hair cells of the wing surface has been made. The basic assumption is that the irradiated stem cell, in giving rise to discrete areas of homozygous tissue, reveals its state of determination at the time of irradiation. These areas become more restricted with time, indicating a progressive or stepwise determination of the final elements of the adult structure. Some regulation may occur within a single disc, but mixtures of discs inevitably sort according to their origins.

Prolonged culture of imaginal discs in adult flies: If imaginal discs are placed in adults, there is no pupal ecdysone to effect differentiation. The disc cells remain in the host throughout

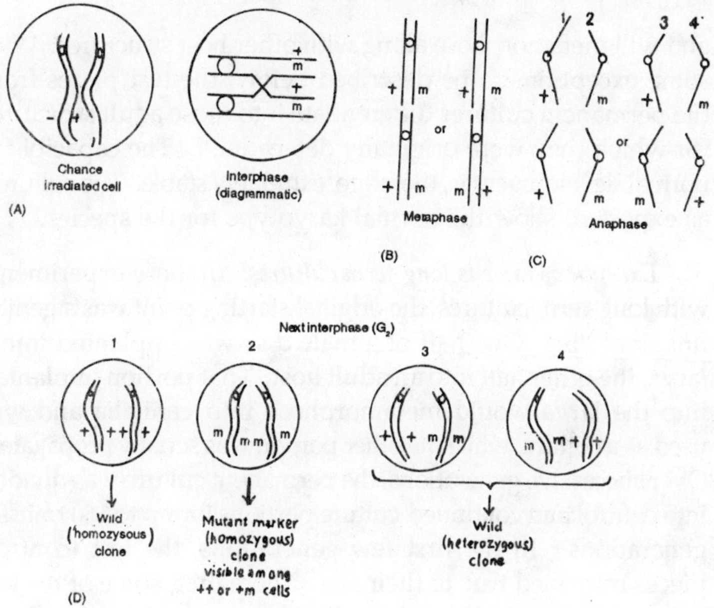

Fig. 6.13. Mechanism of X-irradiation induced somatic crossing over. The somatic crossovers induced are random and depend on: somatic pairing during mitotic interphase, and that the affected nucleus is in the four-stranded G portion of interphase. Two tandem orientations at metaphase (B) are possible, leading to four heterozygous anaphase configurations (C, 1-4). In D, heterozygous types 3 and 4 are identical, type 2 is mutant, and type 1 is genetically homozygous wild type and appears identical with 3 and 4.

its lifetime and continue to proliferate as apparently undifferentiated imaginal disc cells. The adult host fly lives about one month, and therefore, proliferating cultures must be transferred to new hosts about every two weeks. Such unlimited growth has been continued for at least six years and over 160 transplant generations. Many subcultures may be made from the original culture at each transfer.

When such long-term cultures are transplanted back into larvae instead of adult hosts, they will be subject to β-ecdysone and will metamorphose along with other host structures. With some exceptions to be described below, the test pieces from the permanent cultures differentiate into those adult structures for which they were originally determined. The capacity for normal development is, therefore, extremely stable. The cultures, as expected, show the normal karyotype for the species.

Exceptions arise is long-term cultures: In these experiments with long-term cultures, the original starting point was a genital imaginal disc. One half of a male disc was implanted into a larva, the other half into an adult host. That portion implanted into the larva would metamorphose into genitalia and was used as a control, while the other portion was serially propagated. Over successive generations, the permanent culture was divided into control and continued-culture portions for up to 160 transfer generations. In the first few generations, the test (control) pieces remained true to their fate. Thereafter, some of the test pieces developed at variance with their normal fate into head organs or leg parts. The developmental path of the cells had changed. This phenomenon has been called transdetermination. For each generation of the transfer culture, there existed a distinct frequency of transdetermination in a specified direction. The first transdetermination led with about equal probability to head or leg structures. From these states, a transdetermination to wing structures could develop. A direct change from genital to wing structures was not observed. Finally, rudiments of thorax could arise from transdetermined wing cells. It seems impossible for transdetermined thorax cells to revert back to genital cells. This sequence described for the genital disc has been shown to occur in other imaginal discs. Transdetermination changes take place after prolonged culturing, each disc has its own series of changes which follow in sequence.

No ready explanation of transdetermination is available. It has been found that a positive correlation exists between the rate of proliferation of cells in serial culture and the frequency of transdetermination. Cell proliferation could conceiveably lead to changes in repression or activation of genes through chance changes in cytoplasmic synthesis of histones or other regulator molecules.

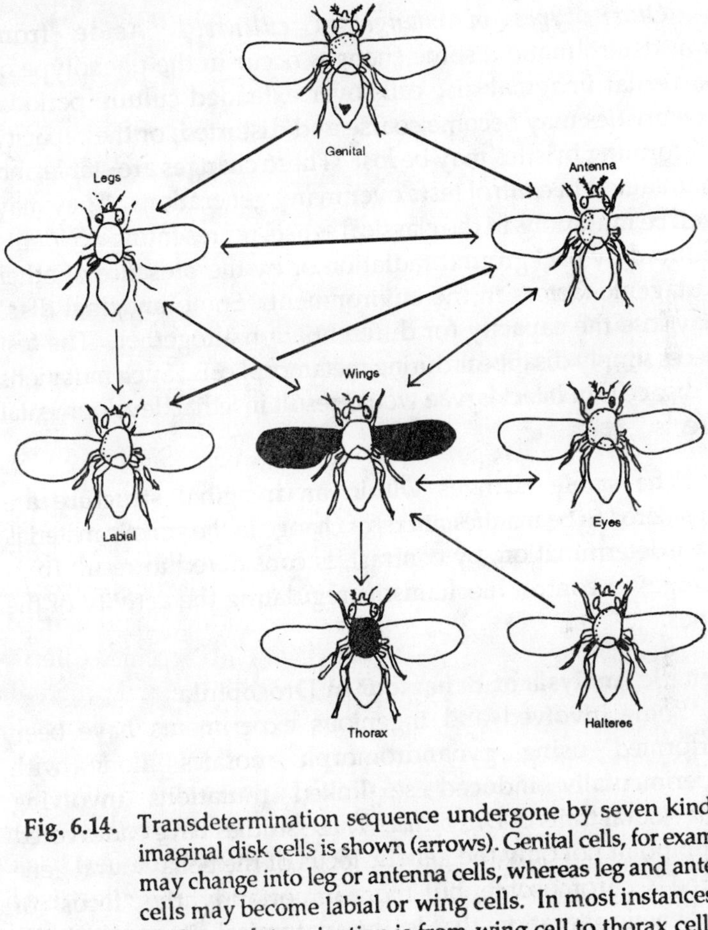

Fig. 6.14. Transdetermination sequence undergone by seven kinds of imaginal disk cells is shown (arrows). Genital cells, for example, may change into leg or antenna cells, whereas leg and antenna cells may become labial or wing cells. In most instances the final transdetermination is from wing cell to thorax cell; the change to thorax appears to be irreversible.

Serial culture of imaginal cells demonstrates the degree of stability of the differentiated state. The departures from the normal fate of the original disc cells (transdetermination) indicate that variation is indeed possible, for changes may occur in the differentiated state. Aside from consideration of possible control mechanisms acting to preserve the original fate of the cells, transdetermination provides yet another example of pluripotence among embryonic nuclei.

Other types of change in culture: Aside from transdetermination, some changes occur in the phenotype of particular imaginal disc cells over extended culture periods. The bristles may become coarse and distorted, or the capacity for forming bristles may be lost. These changes are stable and maintained in control tests over many generations. They may well be mutations in the classical sense, representing accidents induced by background radiation or by the presence of other mutagenic factors in the environment. Some imaginal discs may lose the capacity for differentiation altogether. The test pieces simply disappear during metamorphosis. Such mutations in disc cells in intact larvae would result in lethal developmental effects.

The stable changes within an imaginal structure are considered to be manifestations of change in the genetic material. Transdetermination, by contrast, is considered to result from changes in control mechanisms regulating the activity of the genes.

Genetic Analysis of Behaviour in Drosophila

Some involved and ingenious experiments have been performed using gynandromorph mosaics along with experimentally induced sex-linked mutations involving behavioural traits of *Drosophila*. These studies have endeavored to pinpoint not only the genetic locus of the behavioural gene in the X chromosome, but more importantly, the "focus" of the genetic alteration, that is, the anatomical site at which the mutant gene exerts its effect in the adult. The focus could be a group of muscles or perhaps a single cell within the central nervous system. It could also lie in a more remote organ whose malfunction influences behaviour indirectly. The utilization of mosaic flies is a powerful tool for locating such foci. Mosaics are composite flies in which some tissues are mutant and some normal, making it possible to identify which part must be mutant for the altered behaviour to be manifest. The study of mosaics has produced a unique kind of fate map for normal *Drosophila* development as well.

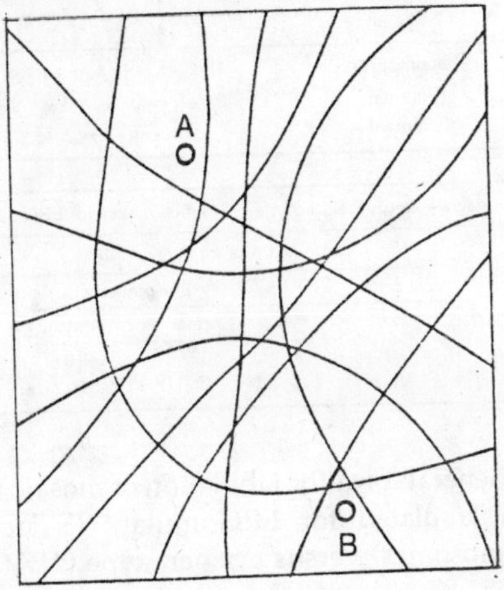

Fig. 6.15. Principle of mosaic mapping. Mosaic boundary lines fall in random orientation on the blastoderm. The farther apart two sites are, the more likely it is that the boundary will fall between them. In that event, the corresponding body structures derived from these sites will be of different genotypes. By observing the proportion of mosaic flies in which two body structures differ in genotype, a measure of the distance between their embryonic sites is thus obtained. One unit of distance (sturt) is defined as a distance such that the boundary lines pass between the sites in 1 per cent of all mosaics. Distance units are called sturts after A. H. Sturtevant, who first recognized the possibilities of mosaic mapping as a means of fate mapping. In the figure, A and B represent two locations on the blastoderm. The lines are a summary of the boundaries passing between the two genotypes in a number of mosaic flys.

Production of mosaics: The method of choice for generation of mosaic flies involves the use of an unstable ring chromosome (X chromosome). During the first division of a female zygote, one ring X may be lost, leaving a mosaic of male (XO) and female (XX) tissue. This loss also produces a mosaic for any sex-linked recessive genes present in the stable X chromosome. After the loss, the remaining ring X stabilizes. The boundary

Surface landmark A	Surface landmark B	
	Normal	Mutant
Normal	a_{11}	a_{10}
Mutant	a_{01}	a_{00}
a, proportion	1) $a_{11} + a_{10} + a_{01} + a_{00} = 1 = 100$ sturts	
1, normal	2) $a_{11} = a_{00}$; $a_{01} = a_{10}$.	
0, mutant	Dist $(\overline{AB}) = a_{10} + a_{01}$	

(A)

Electroretinogram versus eye genotype - 217 flies mosaic for ERG defect genes

Combined data for all eyes

ERG	Eye genotype		
	Normal	Mutant	
Normal	169	0	Distance $\overline{AB} = 0$
Mutant	0	203	

(B)

Fig. 6.16. (A) General form for tabulation of mosaic mapping data. (B) Data tabulated for ERG mutant (5 loci involved on X chromosomes) versus eye genotype (217 flies).

between the male (mutant) tissue and the female (normal) tissue depends on the *random* orientation of the first cleavage spindle, while one ring X is lost. The daughters of the male and female nuclei remain clustered and also remain together (male with male, female with female) when they migrate to the blastoderm. The farther apart any two sites on the surface of the blastoderm are, the more likely it is that the boundary between XX and XO cells will pass between them, so that they will be separately located in male and female sections of the adult fly. Using this technique, a fate map can be projected for the positions of the adult structures on a diagram of the embryonic blastoderm. To complete the map many mosaic flies must be examined, with differently located mosaic boundaries.

Similarly, using sex-linked behavioural mutants, the section of the adult fly in a mosaic can be scored as behaviourly mutant or normal. In the same way that the fate map was put together, a map of the foci of various behavioural mutants can also be constructed. In practice, one combines a recessive behavioural gene with a surface marker, such as eye colour,

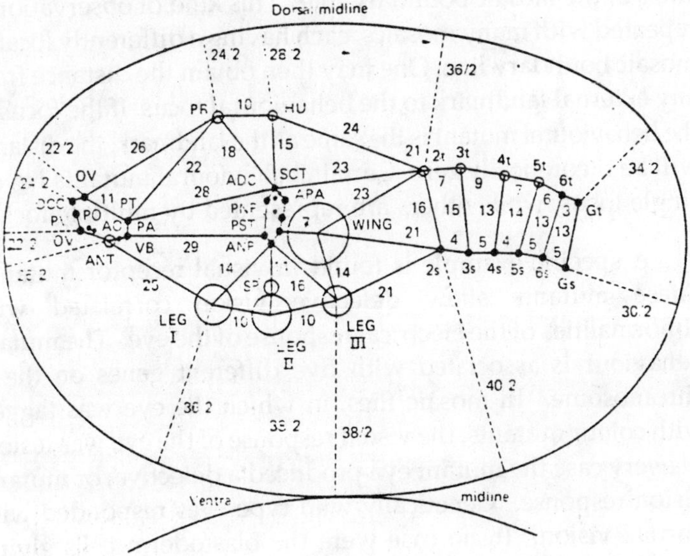

Fig. 6.17. Fate map of the *Drosophila melanogaster* blastoderm for adult external body parts, constructed by mosaic mapping. The map is a projection of the right half of the blastoderm as seen from inside the egg, showing sites of cells that will eventually develop into the indicated parts of the right side of the fly. A similar map, with mirror symmetry, corresponds to the left side. Distances between pairs of sites are indicated in sturts (i.e., the probability in percentage points that the boundary line in a mosaic will fall between the two sites, producing an adult in which the two landmarks are of different genotypes). Dotted lines indicate distances to the nearest midline as obtained by halving the distance measured between two homologous contralateral sites.

body colour, or bristle shape, on the normal X chromosome. Mosaics will be produced with the loss of one ring X chromosome. One then asks whether or not the behaviour exhibited by a mosaic fly corresponds to the genotype of a particular surface landmark. If one landmark is mutant and the other normal, it means that the behavioural focus and the imaginal disc for the structural landmark are on different sides of the mosaic boundary line. This kind of observation is repeated with many mosaics, each having a differently located mosaic boundary line. One may then obtain the distance from any external landmark to the behavioural focus. If the focus of the behavioural mutant is the same as the landmark, the distance will be theoretically zero. Certain behavioural mutants have a single focus, while others are represented by multiple foci.

A specific example is found in visual receptor mutants. These mutants show defective vision correlated with abnormalities of the electrical response of the eye. The mutant behaviour is associated with five different genes on the X chromosome. In mosaic flies, in which the eye was tagged with colour mutants, the visual response of the eye was tested. In every case the mutant eye produced a defective, or mutant, vision response. Genetically wild type eyes responded with normal vision. In no case were the blastoderm cells giving rise to the eye separated by the mosaic boundary from those cells responsible for the visual defect. The focus of the behavioural mutant was, thus, close to, or identical with, the eye itself. The result precludes the possibility that some remote organ is responsible for the visual defect.

More remote foci: Another mutant illustrates a case where the behavioural focus is found outside the organ manifesting abnormal behaviour. The semidominant hyperkinetic gene causes shaking of the legs of flies under ether anesthesia. The shaking effect is more pronounced in homozygous or hemizygous (monosomic) forms. Six hundred mosaic flies were produced in which the XX parts were female, normal body colour, eye colour, and bristles, and heterozygous for the Hk gene. The XO parts were male, yellow body, abnormal eye colour, forked bristles, and hemizygous for the Hk gene, and thus showed the effect more strongly than the heterozygous female parts. The following observations were made :

1. The shaking response of each leg is independent of other legs.

2. The shaking response and the cuticle genotype of the leg are most frequently the same, but not invariably.
3. The independent behaviour of each leg indicates that each has its separate focus.

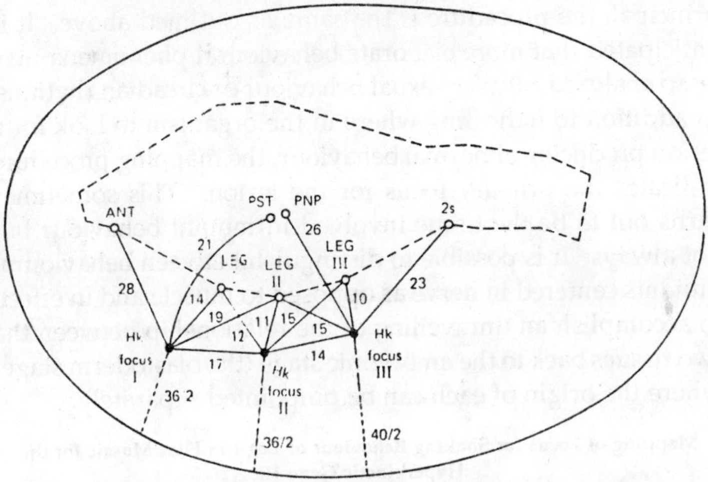

Fig. 6.18. Fate map sites of the behavioural foci for the hyperkinetic mutant Hk. There is a separate focus for each leg: they fall in the region of the blastoderm which, according to embryological studies, gives rise to the ventral nervous system. Flies mosaic for the Hk character were produced by crossing males having an X° chromosome marked with y. cho, Hk, f, ca. to virgin X_R females, and mosaic flies were scored for various body landmarks and the shaking behaviour of each leg. See text for details.

By computing the percentage of mosaics in which the shaking character was independent from the genotype of the leg cuticle, the arbitrary distance between the shaking focus and the leg cuticle site (leg imaginal disc) can be determined. The shaking focus for all three legs can be similarly measured. Each of the foci lies close to its corresponding leg, and all are located at positions ventral to any adult cuticular sites of the leg. The foci are in a region of the blastoderm which has been mapped as giving rise to the ventral nerve cord. Neurophysiological observations confirm that the neurons of the thoracic ganglia behave abnormally in Hk mutants, making these cells the likely focus for the behavioural mutant.

Other foci are more complex and require more complex analysis to locate the two or more foci on the fate map. In principal, the procedure is the same as outlined above. It is anticipated that more elaborate behavioural phenomena may be so analyzed, such as sexual behaviour or circadian rhythms. In addition to indicating where in the organism to look for a lesion producing abnormal behaviour, the mapping procedure indicates the primary focus for the lesion. This sometimes turns out to be the tissue involved in mutant behaviour but not always. It is possible to distinguish between behavioural mutants centered in nerve as opposed to muscle and in effect, to accomplish an unravelling of the relationship between the two tissues back to the embryonic stage (the blastoderm stage) where the origin of each can be pinpointed separately.

Mapping of Focus for Shaking Behaviour of Leg I in Flies Mosaic for the Hyperkinetic Gene Hk

	A. Coxa of leg I		B. Coxa of leg II		C. Coxa of leg III	
Shaking of leg I	Normal	Mutant	Normal	Mutant	Normal	Mutant
Normal	277	50.5	261.5	69	250.5	82.5
Mutant	33	223.5	44	215.5	46.5	215.5

	D. Presutural bristle		E. Antenna (segs 1, 2)		F. Humeral bustles	
Shaking of leg I	Normal	Mutant	Normal	Mutant	Normal	Mutant
Normal	251	84.5	253	82	241	84
Mutant	42	222.5	84.5	179.5	66.5	197.5

G. Shaking of left leg 1	Shaking of right leg I	
	Normal	Mutant
Normal	110.5	49
Mutant	59.5	75

1 Each matrix gives number of each combination observed among 300 mosaic flies examined. Data for right and left sides are combined making a total of 600 sides. "Normal" shaking refers to the small amplitude typical or heterozygous flies. In cases where a body structure is mosaic or where amplitude of shaking is intermediate, a value of 1/2 is assigned to each of

the two corresponding categories. Those rare cases (a few percent) where both structure and shaking amplitude are mosaic are omitted.

Calculation of map distances

A. Shaking of leg I to coxa of leg I = 83.5/600 = 13.9 sturts.

B. Shaking of leg I to coxa of leg II = 113/600 = 18.8 sturts.

C. Shaking of leg I to coxa of leg III = 129/600 = 21.5 sturts.

D. Shaking of leg I to presutural bristle = 126.5/600 = 21.1 sturts.

E. Shaking of leg I to antenna = 166.5/600 = 27.8 sturts.

F. Shaking of leg I to humeral bristles = 160.6/600 26.8 sturts.

G. Shaking of left leg I to shaking of right leg I = 108.5/300 = 36.2 sturts.

Distance of leg I shaking focus of midline = 36.2/2 = 18.1 sturts.

Fig. 6.19. Data for the mapping of focus for shaking behaviour of Leg I in flies mosaic for the hyperkinetic gene Hk.

Cancers in Drosophila

Genes and tumors: Malignant neoplasms which destroy the host are not found in adult flies, nor in any other invertebrates. Malignancies have been uncovered in the larvae, possibly because cell divisions are still going on in the larvae and not in the adults. The malignancies to date are of two kinds. One is a neuroblastoma, which invades and destroys the larval brain. This tumor can be subcultured in the adult abdomens for many generations, where it remains malignant and invasive. The second neoplasm is a lethal though noninvasive tumor of the imaginal discs. These larval discs are disorganized and fail to metamorphose into epidermal structures on implantation into normal metamorphosing larvae. When transplanted to adult hosts, the discs grow rapidly into very dense structures and cause the death of the host. Both of these growths show some changes associated with neoplasms in vertebrates, namely, reduced cell contacts, nuclear deformation, and surface membrane changes. Both of these tumors appear to arise from mutations. It appears likely that a large number of mutants which might produce neoplasms can be produced in *Drosophila*. This leads to the possibility of study of the genetic change needed to produce a particular malignant cell property. A system of this kind should be invaluable for an understanding of cancers in other systems in which the genetics is not so thoroughly known.

Gene Activity During Development

The organization of the genome: In 15 other systems examined (among them *Xenopus* and the sea urchin) the DNA of the genome is typically organized into short reiterated nucleotide sequences interspersed with longer single-copy sequences. The short repetitive sequences average about 300 nucleotide pairs in length, compared with 1000-2000 nucleotide pairs in length in the unique-sequence fraction.

Measurements of reassociation kinetic of *Drosophila* DNA sheared to different lengths and reannealed in various combinations indicates that there are few short sequences (about 300 nucleotide pairs long) present in *either* fraction of *Drosophila DNA*. The single-copy sequences appear to be interspersed with reiterated sequences of 5500 nucleotide average length. It is estimated that 2800 of these reiterated sequences are interspersed with single-copy DNA throughout the genome (excluding the satellite DNA, which is highly reiterated and no included in these studies). The replacement in *Drosophila* of short repetitive fragments by longer sequences may be of importance when considering the possible regulatory role ascribed to reiterated, interspersed DNA encountered in other species.

RNA and protein synthesis: The mature egg contains a supply of ribosomes and other classes of RNA supplied by the nurse cells. Mutations involving the nucleolus, found in association with the sex chromosomes, lead to early embryonic lethality unless the zygote contains sufficient mRNA cistrons supplied by the male sex chromosome (X or Y). New RNA synthesis begins at the time of blastoderm formation and coincides with the appearance of nucleoli in the embryonic cells.

Temperature mutants: Gene function can be studied at several different levels. The first level is the transcriptional level; the second, the translational level; the third, the period of protein activity; and the fourth, the morphogenetic effect of the protein activity. Using temperature mutants, which grow normally at lower temperatures (17° C) but are killed by temperature tolerated by wild type flies (30° C), one can study the third and fourth levels of consequences of gene activity. Temperature mutants manufacture protein products which are thermally denatured at lower than normal temperatures and rendered non-functional. These non-functional molecules cause developmental defects or death at times when they would normally be called upon to function. The functioning time of a protein may be considerably removed from the time of transcription or translation of the mRNA concerned with the synthesis of the molecule. A single "temperature" locus may be involved at more than one stage of development. This kind of analysis indicates critical periods of activity for gene products, but does not reveal when the particular gene was transcribed or the message translated.

RNA synthesis during pupation and metamorphosis: The salivary glands and certain other larval tissues contain giant polytene chromosomes. These chromosomes contain many parallel copies of chromosome strands and are also distinctively banded. The bands of the salivary chromosomes correlate with genetic mapping data, as well as with information of chromosome deletions, inversions, and so forth, to produce a chromosome map showing the linear arrangement of genes.

During RNA synthesis, the bands in active regions alter their physical state and become spherical, creating a chromosome puff. Puffing then becomes an indication of transcription at a particular chromosomal region, on a specific chromosome. The puffing patterns change with time during development. some puffs are present at all times during development. At any rate, the chromosomes of the same organ at the same developmental stage provide a constant puff pattern. By comparing puff patterns from different stages and from various tissues, one can determine that the transcription patterns change both with time of development and with the tissue under observation. The puff pattern is altered during development by the hormone ecdysone, controlling pupation and metamorphosis. The effects of the hormones are mediated directly through the nuclei by the individual chromosomes.

Changes in the puffing patterns of the giant dipteran chromosomes are compatible with the theory of selective activation of genome transcription as a basis for control of developmental events.

7

Development of Chironomus Thummi

The Midge Chironomus Thummi

The larvae of chironomids, along with those of drosophilids and sciarids, are widely used in cytological and biochemical studies of cell differentiation. The description of the purring pattern in the polytene chromosomes of these larvae at various stages of metamorphosis has enabled a concept to be formulated regarding changes in the spectrum of the most active sites on chromosomes in the course of morphogenesis. Along with the results of biochemical and cytogenetic studies, these data provided the basis for current views about gene activity in the course of cell differentiation.

Studies of the cell protein spectrum and changes in this spectrum during dipteran development have confirmed the idea of differential gene activity and enabled the functional significance of several puffs to be elucidated, in particular those coding for tissue-specific proteins.

In diptera, as in many other organism, RNA templates with a long half-life have been found; such RNA species are synthesized a long time before the genetic stage at which they are translated. It should be remembered that both transcription and replication of DNA in various organs of diptera are discontinuous in the course of larval development; this activity drops or may even cease completely during molts. Therefore, just as with the general problem of regulating metamorphosis, great attention is being paid to hormones since they play a major role in controlling chromosomal activity. Compared with *Drosophila*, chironomids have only comparatively recently been introduced as experimental species which can be reared under laboratory conditions; this is only possible for a few species. These include *Chironomus* thummi Kieff. (subspecies *thummi* and *piger), C. tentans*, C. *pallidivitatus*, and some others. The large size of chironomid larvae, as well as the high degree of chromosomal polyteny in a number of larval tissues, make these species particularly promising for developmental biology.

Chironomid larvae are aquatic organisms with integuments that are easily permeable to many compounds. They are very suitable for studying the effects of various substances on the whole animal or its individual organs. It is important to emphasize that under vital conditions chironomids can be studied easily. This is because their skin is translucent and the state of the cells in almost all the organs can be examined directly under the microscope. The presence of polytene chromosomes with a markedly higher degree of polyteny in the larval organs of chironomids than in drosophilids makes them an indispensable species for studying many problems related to chromosomal structure and function. Another important problem lies in elucidating the relationship between chromosomal activity and cell physiology during such critical periods as larval molts, after molt, and intermolt growth.

Unfortunately, the life cycle of chironomids has not been fully studied and more detailed information regarding the embryonic stages is particularly necessary. Data on the developmental physiology of chironomids are very scarce; there are few genetic and cytogenetic investigations. This is because it is difficult to obtain progeny from individual pairs, which cannot be traced during swarming. However, methods designed to mate males and females outside swarming have been described: this should facilitate future genetic and cytogenetic studies. The development of methods for rearing chironomids on a laboratory scale may in the future provide the basis for rearing vast numbers of them. This is economically promising because they make excellent fish food. This section contains basic information on the main features of the life cycle of *Chironomus thummi* kieff, in laboratory culture under conditions of artificial rearing. Up-to-date studies on chironomid development place great emphasis on analyzing the larval stage and metamorphosis, so there is plenty of information on these. We deal primarily with problems of the morphology and physiology of these developmental stages. Little work has been done on the embryogenesis of chironomids, and this is discussed briefly.

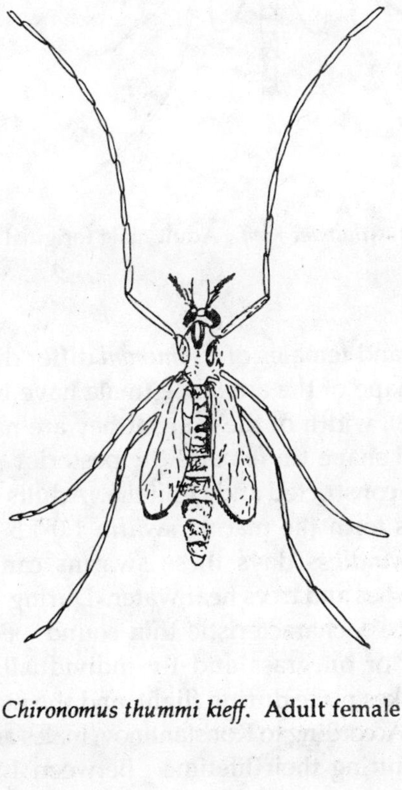

Fig. 7.1. *Chironomus thummi kieff.* Adult female

Fig. 7.2. *Chironomus thummi Kieff.* Adult male (original sketches by Valleva).

Reproduction

Adult males and females of C. *thummi* differ distinctly in the length and shape of the antennae (male have large, wide plumed antennae), width of the wings (they are narrower in males, abdominal shape (in females the posterior part of the abdomen is more constricted and genitalia. Adults live only 4 or 5 days. Males form the mating swarm 1.0-1.5 days after emerging. On windless days these swarms can be easily observed over bushes and trees near water. During swarming the males produce a characteristic thin sound. Females sit either on bushes or on grass and fly individually into the swarm; mating takes place during flight, and the mating paris leave the swarm. According to Konstantinov, males and females mate only once during their lifetime. Between 1.5-2.0 days after mating the females lay egg masses on water. Oviposition takes place either at sunrise or sunset. The egg mass is invested in a gelatinous matrix. The number of eggs in the egg mass of a *C. thummi* laboratory specimen varies from 300 to 900. The length of the egg mass is from 7 to 20 mm. The eggs in the mass are arranged regularly with 10-17 eggs in each row; the gelatinous matrix is divided into corresponding segments. The central fibers consists of thicker gelatinous material, and their shape is significant as a taxonomic trait. The egg mass is attached to any solid object on the water surface by a gelatinous cord. The eggs of *C. thummi* are elongated (length 0.3-0.5 mm, width 0.12-0.14 mm). They have a dark-brown pigmentation depending on the colour of the yolk. The chorion is colourless, and the micropyle is located on the rounder anterior pole. The embryonic development of chironomids at 20-22° C lasts 2.5-3 days. A characteristic feature is the early segregation of primary germ cells.

Fig. 7.3. ***Chironomus thummi Kieff*. Adult male.**

When embryonic development has been completed, the larvae hatch from the egg but do not leave the gelatinous matrix for 24 h after hatching. The larvae which have left the gelatinous matrix are capable of building specialized structures in which they live. These are tubes consisting of small particles of mud and detritus glued together by a salivary gland secretion. Each tube is open on both ends, and water is drawn through the tube by anteroposterior undulation of the larva's body. Particles of food suspended in the water are deposited on the sticky walls of the tube and may be eaten by the larva. In addition, the larvae can emerge from their tubes, spread salivary gland secretion with their anterior forelegs, and then withdraw into the tube again, dragging along the secretion with any adhering particles.

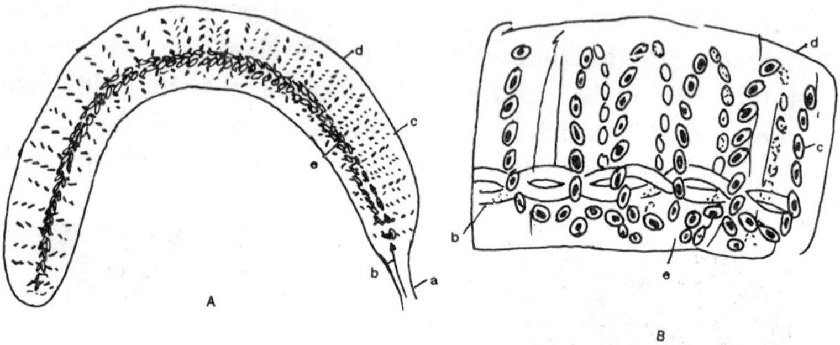

Fig. 7.4. Egg mass of *Chironomus thummi*. (A) General view; (B) part of the egg mass at higher magnification, a) Gelatinous stalk for attachment; b) central fibers of gelatinous matrix; c) eggs; d) gelamous envelope; e) gelatinous matrix.

The larvae are microphagous, feeding on algae, protozoa, small invertebrates, bacteria and detritus. The larval stage is the longest of the chironomid life cycle and lasts 35 days on average. However, the duration of larval development depends on environmental conditions, for example in winter the larval stage can continue without metamorphosis for several months.

The pupae of *C. thummi* do not feed; when mature they leave their tubes and move to the surface of the water. The ascent of thė pupa in water is aided by air accumulated under the pupal skin around the adult's thorax. The adult emerges from a dorsal slit in the thorax of the pupal skin.

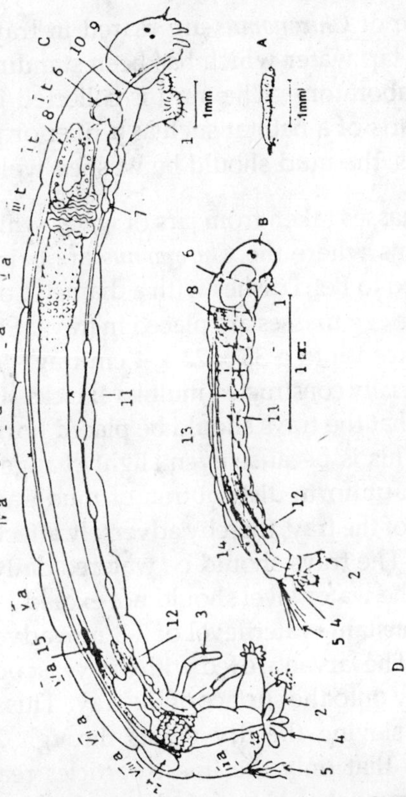

Fig. 7.5. Larval stages of *Chironomus thummi*. A, B) Ist Instar larvae (different magnifications); C) 4th instar larva; abdominal segments; 1, 2) appendages with a circle of hooklets (for locomotion) on the ventral side of the first and last body segments; 3) two pairs of appendages (tubuli) of VIIIth abdominal segment; 4) four anal papillae; 5) bristle tuft; 6) brain; 7) ventral nerve cord; 8) salivary gland; 9) pharynx; 10) esophagus; 11) gut; 12) Malpighian tubules; 13) aorta; 14) heart; 15) ovaries.

Culture

A procedure for rearing *C. thummi* under laboratory conditions has been developed by Konstantinov. We have used this method with some modifications throughout 20 years of continuously rearing this species.

The larvae of *Chironomus* are reared in trays filled with moist mud or tap water which has been standing for several hours in the laboratory. The mud is collected in summer in the shallow parts of a habitat such as a river or pond. Before filling the days, the mud should be washed well and boiled.

The egg masses taken from jars of water which have been placed in rooms where the *Chironomus* are being reared are then transferred to Petri dishes with a diameter of 11 cm for 2-3 days; then the egg masses are placed in trays. Six egg masses are usually placed into a 35 x 23 x 4 cm tray. The trays are placed into specially constructed multilevel racks. It is important to emphasize that the trays should be placed horizontally into these racks. This is because even slight changes in position result in a non-uniform distribution of mud and water over different parts of the tray, thereby adversely affecting the larva development. The trays should be watered daily to keep the mud wet, but the water level should not exceed 1-2 mm above the mud. A constant water level of 2-3 mm adversely affects development. The larvae are fed with dry yeast powder which is spread evenly onto the surface of the tray. This can be easily performed by sieving the dry yeasts through 2-3 layers of cheesecloth, so that only the finest particles reach the mud surface. The larvae should be fed every other day. It is better to spread the food before watering the trays. The dry yeasts should be heated at 50° C in a thermostatically controlled oven before use.

The mud should be washed and boiled after each generation. Larvae are very sensitive to oxygen deficit during development. If this happens, the tubes extend high above the surface of the water and the larvae leave the tubes for a long time. The racks with the trays should be placed in a separate room or in a cage (180 x 80 x 80 cm) and tightly covered with fine mesh gauze. Large jars of water are placed on the floor near the racks. The females lay eggs into these jars. Egg masses are collected daily.

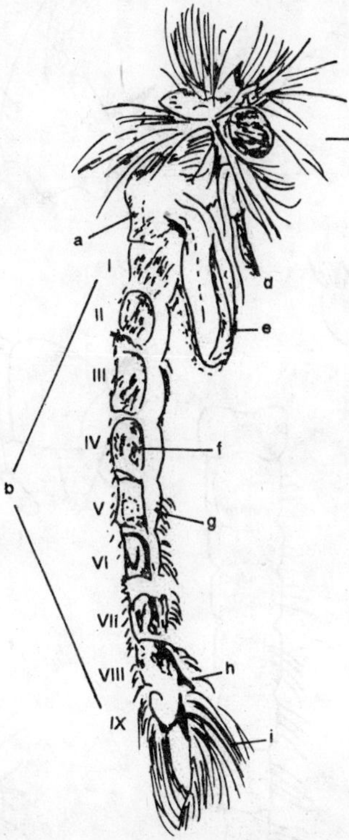

Fig. 7.6. Pupa of *Chironomus thummi.* a) Cephalothorax; b) I-IX, abdominal segments; c) thoracic horns; d) leg sheaths; e) wing sheaths; f) shagreen; g) bristles; h) spur of segment VIII; i) swim fin.

The larvae can also be reared in water. This method of rearing is used to maintain laboratory stocks of *C. thummi* and other chironomid species — *C. tentans* and *C. pallidivitatus.* The water should be well aerated when using this method.

The larvae of chironomids usually develop asynchronously even if they originate from the same egg mass. To synchronize larval development, the aeration and feeding conditions should be strictly standardized. The population density is of fundamental importance. By changing it one may obtain larval

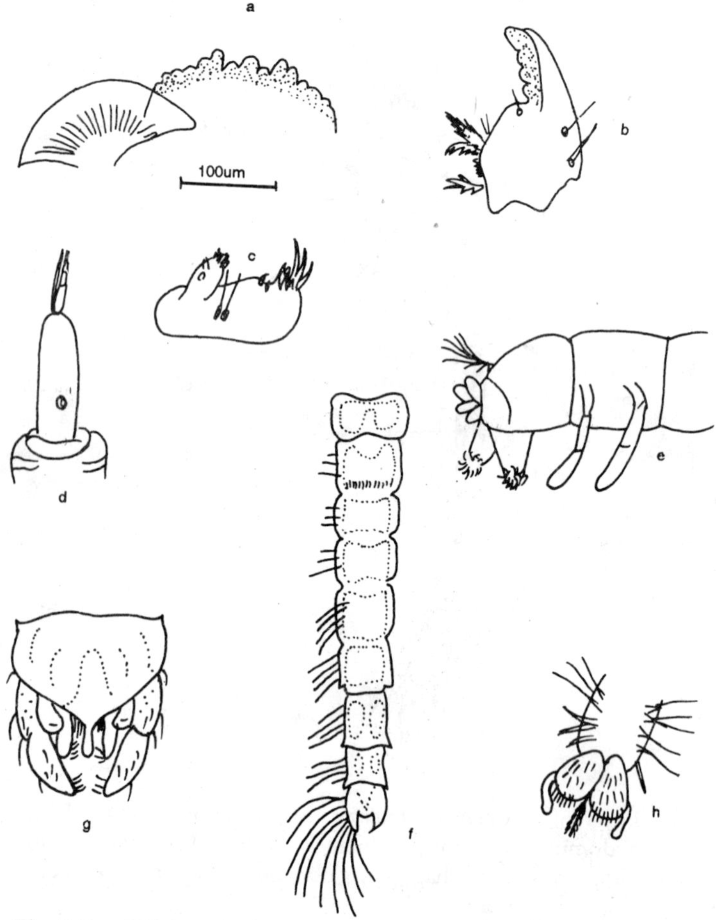

Fig. 7.7. Diagnostic features of *Chironomus thummi.* a) Submentum with submental plate; b) mandible; c) maxilla; d) antenna; e) appendages of VIIIth abdominal segment; f) shagreen pattern and characteristic features of the pupal abdominal segments; g) hypopygia of adult male; h) tip of adult leg.

population developing at different rates and with a different degree of asynchrony. We have obtained the optimal uniformity of larval development at a population density of 0.6 larvae/ cm^2. A drop in the population density by about one-third did not bring any marked improvement in the synthrony. Conversely, a four-to sixfold increase in the population density resulted in a marked desynchronization and a decreased rate of larval development in the population.

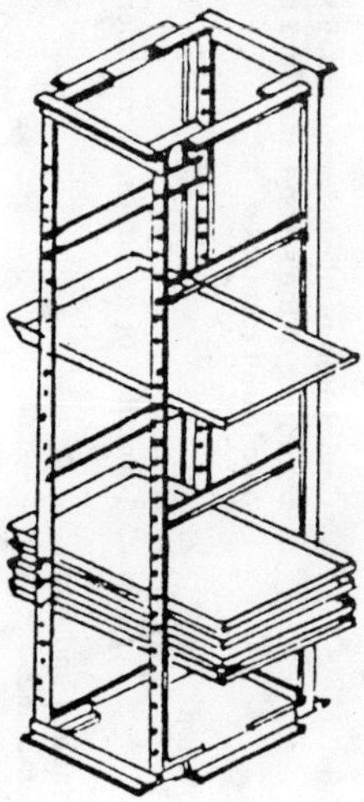

Fig. 7.8. Rack of trays for rearing chironomids.

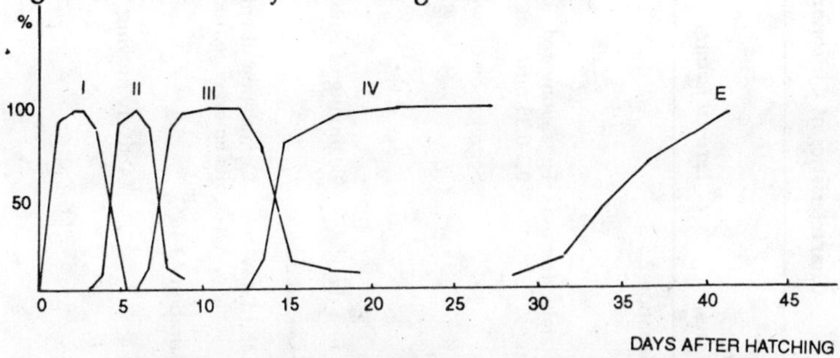

Fig. 7.9. Duration of larval development of *Chironomus thummi* under laboratory conditions (15-25° C). Abscissa shows time after larval hatching from the gelatinous envelope; ordinate gives the number of a given instar as a percentage of the total. I-IV, Larval instars; E, emergence of adults

C. thummi should be reared under a regime of 15 h of light of 9 h of dark at a standard temperature of 20-22° C.

Normal Development

The general characteristics of the normal developmental

TABLE 7.1

Characteristics of Chironomus thummi Development Stages

Stage	*Time after oviposition (hours, days)*	*Diagnostic features* *External features*	*Internal features*
Embryonic development at 20° C (arbitrary stages):	Total duration 2.5-3 days		
1	Immediately after oviposition	Eggs elongated, dark brown, length 0.3-0.35 mm	Completion of the maturation divisions
2	2.5 h	Segregation of two primary germ cells	Four nuclei embryos
3	10-15 h		Completion of blastoderm formation and beginning of the second and third embryonic leaflet formation
4	End of first day	Beginning of embryonic rotation	
5	End of second day		Formation of mouth organ rudiments
6	End of third day	Completion of embryo formation and its emergence from egg membranes	
Postembryonic development at 23-25° C:	Total duration 43-46 days		
Ist instar	Days 3-4	Actively moving larva feeding on mucus	Head ganglion in the head capsule

	Days 4-8	Larvae immediately after leaving the egg mass 0.88 + 0.07 mm in length with width of the head capsule 0.104 + 0.002 mm. Mouth parts are incompletely formed as compared with later larval instars: only one pair of bristles on labrum, frontal and oral fields are smooth, premandibulae devoid of bristles. Triple median notch in the submentum, antennae are short (44 m), relation between the first and second antenna segments is 0.77. No appendages of the VIII ventral segment. The base of bristle tuft is low.	Hemolymph is colorless due to low hemoglobin concentration. Tracheae are not filled with air, corpus allatum consists of two cells.
IInd instar	Days 8-10	Larval length at the beginning is 1.73 + 0.08 mm. Head capsule width 0.197 + 0.003 mm. Formation of mouth parts completed. Submentum acquires the shape characteristic of the species. Appendages on the VIIIth ventral in segments appear	Head ganglion displaced to the thoracic segment. Hemolymph acquires pink color. Tracheae are filled with air. The number of cells corpus allatum is 4-12.
IIIrd instar	Days 11-17	Larval body length at the beginning of the instar 3.32 + 0.08 mm. Head capsule width 0.326 + 0.008 mm. Appendages on ventral segment undergo elongation	Pink color of hemolymph intensifies. The number of cells in corpus allatum is 12-40
IVth instar (phase 1-5)	Days 17-36	Larval body length from 6.74 + 0.24 mm at the instar beginning to 12.32 + 0.48 mm at the instar end. Head capsule width 0.523 + 0.01 mm	Intensive gonad development. Hemolymph is dark red in color. Progressive development of imaginal disks proceeds in the course of the instar The number of cells in corpus allatum is 40-180

Phase 6-9 (prepupa)	Days 36-38	Swelling of thoracic segments. Body contracts and tapers, false legs decrease in size	Imaginal disks continue to form Characteristic large cells associated with the epidermis appear on the ventral body side. Metamorphosis puffs in salivary gland chromosomes appear. Formation of oocytes begins in female gonads, in males spermatids and spermatozoa are abundant
Pupa	Days 38-41	Change of body pigmentation from red to dark grey. Fusion of the head and three thoracic segments in cephalothorax. Formation of pupal integuments. Formation of prothoracic horns	Gradual lysis of larval organs (salivary glands, intestine, fat body, and others, with the exception of Malpighian tubules and neural chain). Formation of adult organs from the imaginal disks completed.
Adult	Days 43-46	—	—

stages of *C. thummi* are given in Table 7.1. This table shows that the average lifespan of *C. thummi* under laboratory conditions at 20-25° C is 45-48 days. The larval period of development last 35 days on average, and three molts divide it into four larval instars. Since larval development is asynchronous, the data on instar duration can be conveniently expressed as a diagram. The mean duration of a larval instar in this diagram is taken to be equal to the length of the line corresponding to the 50% diagram cross section at any given instar (the prepupa is included in the IVth larval instar). It follows from that the mean duration of the first larval instar is equal to 3.5 days; of the IInd instar, three days; of the IIIrd, seven days; of the IVth (including the prepupa), 21 days. Prepupae live 4-5 days.

As previously mentioned, the population density greatly affects the rate of larval development. The emergence of adults in populations with a density of 0.2-0.6 larvae/cm^2 begins 21 days and ends 35 days after oviposition. In populations of a higher density, adults begin to emerge 35 days after oviposition although the majority of the population does not complete metamorphosis for at least another month.

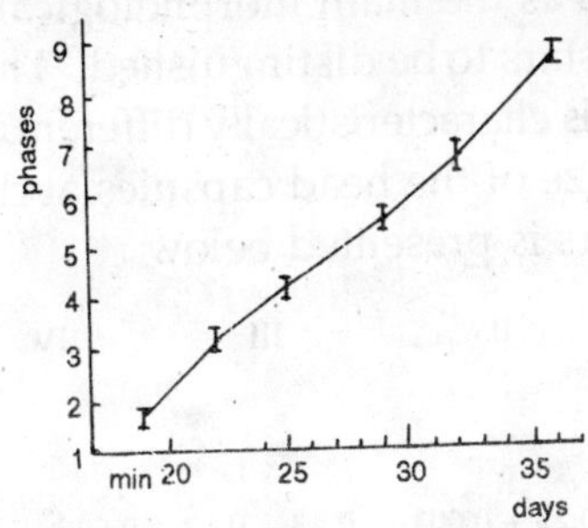

Fig. 7.10. Duration of the different phases of the 4th larval instar of *Chironomus thummi*. Abscissa, time (in days) after hatching ordinate, 4th instar phases; rh, red-head phase.

These data show that the duration of the last two larval instars, particularly that of the IVth instar, is quite long. The larval development becomes particularly asynchronous during this period. Therefore, the use of data on the mean duration of an instar and of diagrams appears to be more suitable for cytological and developmental biological studies compared with the use of estimates of a minimal instar duration.

Characteristic Features of the Larval Instars

The animals grow rapidly during the larval period, and the larvae of different instars differ markedly in size. The mean length of the body and salivary glands (in millimeters) at the beginning of each larval instar (after the molt) and by the end of larval development is shown below :

Larval instar	I	II	III	IV	End of IVth instar	References
Body-length l	0.88 ± 0.07	1.73 ± 0.08	3.32 ± 0.08	6.74 ± 0.24	12.32 ± 0.46	Our data
	-	2.9	3.7	6.9	15.5	[46]
Length of salivary gland	0.08 + 0.01	0.15 + 0.01	0.29 + 0.01	0.60 + 0.05	1.07 + 0.08	

The length of the larvae, however, cannot serve as a reliable criterion of instar determination since the length varies, depending on developmental conditions such as temperature, feeding conditions, and oxygen supply. The width of the head capsule serves as the main morphological criterion and enables different instars to be distinguished. The mean width of the head capsule is characteristically different for each larval instar. The mean size of the head capsules at different larval instars in millimeters is presented below :

Larval instar	I	II	III	IV	References
Head capsule					
Length	0.15 ± 0.008	0.25 ± 0.003	0.43 ± 0.12	0.69 ± 0.12	Our data
Width	0.10 ± 0.002	0.20 ± 0.003	0.33 ± 0.008	0.52 ± 0.01	Our data
	0.125	0.20-0.25	0.33—0.38	0.48-0.68	[44]

Since the width of the head capsule increases only after the molts and shows practically no changes during the intermolt periods, the measurement of this characteristic enables the larval instar to be reliably determined. One should bear in mind, however, that the absolute size of the head capsule may vary somewhat depending on the conditions for larval development. Despite this, the size ratio for different larval instars remains unchanged. This ratio should be used for determining larval instars.

Although the increase in the larval size during larval development is a characteristic trait, other morphological and physiological characteristics can also be used to identify individual larval instars. These include certain changes in the body structure. During the first larval instar there are no appendages (tubuli) on the VIIIth abdominal segment. They appear after the first molt and then become enlarged; bristle tufts undergo elongation. The relative length of the antenna feelers decreases with age, and there is a change in the length ratio of antenna segments and in the mentum shape. The intensive development of the imaginal disks begins during the IVth larval instar.

Body pigmentation of the larva changes during this period, too. This is related to an increased hemoglobin concentration in the hemolymph. The first instar larvae appear colourless, but during the second and third larval instars they gradually acquire a red colour, and during the fourth larval instar they assume the dark-red colour characteristic of chironomids. It is believed that hemoglobin is absent during the first larval instar, but our own observations (Kolesnikov, unpublished) suggest that hemoglobin is present during the first larval instar, but that its concentration is too low to colour the larvae.

The tracheae begin to fill with air after the first molt.

After presenting this short list of differences between the larval instars, it should be stressed that both in C. *thummi* and in other chironomid species the early larval instars have been far from sufficiently investigated. The most detailed information is available for the last larval instar which is more accessible to

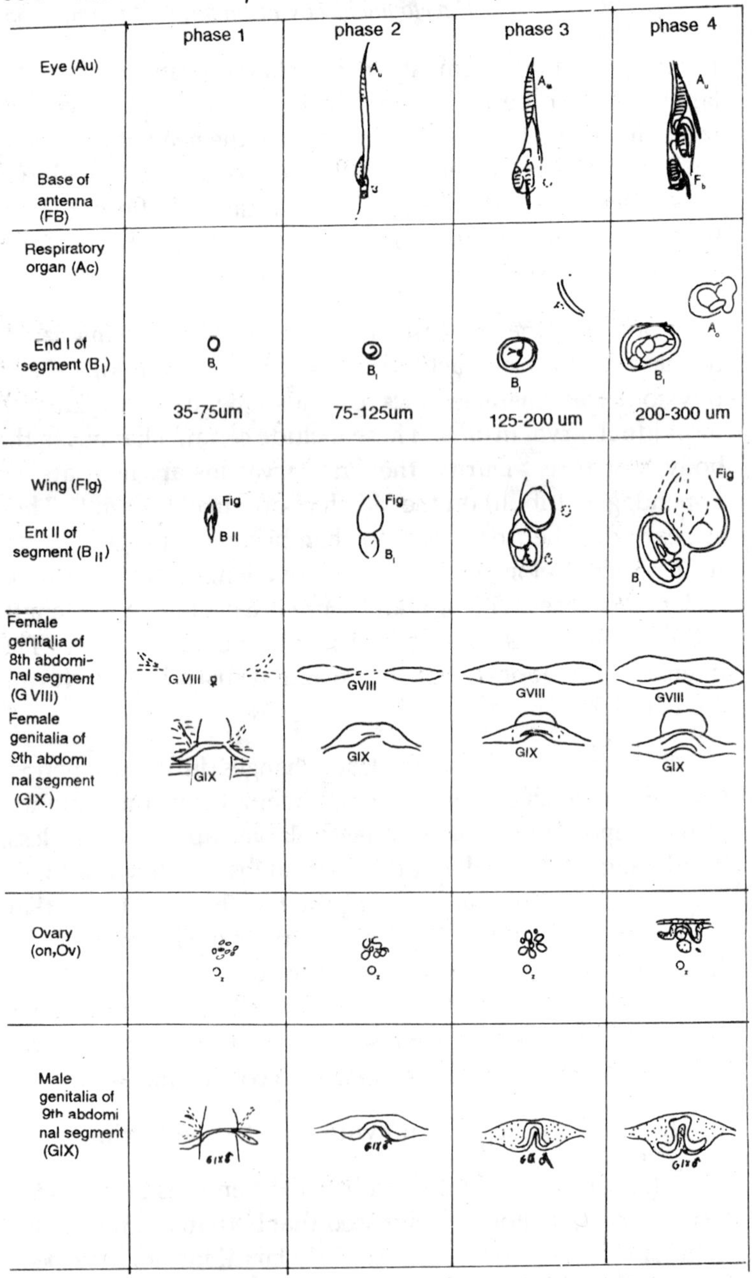
phase 1
phase 2
phase 3
phase 4
Eye (Au)
Base of antenna (FB)
Respiratory organ (Ac)
End I of segment (B I)
35-75um
75-125um
125-200 um
200-300 um
Wing (Flg)
Ent II of segment (B II)
Fig
B II
Female genitalia of 8th abdominal segment (G VIII)
Female genitalia of 9th abdominal segment (GIX)
G VIII
GVIII
GIX
Ovary (on,Ov)
Male genitalia of 9th abdominal segment (GIX)

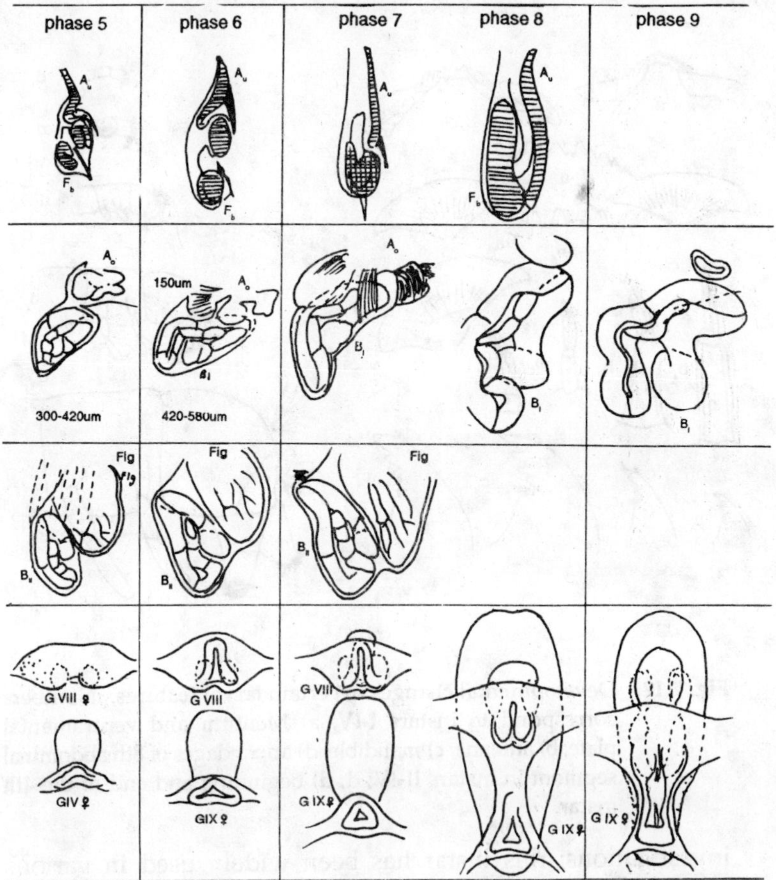

Fig. 7.11. Development of the imaginal disks in *Chironomus thummi*.

Fig. 7.12. Developmental changes in certain larval features. Numbers correspond to instars I-IV. a) Mentum and ventramental plate; b) antenna; c) mandible; d) appendages of 8th abdominal segment in instars II-IV; d, d) beginning and end of the 4th instar.

investigations; this instar has been widely used in various cytogenetic and biochemical studies. As the IVth larval instar lasts for a relatively long period, it is necessary to identify more accurately the consecutive development phases during this period. Such identification is also essential for experimental purposes since analyzing many developmental aspects requires a large number of synchronously developing animals. As the natural development of the larvae is rather asynchronous, it is necessary to select larvae at the specifically required developmental phases using reliable criteria. Essential criteria for distinguishing consecutive developmental phases of the IVth instar larvae are described in the paper by Wulker and Gotz. They compiled a detailed description of the formation of imaginal disks corresponding to different organs and, using these criteria, divided the IVth larval instar (including the prepupal stage) into nine phases. The map showing the location of imaginal disks in a *C. thummi* larva is presented in, while the scheme of consecutive stages in their formation during the development of the IVth instar larva. The system for determining the individual phases of the IVth instar lava according to Wulker and Gotz is suitable for practical use and can be easily applied to selected larvae corresponding to each of the IVth instar phases. Operationally the long-term practice was to divide the IVth instar into early, middle, and late periods using mainly such characteristics as the duration of development, larval body size, shape of imaginal disks of the forelegs, and the set of puffs; there is no need to emphasize that this division was rather arbitrary and not very reliable.

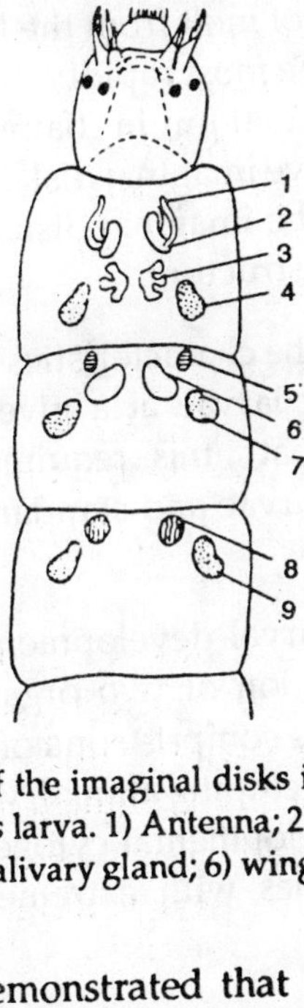

Fig. 7.13. Location of the imaginal disks in the thoracic segments of a *Chironomus* larva. 1) Antenna; 2) eye; 3) respiratory organ; 4) Ist leg; 5) salivary gland; 6) wing; 7) 2nd leg; 8) haltere; 9) 3rd leg.

It has been demonstrated that the size of the imaginal disks may vary significantly under different rearing conditions and depending on the physiological peculiarities of the larvae. For example, the size of the imaginal disks is significantly affected by the population density. As can be seen from at a population density of 0.2-0.6 larva/cm^2, the diameter of the foreleg imaginal disks at the redhead stage (first 12 h after the molt from the IIIrd to the IVth instar) was about 80 μm. This corresponds to the second developmental phase according to Wulker and Gotz rather than the first phase, as could have been expected for the beginning of the IVth instar. Upon a significant increase in the population density (2.5-3.8 larvae/cm^2) the diameter of these imaginal disks was found to be 50 μm, which completely corresponds to the value listed by Wulker and Gotz for the first phase of development. The rate of larval development also affects the size of the imaginal disk. If the size of the imaginal disks at the redhead stage is compared to the larvae developing from one egg mass which take a different time to reach the stage of molt from the IIIrd instar to the IVth instar, one notes that the most rapidly developing larvae have imaginal disks of about 80 μm in diameter, while the larvae completing the molt have imaginal disks with a diameter of 50 μm. The diameter of the imaginal disks always corresponds to their morphological structure.

Thus, when using the characteristics of the imaginal disks as criteria for selecting larvae at a given stage, one should always remember that this requires keeping a certain population density of larvae and standardizing other rearing factors.

While describing larval development, we would like to mention the differentiation of two organs, the gonads and salivary glands. Gonads comprise one of the most important characteristics of the developing animal, while salivary glands are widely used in developmental cytogenetics owing to the presence of chromosomes with extremely high degrees of polyteny.

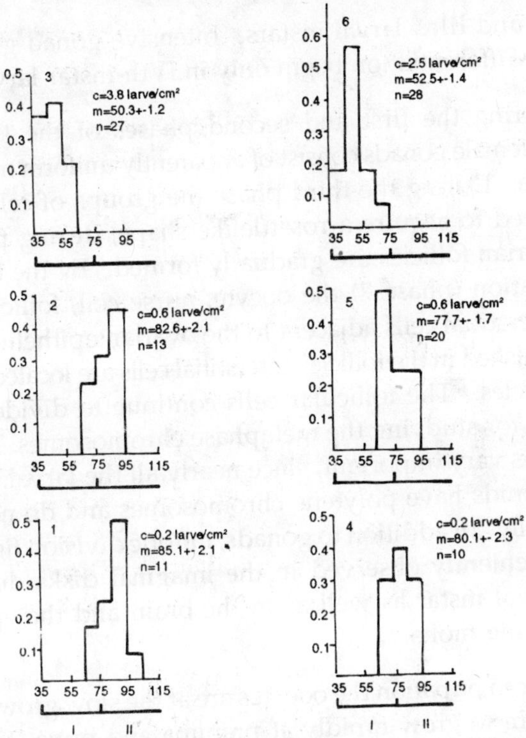

Fig. 7.14. Size changes in the 1st leg imaginal disk in the 4th instar. Different population sizes of *Chironomus larvae* at the redhead stage were used. Abscissa, diameter of imaginal disk (µm); ordinate, frequency of occurrence. I, II) Developmental phases. c, population density; m, mean disk size; n, number of larvae examined. 1-3 and 4-6 are the results of two independent experiments.

Gonads

As previously mentioned, the primary germ cells in *C. thummi* segregate at the posterior egg pole at the stage of the tetranuclear embryo. The two primary germ cells undergo two divisions and the resulting group, consisting of 8-16 cells, migrates inside the cleaving embryo. These cells do not change until the larva hatches from the egg; this is followed by the beginning of gonad growth. It proceeds very slowly between the Ist and IIIrd larval instars. Intensive gonad growth and oocyte differentiation begin only in IVth-instar larvae.

During the first and second phases of the IVth instar larvae, female gonads consist of apparently uniform multiplying oogonia. During the third phase the groups of cells become organized to acquire a rosettlelike shape; during phases 4-6 the ovarian follicles are gradually formed. By the beginning of pupation (phase 7) the oocyte, nurse cell, follicular cells, and interstitial cells adjacent to the ovarian epithelium can be distinguished in the follicle. Interstitial cells are located between the foilicles. The follicular cells continue to divide and are suitable for studying the metaphase chromosomes. The latter feature is very important, since nearly all the larval tissues in chironomids have polytene chromosomes and do not divide mitotically. In addition to gonads, mitotic division figures can be conveniently observed in the imaginal disks during the IVth larval instar as well as in the brain and the epidermal cells during molts.

Prior to pupation the oocytes are at the slow growth stage; they begin to grow rapidly at prepupa and pupa stages and cease at the adult stage. Prior to pupation the ovary can be clearly distinguished from the testis, owing to its more elongated shape and larger size; it is located in the VIth abdominal segment. Examination in vivo enables oocytes of different sizes to be distinguished.

By the end of larval development the testis is markedly smaller than the ovary. It has an oval shape and is also located in the VIth abdominal segment.

During phases 1-4 of the IVth instar larva the testes contain mainly spermatogonia which are clustered in rosettelike groups. Spermatocytes appear during phases 5-6. Prior to pupation the testes may contain spermatocytes and the pachytene, diplotene, and metaphase I stages, as well as spermatids and spermatozoa. In the prepupae metaphase II and telophase II stages, as well as spermatids and spermatozoa, are the most frequent stages.

Salivary Glands

The salivary glands complete their formation by the time of larval hatching. The salivary gland of *C. thummi* comprises a flat sac whose wall consists of 28-36 glandular cells. The cells are arranged as one layer around a central gland lumen. The number of cells in the gland does not change throughout larval life. There are two salivary glands and their two ducts unite in the narrow chitinized duct, which carries secretion to the mouth of the larva. The imaginal disk of a mature salivary gland is located at the source of the duct within the gland. The glandular cells produce glycoprotein secretion, which is formed in Golgi bodies and then released to the central secretory lumen of the gland.

The gland is differentiated into lobes. The four cells forming the special lobe produce granular secretion of a particular type. Electrophoretic study of the secretion proteins produced by this lobe has shown, the presence of an additional fraction with a molecular weight of 160 kD which was lacking in the secretion synthesized by the cells of other lobes. The secretion of the special lobe also differs from that of other lobes in carbohydrate composition and the intensity of carbohydrate synthesis.

It is now established that the genes coding for the secretary proteins of chironomids are located in the Balbiani rings. In *C. thummi* the Balbiani rings (BR) are located in the short chromosome IV. In the cells of all lobes, two Balbiani rings are present: BR1(c) and BR2(b). In the special lobe an additional Balbiani ring BR3(a) is present. It is most probably responsible for the synthesis of sp. 160, the additional fraction of secretory protein referred to above.

The salivary gland cells contain extremely large polytene chromosomes which are very suitable for cytogenetic analysis.

The formation of polytene chromosomes in salivary gland cells begins quite early during their differentiation. At the beginning of the first instar the salivary gland cell nuclei look like typical interphase nuclei, but their content of DNA attains 32-64 C. Two cycles of DNA replication take place during the Ist larval instar, 2-3 during the IInd larval instar, and 1-2 in larvae of the IIIrd and IVth instars; 11-12 DNA replication cycles proceed during larval development, and the completely formed polytene chromosomes are between 4000 and 8000 C. An increase in the quantity of DNA templates by endoreplication in salivary gland chromosomes provides the genetic basis for gland cell growth. The process of endoreplication proceeds in relation to body growth.

A total of 336 puffs has been identified in the giant polytene chomosomes of *C. thummi* salivary glands. As the puffing pattern in salivary gland chromosomes is quite stable during the intermolt period of the last two larval instars (studies of puffing are impossible during earlier larval instars owing to the low polyteny of the chromosomes).

The larval molts are important periods in insect larval development. It is in known that molts are accompanied by significant changes in the concentration of hormones such as ecdysone and juvenile hormone in the hemolymph, leading to characteristic changes in the physiological state of the larval tissue (for review see). These changes are primarily observed in the cells of the epidermis, forming the cuticle. However, as already mentioned, synthesis of DNA, RNA, and protein in a number of larval organs decreases or even stops completely during the molt.

The molting physiology, however, is far from being completely understood even for such classic experimental animals of invertebrate physiology as butterflies, ticks, and crustaceans. This process has been investigated in *Chironomus* to only a small extent, although this species is one of the few experimental models to provide an opportunity for comparing the morphological and biochemical characteristics of molting larvae with visually observable changes in chromosomes. The fragmentary data available suggest that further work in this direction could be quite promising. Using the salivary gland as a model, one can expect to observe interesting correlations between changes in the puffing pattern and a number of physiological characteristics reflecting the cavity of the gland.

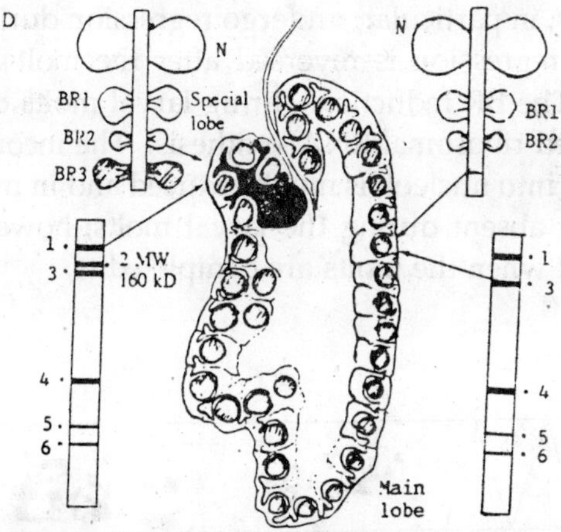

Fig. 7.15. Salivary gland of *Chironomus thummi* 4th larval instar. A) Whole mount and Feulgen preparations. Arrows indicate lobe boundaries. Letters and numbers indicate cells in different lobes. B) Cytochemical differences of the secretion from the special lobe and other lobes of the salivary gland. Paraldehyde—fuchsin staining. SL, special lobe. C) Differences in the electrophoretic patterns of secretory proteins from the special (SSL) and the main (SML) lobes. Polyacrylamide gel-electrophoresis. D) Sketch illustrating differences between the special lobe and other lobes of the salivary gland in the pattern of Balbiani rings and electrophoretic patterns of secretory proteins.

Studies of putting during the larval development of C. thummi have shown that, as in other Diptera, the most significant changes in the purring pattern take place during the metamorphosis molt when metamorphic puffs appear and some of the larval puffs are reduced.

It has been demonstrated that larval molts are also associated with considerable changes in the puffing pattern. The main puffs, previously regarded as typical only for the metamorphosis molt, also appear during these periods. Only the late puffs of metamorphosis do not develop during larval molts. A characteristic feature of the puffing pattern during larval molts is the absence of regression of the main larval puffs, with the exception of tissue-specific puffs which become inactive. The Balbiani rings, in particular, undergo regression during larval molts. This regression is reversed after the molt has been completed . The BR reduction during larval molts correlates with decreased ribosomal RNA synthesis. The incorporation of ^{3}H-uridine into nucleoli is much reduced and in many cells is completely absent during the larval molts; however, it is fully restored when the molts are completed.

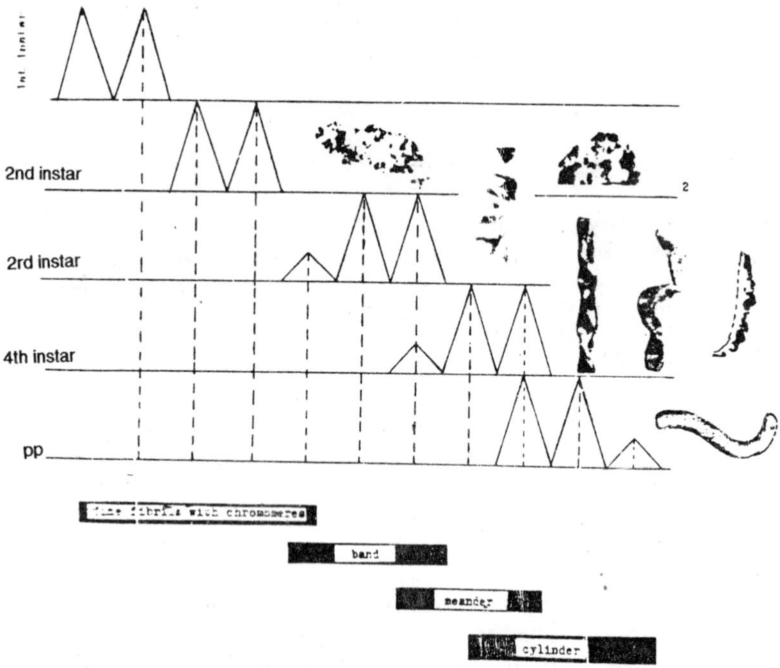

Fig. 7.16. Formation of polytene chromosomes during larval development. Abscissa, degree of polyteny indicated by

cytophotometry of the Feulgen-stained chromosomes; ordinate, larval instars. (For each instar, photometry was conducted at the redhead stage, when DNA synthesis is very low.) PP, prepupa. Photographs of chromosomes with different degrees of polyteny and corresponding plastic models are shown on the right-hand of the figure.

In order to explain the functional aspect of puffing properly we must emphasize that the reduction of RNA synthesis in the BR and the nucleoli is accompanied by a cessation of secretion production. Electron microscopy apparently demonstrates that, as a consequence of ecdysone action, the

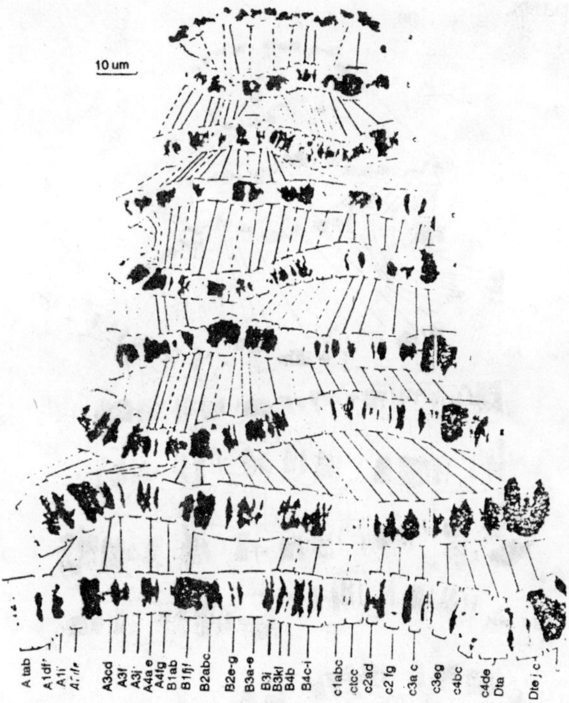

Fig. 7.17. Puffing patterns in the right arm of the Ist chromosome. a-c) 3rd Instar — 3, 5 and 7 days after molting, respectively; d) 4th instar molt; e-g) 4th instar, phases 1 (redhead stage), 3, and 6; h) prepupa, phase 7; i) red pupa. c, centromeric region. In this and the following figures the locations of the main puffs are connected with lines.

Golgi bodies become inactive during molts, secretory granules disappear from the cytoplasm, and the secretion synthesized earlier is actively released into the lumen of the gland. Ultrastructural changes of the cytoplasm of the salivary gland cells during larval molts can also be detected. The cessation of secretion formation is accompanied by the regression of a large area of endoplasmic reticulum (ER) with attached polysomes. This regression predominantly involves the areas of ER located in the apical parts of the cells which are directed toward the lumen. Mitochondria migrate *en masse* from the distal cell regions where they are usually concentrated in the apical parts.

Fig. 7.18. Puffing patterns in the left arm of the Ist chromosome. a-c) 3rd Instar — 3, 5 and 7 days after molting, respectively; d-g)

4th instar, phase 1 (redhead stage), 1-3, and 5-6, respectively; h, i) prepupa phases 8 and 9; j) red pupa. c, centromeric region.

Thus, during the larval molts when the synthesis of tissue-specific BR RNA and ribosomal RNA is reduced, the salivary gland cell cytoplasm undergoes-significant ultrastructural changes. These changes include the disappearance of secretory granules, changes in the shape of the Golgi bodies, migration of mitochondria from the distal zones to the zone of microvilli turned towards the lumen, and the formation of vast regions of destroyed rough endoplasmic reticulum [2,35,39].

After the molt has been completed, the normal cytoplasm ultrastructure recovers along with the normal puffing pattern and the resumption of RNA synthesis in the Balbiani rings and the nucleolus.

Periodic reduction in the activity of the BR and the nucleolus, and simultaneous activation of the main set of metamorphic puffs during larval molts, can be explained by an increased titer of ecdysone in the larval hemolymph during these periods. Changes of the puffing pattern and cytoplasmic ultrastructure can be experimentally induced by injecting ecdysone into the intermolt larvae.

The described transformations of the puffing pattern and cytoplasm ultrastructure can be seen as the periodic reprogramming of terminally differentiated cells during larval molts. A similar periodic switching on and off of the tissue-specific genes and corresponding changes in the cytoplasm ultrastructure during larval molts is described by Suzuki in the silk gland of the silkworm.

An almost complete arrest of DNA replication is another characteristic feature of the molt period.

In order to study the sequential changes of larvae during the molt, one has to find a group of traits enabling larvae to be identified and selected at vital phases, such as the preparation

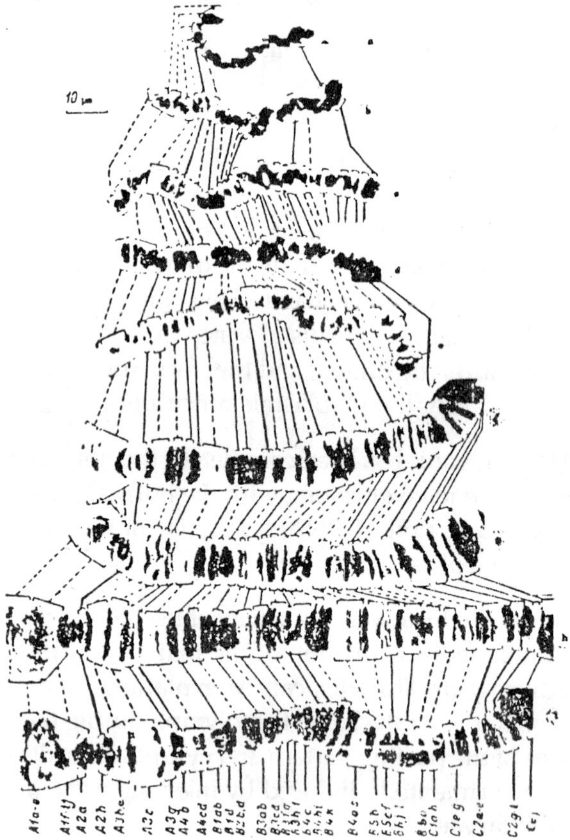

Fig. 7.19. Puffing patterns in the right arm of the 2nd chromosome. a-c) 3rd Instar — 1, 3 and 5 days after molting; d) molt; e,f) 4th instar, phases 1 (redhead stage) and 6; g-i) prepupal phases 7, 8 , and 9. c, centromeric region.

to molt, the molt itself, and immediately after the molt. These traits include the state of the epidermis and cuticle, the position of eye spots on the head capsule, and pigmentation of the head capsule and mouth parts. Some of these trait have been described briefly [13, 18, 46, 64]. We have studied the changes of these characters in the course of the last two larval molts and the metamorphosis molt. During the molt from the IIIrd to the IVth instar, for example, there are marked morphological

Fig. 7.20. Puffing patterns in the left arm of the 2nd chromosome. a, b) 3rd Instar — days 5 and 7 after molt; c) 4th instar molt; d, e) 4th instars, phases 1 and 6; f-h) prepupa phases 7, 8 and 9; i) red pupa.

changes about 1.5 days prior to ecdysis. At this time, the eye pots have already shifted to the center of the head capsule, while the cuticle begins to separate from the epidermal cells. A narrow exuvial space may be detected in the anterior body segments and in the appendage with hooklets on the last body segments, and the thoracic segments become thicker. Twelve hours prior to the molt the eye spot eventually shift to the base of the head capsule and the exuvial space can be seen

Fig. 7.21. Puffing patterns in the right arm of the 3rd chromosome. a, b) 3rd Instar -- days 1 and 5 after molt; c) molt; d-f) 4th instar, phase 1 (redhead stage), phases 1, 3, and 6; g, h) prepupa, end of phase 9; i) red pupa. c, centromeric region.

clearly in all segments. At this time the salivary glands stop supplying secretion, and swell. The larvae, having completed ecdysis, can be easily identified by the bright red colour of the head capsule; hence the description "redhead larvae." The true redhead stage in the IVth instar larva continues for 8-10h, then the head capsule becomes gradually darker, assuming a brown pigmentation by the twelfth hour. Pigmentation of the mentum and mandibula change in parallel with the pigmentation of the head capsule. Use of these characteristics results in more reliable determination of the consecutive stage of larval development and of the first phase, of the IVth larval instar. Correspondingly, the larvae of different stages can be selected for experiments. The latter is quite important since the redhead stage is suitable for various studies, it is particularly suitable for determining the quantity of DNA in the salivary gland nuclei, since the intensity of DNA replication decrease at this time; this stage is also promising for the detection of factors controlling the initiation of DNA, RNA , and protein synthesis during larval development.

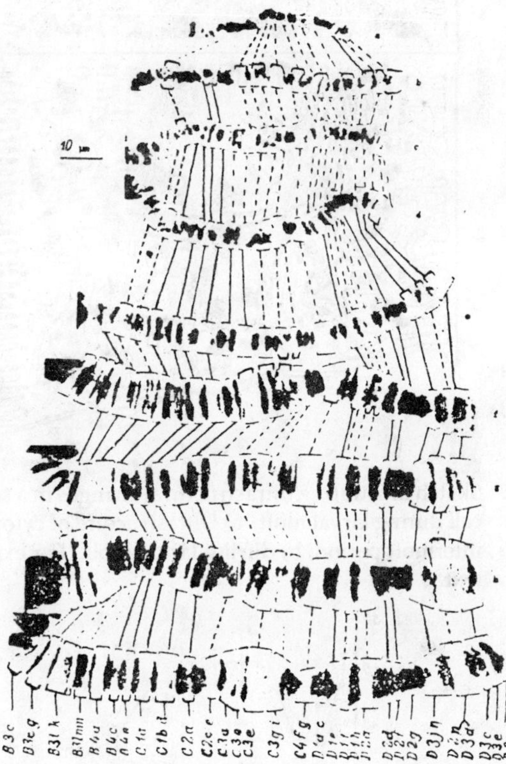

Fig. 7.22. Puffing patterns in the left arm of the 3rd chromosome. a, b) 3rd Instar — days 3 and 7 after molt; c-f) 4th instar, phase 1 (redhead stage) , phases 1, 2 and 6; g, h) prepupal phases 7 and 9 (pupation); red pupa.

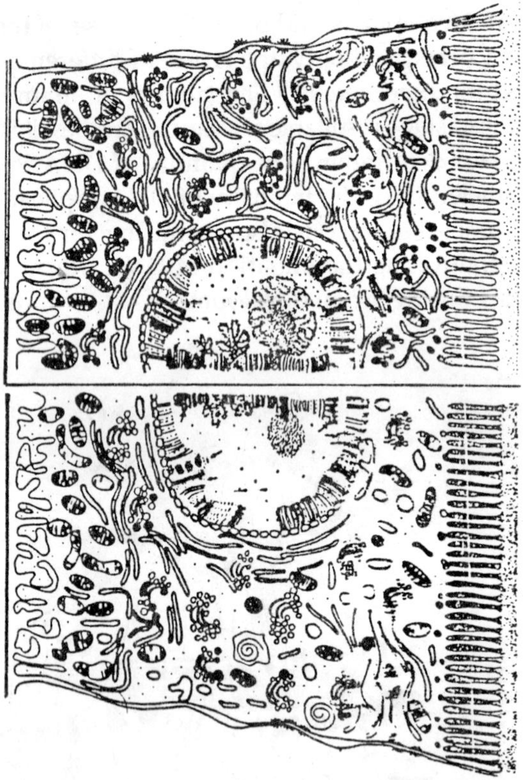

Fig. 7.23. Sketch illustrating ultrastructural changes in a salivary gland cell during larval molt. a) Ultrastructure of cytoplasm during intermolt period; b) during larval molt. For explanation see text.

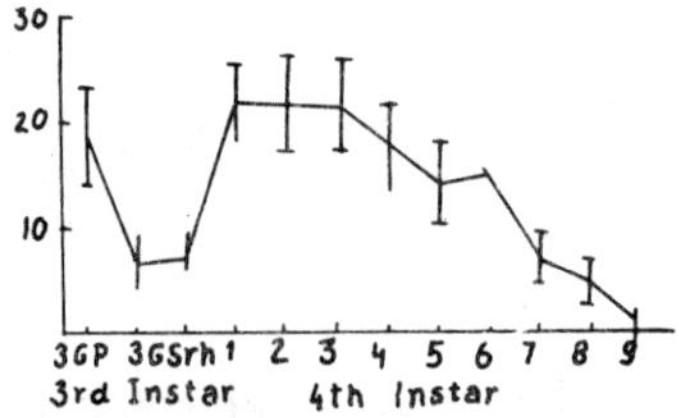

Fig. 7.24. Frequency of ^{3}H-thymidine-labeled nuclei in salivary gland cells of the last larval instars, 3rd Instar. 3GP, 3GS, stages of preparation to molt; 4th instar; 1-9, phases. rh, redhead stage.

The complex of traits, which has been used in the case of the last larval molt, can also be employed during the molt of the second instar larvae to the third instar larvae. It should be borne in mind, however, that the duration of the redhead stage in this case is only a little more than half the molt stage from the third to the fourth instar.

PREPUPA AND PUPA

Prepupa

The transition to metamorphosis begin from the prepupa stage, and the state of the endocrine system changes markedly. The epidermal cells are switched over to the synthesis of the pupal cuticle and the imaginal disks grow intensively. Externally, the transition to the prepupal state is expressed as a marked thickening of the larval thoracic segment, gradual contraction of the head, and shortening of the body. The formation of metamorphic puffs begins during this period. The prepupal phase leads to the metamorphosis molt, resulting in the pupa.

Pupa

The pupa differs drastically from the larva in its morphology. The pupa does not feed, so its wet weight is not much more than half that of the larva. The morphological peculiarities of the pupa are primarily associated with the formation of the cephalothorax following the fusion of the head and three anterior segments. The remaining ten segments form the abdomen and the leg and wing sheaths. Envelopes containing wing rudiments are located laterally on the pupal body and the prothoracic horns are formed on the dorsal side of the anterior segments.

The pupa stage lasts 2.5-3 days, and its pigmentation gradually changes. Immediately after pupation it is red and then becomes dark grey. In the course of pupal life the larval organs, excluding the nervous system and the Malpighian tubules, undergo histolysis. Imaginal disks of the larval organs produce the corresponding organs of the adult animal.

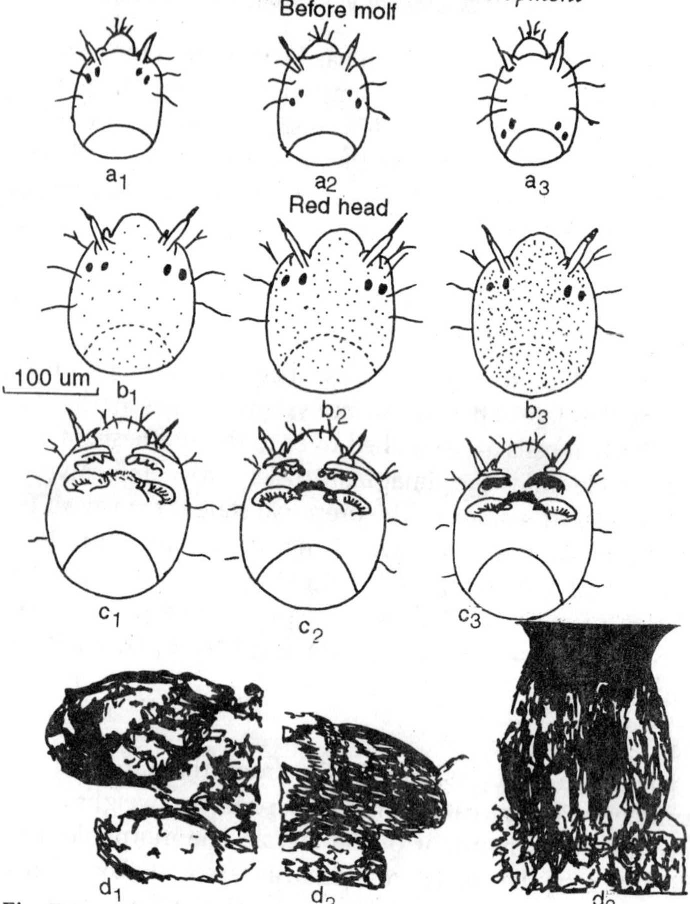

Fig. 7.25. Changes in the position of the eye spots (a_1-a_3) and pigmentation of the head capsule (b_1-b_3) and mouth parts (c_1-c_3) during molt of the 3rd larval instar. a_1) Usual position of the eye spots during the intermolt period up to 36 h prior to molt; a_2) 12-24 h prior to molt a_3) immediately before molt; b_1) red head colour immediately after molt; b_2) light-brown colour 8-10 h after molt; b_3) dark brown pigmentation 12 h after molt; c_1) yellow pigmentation of the mentum and absence of mandible pigmentation after molt; c_2) dark brown mentum pigmentation and yellow pigmentation of mandible 8-10 h after molt; c_3) submentum and mandible are dark brown 12 h after molt; d_1-d_3) photographs of the eye spot of living larvae.

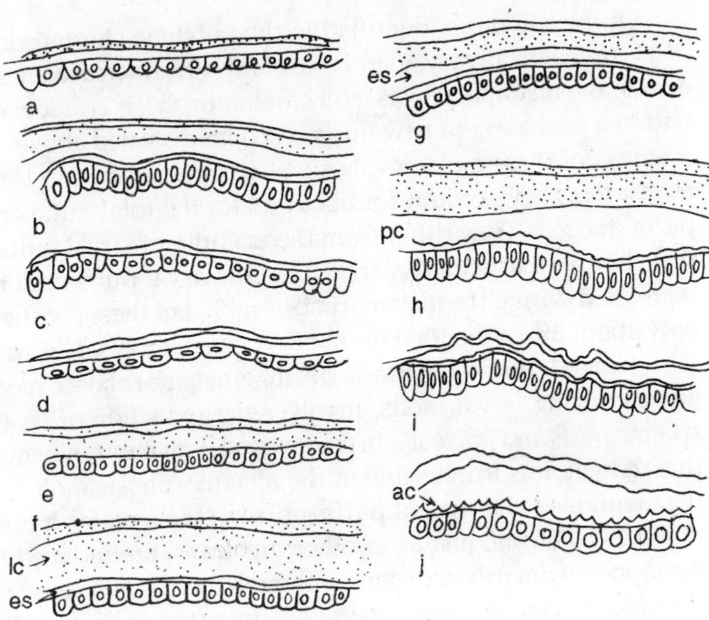

Fig. 7.26. Changes in epidermal cell morphology and the formation of the exuvial space during larval and metamorphic molts. a) Intermolt period of 3rd instar; b) prior to the molt of the 3rd to the 4th larval instar; c) 4th instar immediately after molt (redhead); d) phase 1-2; e) phase 3; f) phase 4; g) phase 6; h) phase 8; i) phase 9; j) pupa. es, exuvial space; lc, larval cuticle; pc, pupal culticle; ac, adult cuticle.

The presence of such complex morphogenetic transformation during metamorphosis makes this period of dipteran development extremely attractive to those studying developmental cytogenetics. Many studies have dealt with the puffing pattern in polytene chromosomes of several larval organs of chironomids and drosophilids during metamorphosis. The authors of these papers agree that the most distinct changes of the puffing pattern take place during metamorphosis when a special complex of puffs is formed. These puffs are, as a rule, larger and synthesize RNA more actively than other puffs (excluding the Balbiani rings in chironomids).

Of the 336 puffs identified in the polytene chromosomes of *C. thummi* salivary glands, around 40% increase in size during the metamorphosis molt. Metamorphosis is associated with the formation of new puffs. As demonstrated above, the main set of these metamorphosic puffs appears during larval molts, too; they account for about 4% of the total number of puffs studied. They differ from the majority of larval puffs in that they are particularly large. Certain larval puffs undergo regression during the metamorphosis molt, but these constitute only about 2% of the total number. A characteristic feature of the changed puffing pattern during metamorphosis molts, just as during larval molts, involves the reduction of tissue-specific puffs and nucleoli. In contrast to larval molts, however, this reduction is irreversible in the metamorphosis molt. All the mentioned changes of puffing during the metamorphosis molt in *C. thummi*, just as in other dipterans, are induced by hormones, primarily ecdysone, although there is evidence that juvenile hormone also plays an important part. The metamorphosis molt leads to a drastic increase in the activity of lysosomal enzymes and, as a result, the salivary glands undergo lysis in the late pupa just as do other larval organs.

The pupal stage ends in the metamorphosis molt leading to the formation of the adult. As previously mentioned, adult chironomids have a short lifespan, and hardly ever feed; they are only occasionally used in experimental studies.

8

Development of Silkworm Bombyx mori

Moths are the close relatives of butterflies. Their antenna are short and feathers like and the body is stout. At rest, this wings are horizontally placed. There are several species of moths but the most species of moth used in semiculture is the *Bombyx mori.*

The scientific biography of the silkworm contains a wealth of observations of tremendous biological significance; therefore it seems inevitable that this economically important insect will become an established laboratory species in developmental biology. The advantages of using the silkworm in this area include primarily the abundance of available genetic information relating to development. However, cytogenetic investigations of silkworm chromosomes, despite the progress in this field, present considerable difficulties because of the large chromosomal number (2n = 56) and their morphological uniformity. The high polyploidy of the silk-gland is as yet insufficiently studied at the chromosomal level; the supposedly polytenic chromosomes of this organ do not appear to possess disk organization.

One unquestionable advantage of the silkworm is associated with the possibility of artificial parthenogenetic and androgenetic cloning which, when combined with the case of biological synchronization, provides the required amounts of isogenic material at any developmental stage from the mature egg to the adult form. Professor Astaurov's fundamental studies in the field of artificial reproduction of the silkworm were reprinted a few years ago, and will only be discussed briefly in this chapter. The cytogenetic mechanism of cloning achieved in the silkworm by the heat-activation procedure is based on the lack of one meiotic reductional division following high temperature treatment. In unfertilized eggs activated by heating, the mitotic division proceeds more or less successfully, and the maternal genotype, whether diploid or polyploid, passes unchanged to the parthenogenetic offspring. Heating the eggs immediately after spermatozoa have penetrated them induces gynogenetic development of the ameiotic type, the development being activated by spermatozoa which introduce the acrosome, centriole, extrachromosomal chromatin, and cytoplasmic self-reproducing structures into the egg.

Various treatments of the unfertilized egg at does lower than those needed for ameiotic parthenogenesis enable meiotic parthenogenesis to be induced. In this case the reduction division reaches completion. The haploid pronucleus divides to yield two genetically identical daughter nuclei which then fuse to form the diploid cleavage nucleus. Such a cytogenetic mechanism leads to the appearance of completely homozygous males which are of great value in various studies. In the case of meiotic parthenogenesis of tetraploids, the role of cleavage nucleus is played by the diploid pronucleus formed in the course of meiosis. Because of this, the resulting descendants (almost always females) are heterogeneous, which enables the isogenic clone to be transformed into a large number of new genetically non-identical clones.

The silkworm was the first animal species in which androgenetic development up to sexual maturity was obtained. Discussions on the general significance of androgenesis in the study of nucleocytoplasmic relations have been given in the Astaurov critical review, while its practical application has been described in studies about the relative significance of radiation damage for the nucleus and the cytoplasm; other

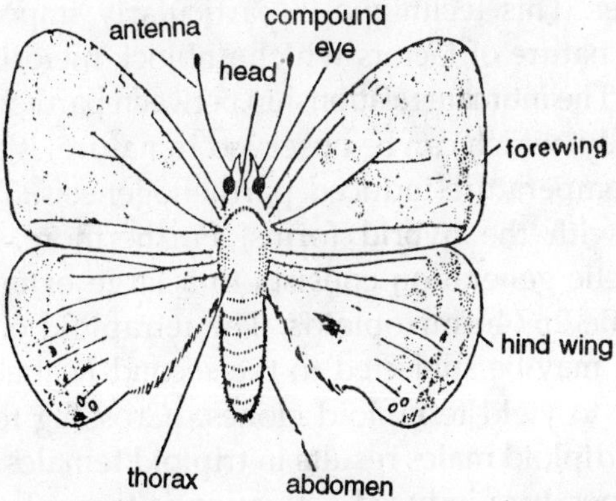

Fig. 8.1. Butterfly.

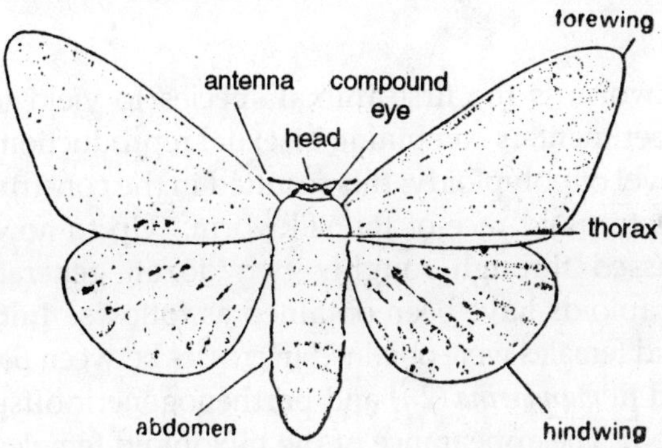

Fig. 8.2. Moth.

papers in this series deal with the interaction of the nucleus and cytoplasm of various species. A technique allowing bipaternal androgenetic hybrids to be obtained has been developed for the silkworm; this opens up new opportunities for studying nucleocytoplasmic relations. Methods allowing one to prevent the activation of eggs following the penetration of spermatozoa, or various physical and chemical treatments, are of crucial significance in developmental studies. The eggs with blocked activation are viable and can be successfully activated later. This technique is particularly important for studying the nature of factors which unblock meiosis or lead to activation. The intimate relationship between parthenogenesis and polyploidy, which can be observed in nature, is also seen in artificial temperature-induced parthenogenesis. The latter occurs best with the hybrid forms. Furthermore, the first parthenogenetic generation consists to a large extent (up to 30%) of female 2n/4n mixoploids. The tetraploid oocytes of such animals may be subjected to the second heat activation step in order to yield tetraploid clones. Crossing tetraploid females with diploid males results in triploid females. In these artificial temperature-induced activation is the only possible way of reproduction since bisexual reproduction results in aneuploidy and the death of the embryos. Triploid parthenogenetic clones of the silkworm have been obtained in this way.

The silkworm is the first animal species to yield to the efforts of experimenters combining bisexual reproduction with a constant level of polyploidy; this resulted in the construction of the new tetraploid race of the silkworm. Up to now this race has passed through roughly two dozen generations. Bisexual tetraploids have been obtained as follows. Initially, triploid hybrid females were obtained in crosses between *Bombyx mori (4n)* and *B. mandarina (2n)* and parthenogenetic offspring were obtained. The appearance of the mixoploid females 3n/6n could be expected among parthenogenetic triploid females just as at the diploid level. Crosses between the mixoploid 3n/6n females and diploid males of *B. mandarina* should, after the fertilization of hexaploid oocytes, yield both male and female tetraploids. Such allotetraploid males have indeed been found; in contrast to autotetraploids they showed partial fertility. Further studies of these experimentally constructed parthenogenetic and bisexual forms may provide interesting opportunities.

The relative ease with which polyploid parthenogenetic forms are obtained and reproduced comprises yet another important biological advantage of the silkworm. It could well be that the silkworm polyploids derived by other techniques, such as centrifugation, colchicine treatment, or heating of freshly inseminated eggs, may be successfully cloned using the temperature-activation procedure.

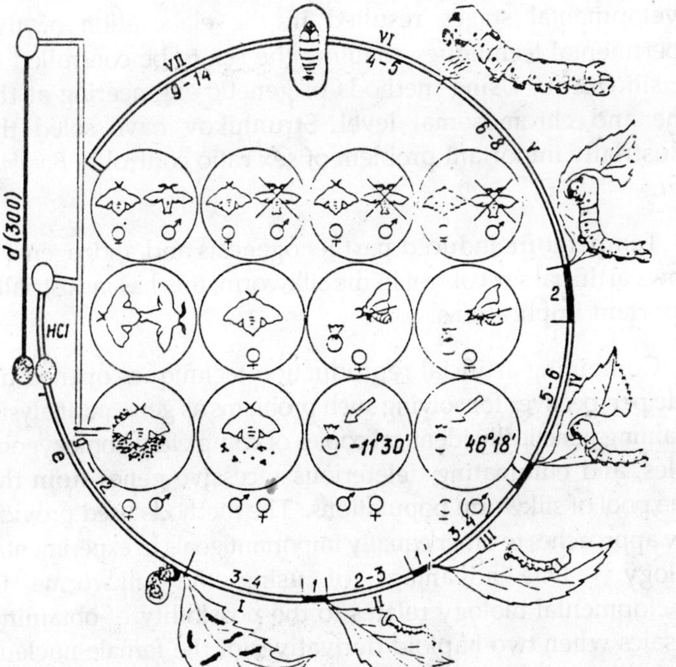

Fig. 8.3. Life cycle of the silkworm *Bombyx mori*. I-V) Larval instars; VI) spinning and prepupa; VII) pupa. Arabic numerals refer to the duration (in days) of corresponding instars, as well as to intermediate molts, etc. If the eggs obtained using one of the four techniques described (vertical columns, see text) are treated with HCl prior to diapause (d), embryogenesis (e) proceeds to the hatching of the 1st larval instar.

Cooling eggs which begin parthenogenetic development results in the appearance of autotetraploid forms. The activation of eggs in the bodies of live moths of different ploidy, followed by insemination at various stages of parthenogenetic ameiotic development, is of considerable methodological significance. This provides ample opportunities for obtaining various types of polyploids combining several genomes of genetically non-identical forms. The modification of cytogenetic mechanisms participating in egg maturation and control of early developmental stages resulted in the elaboration of five experimental techniques enabling the sex to be controlled in the silkworm. Using methods of genetic engineering at the gene and chromosomal level, Strunnikov have soled the industrially important problem of sex ratio control in *Bombyx mori.*

Temperature induced parthenogenesis and androgenesis allows artificial sex control in the silkworm; this has industrially important implications.

Combining artificial reproduction techniques opened up wide perspectives for solving such problems as genome analysis, obtaining genetically identical copies of completely homozygous males, and eliminating deleterious recessive genes from the gene pool of silkworm populations. The methods used provide new approaches to other equally important goals in experimental biology. One advantage of using the silkworms in developmental biology relates to the possibility of obtaining mosaics when two haploid derivatives of the female nucleus are fertilized by two different spermatozoa. Other types of mosaicism have also been described. Normal and double fertilization, antrogenesis, meiotic and ameiotic types of parthenogenesis combined with different levels of ploidy provide opportunities for the construction of a wide variety of experimentally modified cytogenetic structures of developing silkworm embryos: quite a number of these opportunities have already been fulfilled. The initial discovery of artificial parthenogeneis in animals was made by Tikhomirov in 1886, using the silkworm. This stimulated further studies of the phenomenon, initially using marine invertebrates and then other animal species as well. The reversibility of activation-induced changes in the course of artificial and normal activation of the egg, recently described in *Bombyx mori*, may again prove this species to be in extremely suitable model for analyzing the problem of egg activation at the molecular level.

Studies on the physiology and endorinology of metamorphosis, synthesis of silk proteins, radiation-induced and chemical mutagenesis, the use of the silkworm for testing environmental pollution, its application in virology, and other subjects have been described in a recent monograph. The greatest interest has always centered around the function of silk gland. The interests of practical silk farming focused on this system which, owing to its simplicity, attracted the attention of biochemists and molecular biologists, including genetic engineers. The silk fibroin gene has been isolated and cloned by Morrow et al. Important reviews dealing with the synthesis of silk proteins include special issues of *Biochimie Experientia*, and several others.

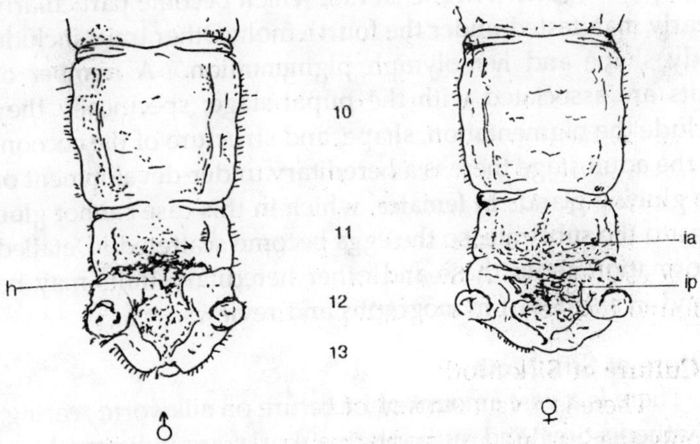

Fig. 8.4. Sexual dimorphism in *Bombyx mori* larvae. 10-13) Segments from the ventral side. h, Herold imaginal bud; ia and ip, anterior and posterior Ishiwata imaginal buds.

LIFE CYCLE

Typically, the life cycle of *Bombyx mori* includes the egg stage, which may last at least 300 days in monovoltinic varieties but may be reduced to 9-12 days by artificially shortening the diapause using hydrochloric acid treatment. It is followed by the larval stage, which is divided into five instars by four molts; the next stage is that of prepupa, which lasts about 2-3 days after the final spinning of the cocoon; the pupal stage then lasts 9-14 days, and is followed by the adult stage.

Genetic analysis has been performed for 328 traits of *Bombyx mori*. The majority of these traits belong to the egg and larval stages; this is a unique aspect of this animal. Mutations affect the following characteristics of the egg : size, shape, pigmentation of the chorion, pigmentation of the yolk, pigmentation of the serosa. Newly hatched larvae are usually dark brown with rather long bristles on all segments of the body. Mutations yielding larvae with reddish-brown pigmentation and short bristles have been described. A marked number of hereditary traits at later larval stages are associated with the pigmentation and spotted pattern of the larvae, which become particularly clearly manifested under the fourth molt. Other traits include body shape and hemolymph pigmentation. A number of traits are associated with the pupal stage; specifically they include the pigmentation, shape, and structure of the cocoon. At the adult stage there is a hereditary under-development of the glueing glands in females, which in this case cannot glue eggs to the substrate, so the eggs become scattered. Detailed information about these and other hereditary traits may be found in a series of monographs and reviews.

Culture of Silk Moth

There is a vast amount of literature on silkworm rearing; only the leading and authoritative works are indicated here. These book, however, frequently either omit details which are important to developmental biology, or mention them only briefly. Therefore, omitting the description of techniques used for feeding and rearing larvae, a more detailed description is given below of manipulations commonly used in laboratory practice.

Obtaining Synchronously Developing Eggs

Unfertilized mature oocytes at metaphase I are extremely easy to obtain from the ovarities by either rubbing or grating them through a suitably sized capron mesh. Eggs obtained in this way are collected in a beaker containing some water, and them washed with water on a piece of cheesecloth stretch over a funnel until the residues of the ovarioles are completely removed from the surface of the eggs.

Freshly inseminated eggs are obtained as follows. Silkworm moths should first be prepared for oviposition. At the day of flight they mate several times with males which emerged on the preceding days and were kept in a cool place at about 15°C. It is not difficult to continue the mating until evening, when even virgin moths begin laying individual unfertilized eggs. At this time, mated females begin intensive oviposition at 25°C, and the moment the eggs are laid may be taken as the time of insemination. Two sheets of thick smooth paper are placed on a table in a dark place. Between 50 and 100 laying females are rapidly placed onto one sheet and allowed to continue oviposition for a certain length of time, depending on the required accuracy of synchronization (e.g., from 30 sec to 5-10 min.). Thirty seconds before the end of the planned period the females are quickly transferred to the adjacent sheet of paper, and egg collection begins again. If only freshly inseminated eggs are required, then the eggs deposited on the first sheet are placed in a refrigerator or are put into an envelope which is then stored in an incubator during the set period. At the required time the eggs are fixed or placed into a refrigerator at 0°C, where the necessary amount of material is gradually accumulated. Later this material is either fixed or subjected to other treatments. It should be borne in mind that cold treatment may lead to very serious artifacts in cytogenetic and other studies, if the eggs are allowed to develop further after chilling.

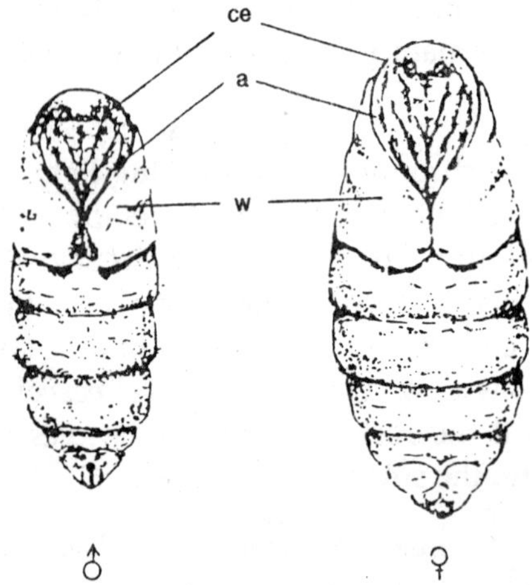

Fig. 8.5. Sexual dimorphism in Bombyx mori pupae. a, antenna; ce, compound eye; w, wings.

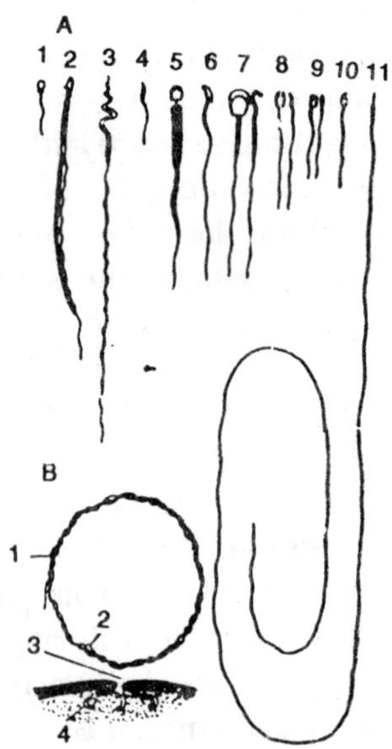

Fig. 8.6. A) Spermatozoa of silkworms compared with those of vertebrates (1-10). B) Rotating spermatozoon of silkworm near the micropylar pole of the egg. 1) Spermatozoon; 2) sperm head; 3) micropyle; 4) micropylar channels.

Large quantities of synchronously developing embryos may be very easily obtained by arivating the eggs extracted from moths using the artificial parthenogenesis method. For this purpose the moths should be taken from a parthenoclone with a high capacity to parthenogenesis. The techniques used for obtaining parthenogenetic and androgenetic eggs have been described in another book of this series.

Larvae and Pupae

Larvae just hatched from the eggs are used. To synchronize a large number of such larvae there must be correct conditions for hatching them from the eggs within a short period of time. For this purpose the hatching occurring in the morning is delayed for a day by a complete exclusion of light followed by bright illumination. The larvae hatched during a certain interval are either fixed or fed to allow development to the necessary stage.

The larvae of subsequent instars are collected immediately after the molts. The procedure consists of placing the required number of larvae on a sheet of paper during their sleeping period. Thereafter, the larvae which undergo molting within a certain period of time are collected. This technique is inapplicable in silkworm farming, but is useful in laboratory practice.

The pupae are similarly synchronized, but in order to see the prepupae one should cut the cocoons with a razor blade immediately after spinning has been completed.

Adults

The adults fly out during the early hours of the morning, and the material may be easily synchronized by the above technique, during the *en masse* emergence from the cocoons or from pupae, if the outer part of the cocoon has been removed.

The flight of moths may be delayed by keeping the cocoons at lower temperatures (12-18°C), but one should bear in mind that this may subsequently affect the quality of the silkworm eggs that moths will lay.

Mating takes place during the morning. Insemination requires about 1 h of mating time, but is frequently prolonged to 2-4 h. A given male may mate several times within one day. Intervals between matings should be 2-3 h, and during this period the males are kept in a paper envelope at 10-12°C. At this temperature the fertilizing capacity of males remains unchanged for 2-3 weeks. The males may also be kept at 4-5°C.

The life cycle as well as the normal and experimental methods used for obtaining silkworm eggs of a conventional monovoltinic variety. The left vertical column in the figure refers to fertilized silkworm eggs obtained by the conventional procedure described above. The second column corresponds to spontaneous or accidental parthenogenesis. The third column corresponds to artificial meiotic parthenogenesis discovered by Sato and Koltzov; it allows completely homozygous males to be obtained, while females are rather rare. The most effective method of obtaining homozygous males in the course of meiotic parthenogenesis has been suggested by Terskaya and Strunnikov. The fourth column corresponds to artificial ameiotic parthenogeneis, discovered by Astaurov using this procedure, one may clone the female genotypes, while the males are virtually absent.

Irrespective of the technique used for obtaining eggs of the monovoltinic silkworm varieties, several days later the egg enters diapause and the next cycle does not start until 10 months later. In the practice of silkworm breeding, the eggs are stored for this long period of time under strictly controlled temperature and humidity conditions. Similar rules should be followed in the laboratory. However, if there is a need to obtain the next generation quickly, the onset of diapause is prevented by treating the eggs with hydrochloric acid as follows.

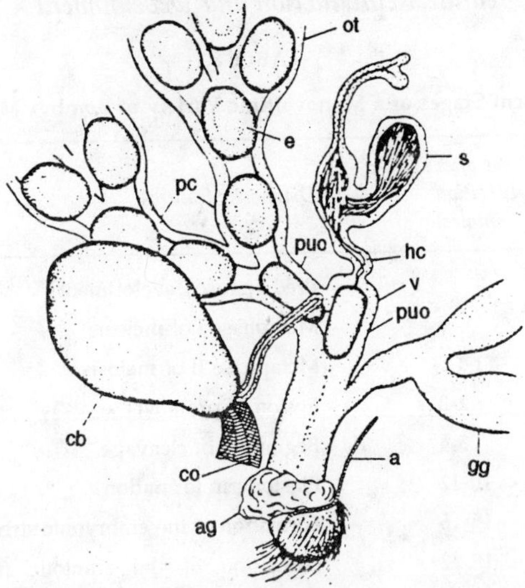

Fig. 8.7. Female genitalia of the silkworm moth. ot, ovarian tubes; e, egg; po, paired oviduct; auo, puo, anterior and posterior part, respectively, of unpaired oviduct; ag, aromatic glands; a, apophyis; v, vestibulum; s, seminal receptacle with helical channel (hc); cb, copulatory bursa; co, copulatory orifice; gg, glueing glands.

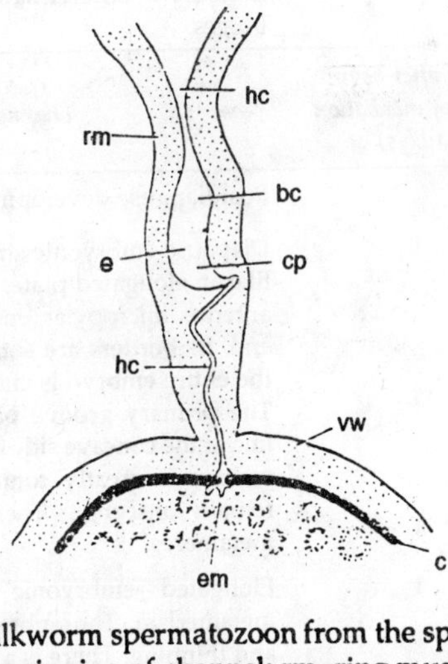

Fig. 8.8. Path of silkworm spermatozoon from the spermatheca to the egg, bc, beginning of channel; rm, ring muscle; e, expansion with cylindrical part (cp); hc, helical channel; vw, vestibulum wall; em, micropyle; c, chorion.

TABLE 8.1

Development Stages of a Monovoltinic Variety of Bombyx Mori L. at 25°C

Stage	*Time after fertilization (hours)*	*Diagnostic features*
		Prediapause development
0	0	Metaphase I of meiosis
1	1-1.5	Metaphase II of meiosis
2	2-3	Fusion of pronuclei
3	3-4	Beginning of cleavage
4	10-12	Blastoderm formation
5	15-16	Formation of the embryonic streak
6	18-22	Formation of the amniotic fold, gradually closing over the embryonic streak which sinks inside; amnion and serosa formed. The yolk gradually segregates to yield /spheres from periphery to the center. Formation of amnion complete. The primary streak appears on the outer side of the embryonic streak; 18 bulges are produced on its inner side from yolk cells; mesoderm differentiation proceeds inside the bulges.

Stage	*Time after beginning of incubation (days)*	*Diagnostic features*
		Postdiapause development
A	0	Diapause embryonic streak. The embryo looks like an elongated plate, which is wider near the anterior micropylar end than at the posterior end. Its borders are somewhat bent inside and the entire embryo is curved along its long axis. The primary groove passes along the middle line of the concave side of the embryo (when the embryo is greatly contracted the groove may remain unseen). There are 18 mesodermal metameres.
B	1	Elongated embryonic streak with distinct metamerism. The embryo undergoes elongation and thinning. There is a distinct metamerism of the ectoderm corresponding to that of the mesoderm. The widened anterior and of the embryo forms two symmetric lateral procuberances (head laminae).

Stage	Time after beginning of incubation (days)	Diagnostic features
C	3-4	Elongated embryo with undifferentiated thoracic appendages. Metameric thickenings of the ectoderm grow more prominently in the anterior region of the embryo than in the posterior one. Seven first pairs of ectodermal protuberances located symmetrically with respect to the midline of the body comprise the rudiments of future head and thorax appendages : antenna, mandibles, two pairs of maxillae, and three pairs of thoracic legs. Appendages develop in the anteroposterior direction. Ectodermal protuberances are poorly marked on the ventral segments. The two depressions of the stomodeum and proctodeum are marked between the head lamina and the posterior segment. Thoracic legs are not differentiated.
D	5	Weakly contracted embryos with marks of ventral appendages. False legs will develop from these appendages at segments III-IV and IX (X). The embryo is somewhat contracted; the head segments are brought into proximity.
E	6-7	Shortening of the embryo and elongation of the stomodeum and proctodeum. The ventral and head neural ganglia are formed, as are the lateral walls of the larva. The anterior head border is somewhat bent inside the egg because of fusion and displacement of the head segments. A pair of ectodermal depressions appear near the base of the labrum; they form the silk-gland duct.
F	8	Beginning of blastokinesis. Embryo reaches maximal contraction. Preparation for blastokinesis is marked by the separation of the embryo posterior end from the membranes. Initial differentiation of primary germinal cells begins.
G	9	S-shaped stage of blastokinesis; the embryo is bent in the shape of a letter S
H	9-10	End of blastokinesis. The lateral walls of the larva are completely formed. The embryo contracted in the course of blastokinesis undergoes quick extension

Stage	*Time after beginning of incubation (days)*	*Diagnostic features*
I	10-11	Stage of caudal elongation. Ventral tip reaches the middle of the egg ventral curvature. Differentiation of organs
K	11	Stage where the embryo bends. Ventral end almost touches the head. The embryo is S-shaped
L	11-12	Helically coiled embryo. The posterior ventral end glides over the head and bend to the egg center, forming a helix. Eggs become lighter in colour
M	12-14	The first instar larvae hatch.

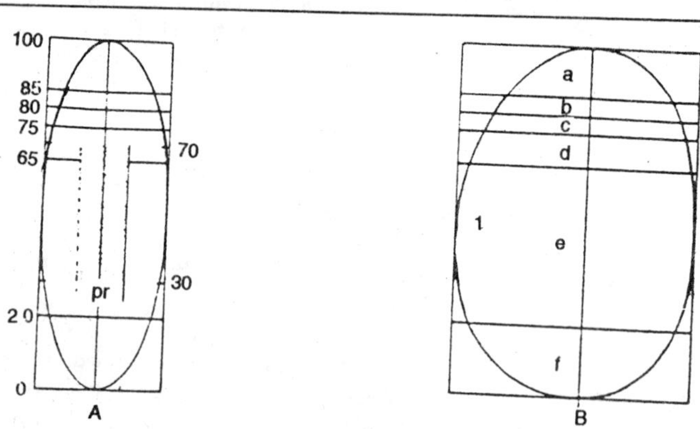

Fig. 8.9. Fate map of the silkworm egg (modified from). A) Ventral view; B) lateral view. Hatched area indicates the determined area of the mesoderm. a,f) Extraembryonic parts; b) head segment; c) jaw segment; d) thoracic region; e) abdominal region; pr) presumptive region of germ cell differentiation; l) ventral side. Relative size of a region is given a number proportional to the maximum diameter of the egg (taken as 100).

The first of the two techniques is used if the next generation must be reared after 11-12 days. A hydrochloric acid solution with a specific gravity of 1.075 at 15°C is prepared. Silkworm eggs aged 20-25 h (at 25°C) are treated with this solution at 46°C for 4-8 min., depending on the exact age of the eggs and the silkworm variety. After treatment the acid is washed away with water and the eggs are left to incubate at 25°C. On the following day, the eggs acquire a rose brownish pigmentation. At this stage they may be placed into a refrigerator (5°C) if it is necessary to delay hatching. This is normally expected after 10 days, but not longer than 3 weeks.

The second technique uses hydrochloric acid with a specific gravity of 1.10 at 15°C. The rose-brownish eggs, aged 42-50 h at 25°C, are placed into a refrigerator at 5°C for 40-60 days. Prior to treatment with hydrochloric acid (48°C for 5-10 min), the eggs are kept for about 3 h at room temperature.

Sexual Dimorphism

Planned crosses frequently necessitate the separation of the males from the females. This can be done very easily using genital disks as criteria at the beginning of the fourth and fifth instars. With a good magnifying glass the male and female larvae can be separated during the third and even the second instar. The paired Herold organs are present in males between the eleventh segment, which carries the caudal spike, and the twelfth segment; they are located on the ventral side under the hypodermis. In the females these segments carry Ishivata disks. Sex identification at the pupal stage presents no difficulties.

Fertilization and Parthenogenesis

The silkworm shows internal insemination. The insemination of eggs outside the body has not been successful because the forward movement of the spermatozoa required to make contact with, and penetrate, the egg does not depend on the spermatozoon itself, but is supported by the female's complex genital system. The spermatozoa resemble very long filaments up to 600-800 μm in length with a diameter of about 1 μm; the head is symmetrical and shaped like an awl. In the female genital tract the spermatozoon rotates, forming a helix with 2-2.5 turns.

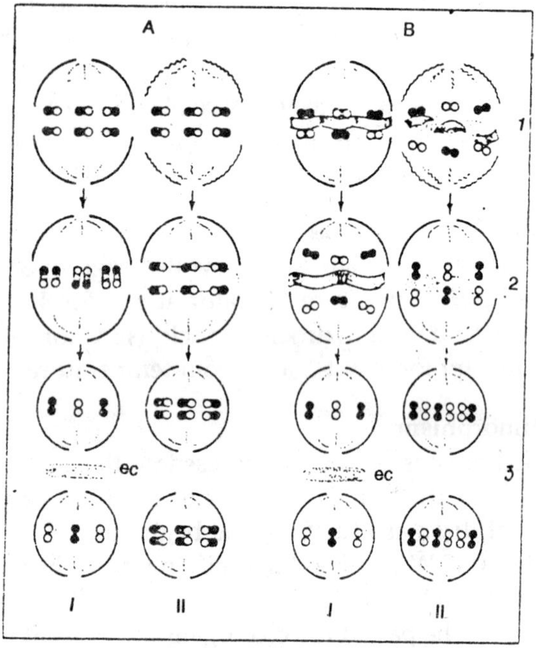

Fig. 8.10. Identification of homologs during metaphase I according to different interpretations of the elimination of segregated chromatin (ec) during the first meiotic division compared to the cytogenetic mechanism of heat-induced inactivation. A) According to Frolova. B) New Scheme, I) Normal; II) following heat inactivation. 1) Metaphase; 2) anaphase; 3) metaphase II.

Artificial insemination in the silkworm has been described. For this purpose the sperm is taken from the freshly inseminated females by opening the dissected copulative bursae containing spermatophores introduced by a male. The spermatophores are then introduced into the genital tract of another female.

Physiological polyspermy is a characteristic feature of the silkworm. The polyspermy for 230 newly inseminated eggs is given follows :

No. of spermatozoa in the egg	0	1	2	3	4	5	6	8	11
No. of eggs	13	30	84	80	17	3	1	1	1

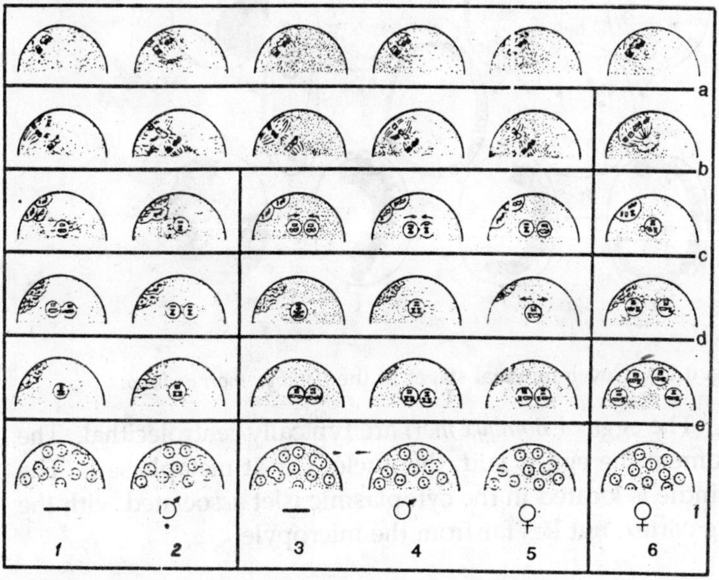

Fig. 8.11. Parthenogenetic variants of *Bombyx mori*.

Freshly laid eggs of *Bombyx mori* have a somewhat irregular ellipsoid shape; the long axis has a length of 1.3-1.5 mm, the middle axis 1.1-1.25 mm, and the short axis 0.5-0.7 mm. The weight of the egg slightly exceeds 0.5 mg. The egg is covered on the outside by a translucent rigid membrane or chorion which protects it from mechanical damage, drying up, and chemical agents. The chorion is wetted by the material secreted by the glueing land, and the egg becomes tightly glued to the substrate the moment it is laid. The outer surface of the chorion shows the imprints of follicle and nutrimentary cells. The micropyle has only one external orifice. The channel passes through the chorion and the vitelline membrane and branches to yield 3-4 canaliculi. This system of micropyle channels follows the spermatozoa to penetrate the egg.

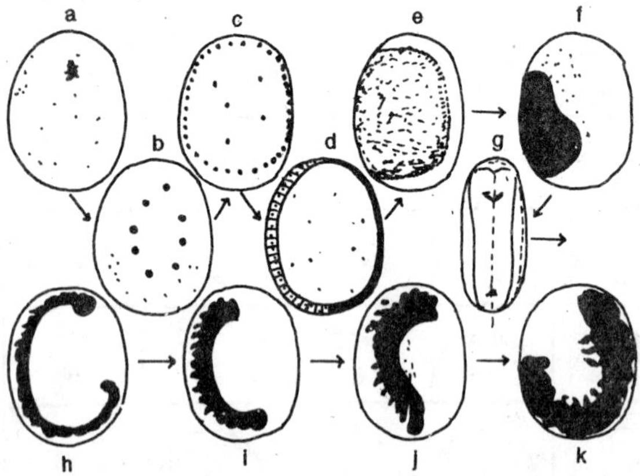

Fig. 8.12. Developmental stages of the *Bombyx mori embryo.*

The eggs of *Bombyx mori* are typically centrolecithal. The moment the egg is laid, the nucleus is at metaphase I. The spindle is located in the cytoplasmic islet associated with the egg cortex, but lies far from the micropyle.

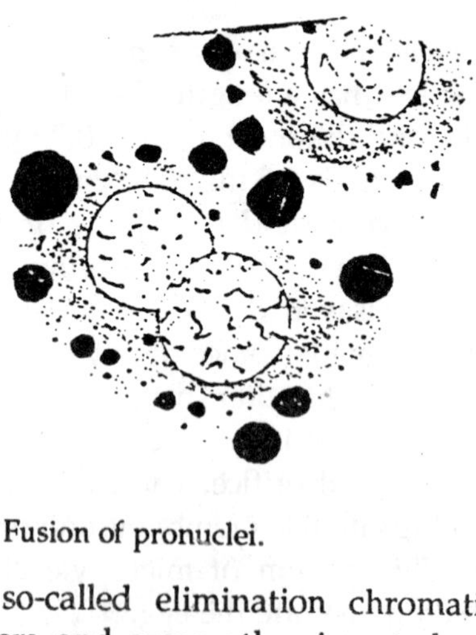

Fig. 8.13. Fusion of pronuclei.

The so-called elimination chromatin characteristic of Lepidoptera and some other insects begins to segregate at anaphase I soon after the egg has been inseminated. A re-

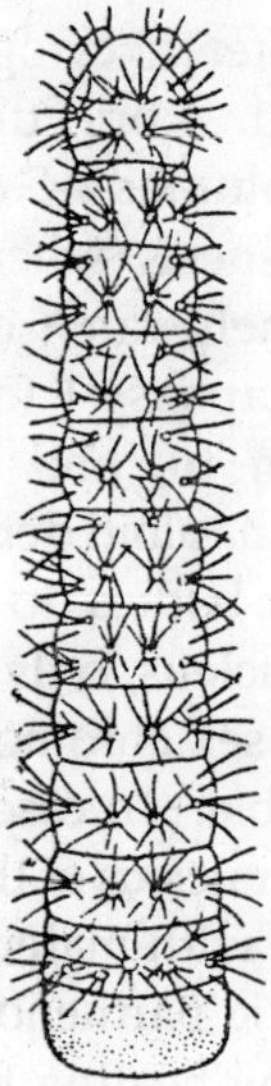

Fig. 8.14. First larval instar, enlarged.

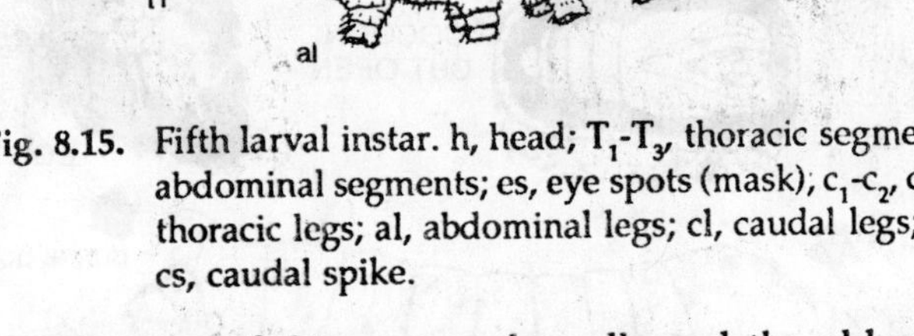

Fig. 8.15. Fifth larval instar. h, head; T_1-T_3, thoracic segments; A_1-A_{11}, abdominal segments; es, eye spots (mask); c_1-c_2, crescents; tl, thoracic legs; al, abdominal legs; cl, caudal legs; s, spiracle; cs, caudal spike.

examination of this process has allowed the old schemes of homologous alignment of chromosomes at metaphase I to be revised and a new mechanism for the artificial ameiotic parthenogenesis, discovered by Astaurov to be suggested. Both mechanisms are presented here.

It has already been mentioned that one important advantage of *Bombyx mori* is that several different cytogenetic variants of parthenogenesis may be obtained. A brief description of them is given below, in this figure, columns 1-5 correspond to the variants of meiotic parthenogenesis when the egg either spontaneously (which is extremely rare), or after treatment with special agents, leaves metaphase I (column a), passes through metaphase II (column b), and after completing maturation divisions yields four haploid derivates (column c). In cases 1 and 2, when the polar bodies do not participate in forming the diploid cleavage nucleus, only the rare absolute male homozygotes carrying ZZ sex chromosomes are viable. The WW constitution is nonviable. In case 3-5 polar bodies participate in forming the diploid nucleus of the parthenogenetic embryo. In such a case females with material genotype may appear. Case 6 refers to ameiotic parthenogenesis when the reduction division is inhibited by heating for 18 min at 46°C (5), while the remaining equational division leaves the maternal genotype unchanged, and this serves as the basis for cloning.

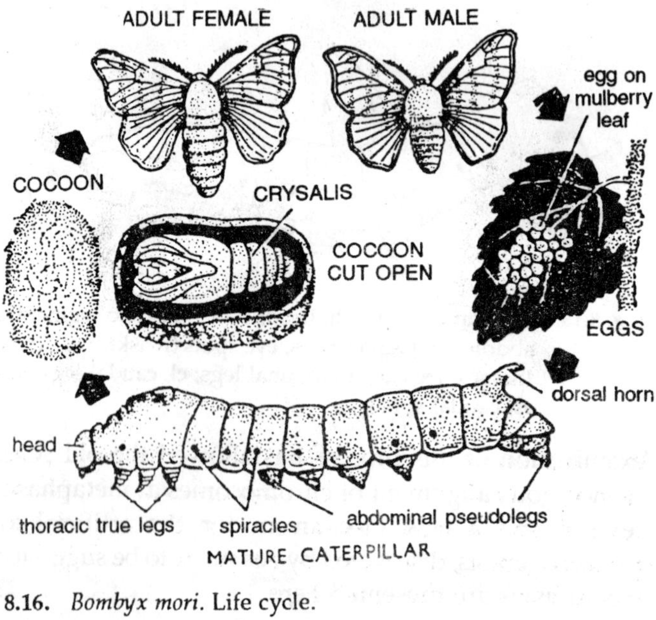

Fig. 8.16. *Bombyx mori*. Life cycle.

Stages of Normal Embryonic Development

A map of the unfertilized *Bombyx mori* egg presumptive regions has been established using the cauterization technique.

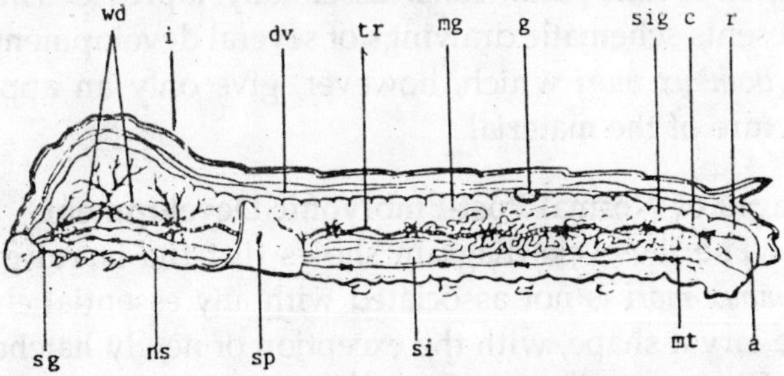

Fig. 8.17. Internal organs of the larva. a, anus; c, colon; dv, dorsal vessel; g, gonad; i, integument; mg, midgut; mt, Malpighian tubule; ns, nervous system; r, rectum; si, small intestine; sg, salivary gland; sig, silk gland; sp, spiracle; tr, trachea; wd, wing disk.

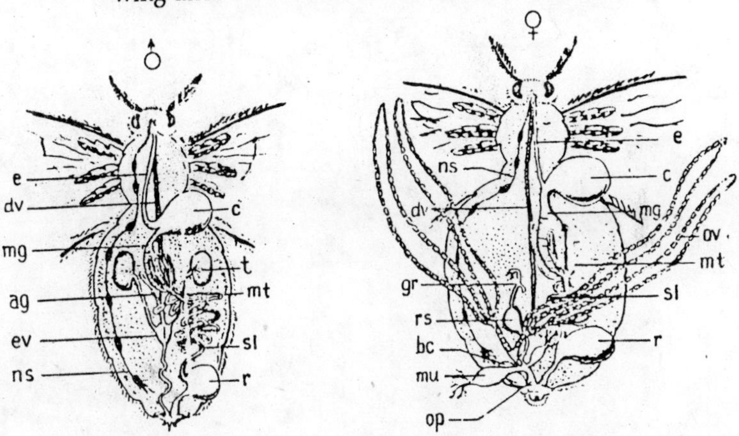

Fig. 8.18. *Bombyx mori*, internal organs of the adult male and female (modified from). ag, accessory gland; bc, bursa copulatrix; c, crop; dv, dorsal vessel; ev, ejaculatory vesicle; gr, glandula receptaculi; mg, midgut; mu, mucous gland; mt, Malpighian tubules; ns, nervous system; e, esophagus; op, ovipositer; ov, ovarioles (note that the relative length of these is 4-5 times greater than shown); r, rectum, rs, receptaculum seminis; si, small intestine; t, testis.

The table of normal development presented here does not pretend to be complete or highly accurate, particularly with regard to time parameters. Essentially, it provides mean data. Presents schematic drawings of several developmental stages of *Bombyx mori* which, however, give only an approximate picture of the material.

Stages of Normal Post-Embryonic Development

The life-cycle diagram shows that the larval growth of *Bombyx mori* is not associated with any essential changes of the larval shape, with the exception of newly hatched larvae which have relatively longer bristles. The outer and inner structure of the larva is shown. Metamorphosis takes place at the pupal stage. The inner structure of the adult male and female.

9

Development of the Honeybee (*Apis mellifera* L)

Honeybee is the most familiar member of the order Hymenoptera. The honeybee usually described in text books is *Apis mellifera,* the common European honeybee. The small sized honeybee, employed for commercial bee keeping (apiculture) in India, is *Apis indica.* Other honeybees commonly found in India are *A. dorrata,* the grant honeybee and A. florae, the little honeybee.

As an experimental animal the honeybee has many advantages. The female, or queen, lays a large number of eggs over a long period extending from spring to autumn. The embryos of one fertilized queen, that is, a queen inseminated by several males, can be obtained for many months and even up to 2-3 years. Although these embryos develop from eggs fertilized by the sperm of several males, the number of males whose sperm is stored in the queen's spermatheca for years is not necessarily large. The spermatozoa of each haploid male are genetically identical. The abundance of larvae and pupae of a similar age in the honeycomb enables sufficient amounts of material to be collected for biochemical studies. Material can be obtained even from a beehive in a town, on a roof, or in an attic, for example.

Normal honeybee males or haploid drones having 16 chromosomes where as the females are diploid and have 32 chromosomes. The drives are extremely suitable for investigating the morphological and biochemical manifestations of recessive mutations. The availability of body pigmentation and eye pigmentation mutations enables genetically marked material to be used. Honeybee embryos have been used in experiments to separate parts of the embryo and for nuclear transplantation. They have been used for studying the nucleus and chromosomes by electron microscopy and determining the amount of DNA in different nuclei. The development of a female larva into a queen, worker bee, or intermediate forms was studied in detail sometime ago, a more sophisticated and up-to-date analysis now seems appropriate. The discovery of a series of multiple alleles of the gene controlling the sex in the honeybee suggests that this species could be particularly interesting for studying the mechanisms of sex determination. The opportunities which the honeybee offers for studying problems of developmental biology remain to be fully realized.

Investigation into honeybee development have an immediate practical aspect since apiculture is a major aspect of the economy. The pollination of field and garden plans by honeybees for crop production is much more important in economic terms than the direct revenue from honey and wax.

Castes and Colony Development

Bees have developed one of the most highly organized societies known to the animal world.

The societies or colonies of the honeybees are *perennial.* At the height of its strength in summer, a good colony of the Western honeybee may consist of 50,000 to 80,000 individuals of which one is the fertile female or the *queen,* a few hundred are the males bees or the *drones,* and the rest are the sterile females or the *workers.* The colonies of the other honeybees consist of small number of individuals, those of the Little honeybee containing only 4,000 to 5,000 individuals.

The Queen

The queen is the only fertile female in a hine, having immensely developed ovaries. She does all the egg laying for the entire colony.

The queen lives for several years. She lays eggs, but cannot collect food, feed larvae or built the nest. The queen's long abdomen contains two large ovaries, each consisting of 110-210 ovarioles. It is generally believed that after the beginning of oviposition the queen does not mate again. The average queen in the high season may lay 800-1500 eggs a day for weeks. A highly productive queen may lay up to 3000 eggs a day, possibly more. She lays about 1,500,000 eggs during her little life.

In the warm nest located in the center of the beehive, the queen lays one egg on the bottom of each empty cell previously cleaned by bees. Hardly ever making any mistakes, she lays fertilized eggs into the worker cells and the queen cell cups, and deposits unfertilized eggs into the drone cells. The ovipositing queen, who has either lost her store of spermatozoa in the spermatheca or did not have any from the beginning, lays only unfertilized eggs into all the cells. Such a queen is usually referred to as a "drone queen."

As a rule, an individual beehive contains the eggs of just one queen. Two queens meeting in a beehive fight until one (usually the older) dies. When the old queen whose egg laying capacity is rapidly decreasing or who unexpectedly dies, is replaced by a new, young and vigorous queen. Some time, the old and new queen may continue to live together peacefully without fighting, but it rarely happens.

Worker Bees

The worker bees belonging to the summer brood live for about a month, while autumn bees live until spring. The number of them in a colony varies from a few thousand to 50,000-80,000, depending on the season and strength of the colony. They collect nectar and pollen from flowers, propolis from resinous buds, and sweet secretions from plant-sucking insects. They bring water, transform nectar into honey, feed all the larvae, queen, and drones, secrete wax from the abdomen, and use it to build the honeycomb. They also clean, heat, ventilate, and guard the beehive; specifically, they prevent worker bees of other colonies from entering their beehive. Worker bees alter the form of their activity according to their age (with marked deviations) and the requirements of the colony. Initially they do a variety of work inside the beehive (household bees) and at the age of 2-3 weeks they become active outside the beehive (field bees).

The two ovaries in a worker bee contain 1-2 ovarioles, but no eggs are formed in them. In special circumstances, for example, if the queen is absent from the colony for a long time, the ovaries of young bees may begin to function. These worker bees are called laying workers; they are capable of laying approximately 20 unfertilized eggs during their life. They lay eggs into any cell, whether or not there are any previously laid eggs already in them; eggs are usually laid onto the cell walls, not on the bottom. These eggs result in the usual drones.

Common (Haploid) Drones

The drones develop from unfertilized eggs in normal colonies only during spring and summer. They vary in number from several hundred to thousands; they are bigger than worker bees, and have no sting. Drones do not participate in the activities of worker bees. They are sexually mature by the twelfth day after leaving the pupae. The amount of spermatozoa in one drone averages 11 million, and the volume of the sperm ejaculate is 1-1.33 mm^3. Drones reared in worker cells (the hunchback brood) are smaller, but their spermatozoa are normal although there are fewer. The haploid drone is hemizygous for all genes. The genotype of all its spermatozoa, with the exception of mutated ones, is identical.

On days when the weather is good the drones fly out of the beehives and gather in special places, sometimes far from the apiaries. Here they find young queens on their wedding flight. The drones mate with the queens in mid-air; and die immediately afterward. During the spring and summer, each colony admits drones into its beehive and feeds them. At the end of the swarming season, particularly when the period of nectar and pollen collection is over, the worker bees no longer feed the drones and do not admit them into the beehive (queenless colonies excepted).

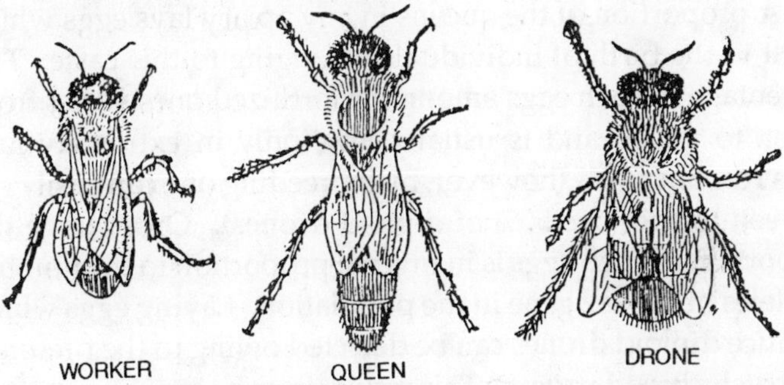

Fig. 9.1 Castes of honeybee.

TABLE 9.1

Dynamics of Development (in days) of Four Honeybee Castes in a Beehive at a Temperature of about 35°C (from, with modifications).

Brood	*Development phase*	*Worker bee*	*Queen*	*Drone* usual (haploid)	*Drone* diploid
Open (cells are open)	Egg (embryo)	3	3	3	3
	Larvae	6-7	5	7-9	0-1
Sealed (cells are sealedwith cups)	Prepupa	2-3	2	4-6	-
	Pupa	9	6	10-14	-
Total		20-22	16-17	24-32	3-4

Note. Variation in time of development is due predominantly to temperature fluctuations. This variation most affects the drone brood, which is frequently located at the border of the heated nest.

Diploid Drones

These drones develop from fertilized eggs, homozygous for any of the multiple alleles of the sex gene. This is the fourth caste of the honeybee discovered between 1951-1963 by

Mackensen and Woyke. It was previously unknown. Members of this caste do not usually reach maturity because nurse bees have an instinct to kill the larvae of diploid drones. Initially these larvae are fed, but in the first hours after hatching they are killed and eaten by the nurse bees.

A proportion of the queens in any apiary lays eggs which result in the birth of individuals belonging to this caste. The percentage of such eggs among the fertilized ones varies from queen to queen and is usually low, only in extremely rare cases reaching 50% (however, by inbreeding one can easily get a queen to produce 50% of diploid drones). On average the proportion of such eggs is in inverse proportion to the number of alleles of the sex gene in the population. Laying eggs which produce diploid drones can be detected owing to the presence of a patchy brood pattern. This results from a certain percentage of empty cells in a good honeycomb that contain pupae or larvae of approximately the same age. The presence of individuals belonging to this caste represents a rudimental mode of sex determination that existed previously.

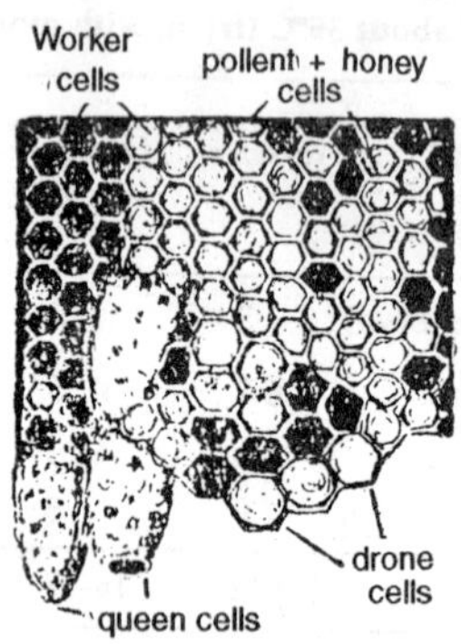

Fig. 9.2. A portion of the honey-comb showing various types of cells.

Living Conditions in the Beehive : Cell and Honeycomb

Before the adult forms emerge from the pupae, development proceeds in the wax cells of the three main castes, which vary somewhat in size for different subspecies and races. The duration of development for the four castes in honeycomb cells is given in Table 9.1. A mass of adjacent worker cells is arranged in strict vertical layers in the honeycomb; the hexahedral cells are located horizontally in the honeycomb. The drone honeycomb consists of large cells with a configuration similar to that of the worker's cells.

The queen cells are oriented vertically with the holes directed downward. They are located individually, usually near the edge of the honeycomb. They do not touch each other, so stay rounded. It contrast to the worker and drone cells, the queen cells are built during the larval growth stage and used only once. The construction of a swarm queen cell begins with a queen cup being made, which has a hole with a diameter close to that of a worker cell, but somewhat narrower than the largest inner diameter of the cup. The empty cup can be left for a long period of time. After the egg is laid into it, the "sown cup" is rapidly transformed into an open queen cell containing a larva, and then into a sealed queen cell with a pupa. If the queen is lost, the colony lays down emergency queen cells for the available young larvae. The number of queen cells varies from 2 or 3 to 100 or more, depending on the quality of the colony, including hereditary factors.

Cells which have been used several times for brood rearing become darker and reduced in volume, owing to the presence of cocoon remnants; consequentially the bees developing inside them are somewhat smaller.

The spaces between the ledges and walls (or the ceiling) in the beehive measure 8-9 mm in width or height. They are left free by honeybees, larger spaces are taken up by the honeycomb or connections, while smaller ones are glued by propolis ("bee-glue," derived from plant resins) and wax (the space between the honeycombs is wider). The discovery of this "bee space" in the 19th century enabled beehives to be developed which could be disassembled with the honeycomb in replaceable frames.

As well as for breeding, the honeycomb is used by the bees for processing or storing food such as nectar, honey, and stored pollen. The worker cells as well as the drone cells store honey. These cells are often longer than normal. Cells with mature honey are sealed by a continuous waxcomb and cells with the pupating brood are covered with porous cups consisting of wax (60% for worker cells and 10-15% for drone cells) mixed with plant fibers and pollen grains.

Cells containing brood are located in the center of the beehive, where they form a nest of varying volume but with a more or less round shape. The temperature in the nest is maintained by the bees at 34-35.5°C. The space under the honeycomb may be free. The aperture is normally located laterally at the lowest level of the beehive, whereas the store of honey is usually placed by the bees in the upper part far from the aperture. In winter the bees collect in a round cluster occupying several adjacent honeycombs close to the stored honey. They maintain the temperature from 6°C at the cluster periphery to 28-32°C in its center.

Colony and Individual

Not only an individual, but the colony as a whole, comprises a biological unit of the honeybee species. The behaviour of an individual worker bee or a queen depends not only on its phenotype and external conditions, but also on the status of the colony in general. The status depends on the phase of the annual and circadian cycle of the colony, heredity, the number of adult household and fields bees, the amount of open and sealed brood in the honeycomb, the quantity of food, the free space for building honeycombs, the design of the beehive, weather conditions, availability of nectar and pollen, presence or absence of artificial food in the beehive, and possible irritating factors.

The number of eggs deposited by the queen depends on the strength of the colony (the number of bees in the colony) and the stage of colony development. For example, while

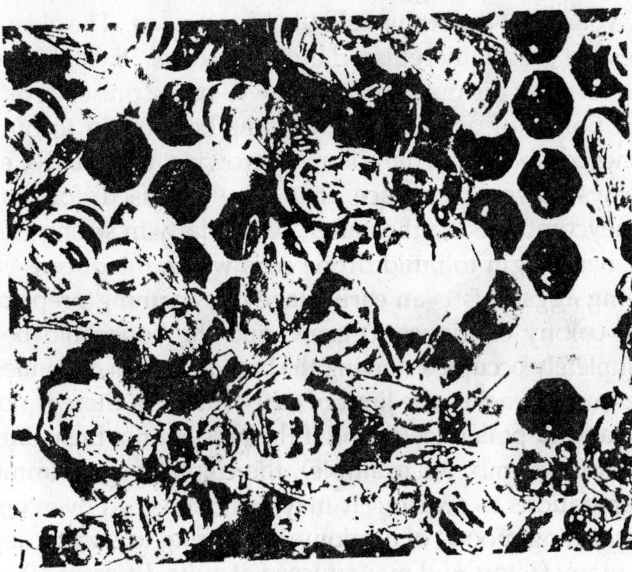

Fig. 9.3. A close-up view of the crowded dance floor of the bees. The antennae of some of the bees touch the abdomen of the dancing bees to determine the rhythm and direction of the dance.

preparing to swarm, the colony enters the "swarming mood", during which the queen temporarily discontinues oviposition. The queen's maximal potential may only be realized in a strong colony.

The Annual Cycle of Colony Development

Basics of Care

By the end of winter (February-March) the queen begins oviposition by forming a nest, which is small at first. Initially the queen only deposits eggs into the worker cells. Worker bees spend the first warm spring day "flying around," emptying their colons, the content of which by the end of winter weights up to a third of the body weight (healthy bees do not defecate in the beehive). In spring the queen gradually increases the intensity of oviposition and the volume of the nest, and the appearance of young bees increases correspondingly; therefore, the volume of the beehive should be increased. The old autumn bees are gradually replaced by young ones. Except in winter, bees usually die outside the beehive. In the presence of nectar and pollen or artificial feeding, the bees begin to build the honeycomb. When the colony is growing, the beehive should always have free space that may be used for building honeycombs. Soon thereafter, particularly in strong colonies, the bees begin to build drone cells which the queen uses for laying eggs; this is an early stage of swarming preparations. The colony continues to grow, and the household bees are completely occupied nursing the brood; growth continues until the queen reaches the limit of oviposition. If there is no large nectar and pollen yield, and if the bees are not busy (building the honeycomb, for example), the colony begins immediate preparations to swarm. It may not be easy to overcome the "swarming mood" of a colony. Preparing to swarm depends on many factors and may proceed at quite different rates. The cups are built well in advance, but not necessarily used. The queen does not deposit eggs in these cups immediately after they are built, and not simultaneously either. She reduces the rate of oviposition and then stops it altogether. When the queens in the queen cells are ready to emerge, the bees, on a clear day, will fly out of the beehive *en masse* along with the old fertilized queen and settle nearby, for example, on a tree branch. Reconnaissance bees inform the swarm about suitable sites and, after a while, the swarm rises up again and flies slowly to the selected site, sometimes several kilometers away, and settles there. The apiculturist should collect the settled swarm without letting it escape, and place it into a free beehive in the evening. After the first swarm the colony may release a second, and sometimes even subsequent swarms, all of which include virgin queens. The colony which released the swarm also remains with a virgin queen. The virgin queen flies out to mate at the age of 6-8 days. In 1 or 2 days she mates with roughly a dozen drones. If the mating does not take place for a month, the queen begins laying unfertilized eggs and becomes a "drone queen". Oviposition may be induced artificially by CO_2 treatment.

Swarming is a form of reproduction in the honeybee colony. The proportion of swarming colonies at the apiary depends on heredity and seasonal conditions, but mainly on the system of apiculture, in other words, the apiculturist. Modern apiculture uses artificial swarming rather than natural swarming, which results in losses.

During the swarming season 2-year-old, or sometimes even 1-year-old, queens are replaced by young ones. When necessary, the colonies themselves conduct the replacement of the old or unsatisfactory queen (she is classed as such if, for example, one of her legs is damaged) by rearing the young queen and replacing the old one independently on swarming. In many regions of the USSR, the end of the swarming season coincides with the main period of nectar and pollen collection. This mobilizes the colony for nectar collection and sometimes results in the arrest of swarming preparations. During nectar collection the honeycomb is needed not only to hold ripe honey, but also to store nectar and to process it. The end of the honey-collection period, particularly if it has been done quickly, enables the colonies to prepare for winter. The workers drive the drones out of the beehive and may even throw out the drone brood. If some nectar collection or artificial feeding carries on, only workers cells continue to be built. The rate of oviposition by the queen slows down. The artificial feeding of colonies, which is carried out before the winter season to increase the stores or replace honey with sugar, should not be started too late. When the bees transfer food into the honeycomb, the queen's egg laying is stimulated. It is important that the young bees resulting from this stimulation reach field age and have time to fly around before winter. Optimally, before winter, the main proportion of the colony should comprise young bees which have not participated in the nectar and pollen collection or even in the transfer of artificial food into the honeycomb. These bees should, however, winter after flying around for a time during which they are able to empty their intestines. Honeybee colonies may be left to winter in the open after making additional thermal insulation from the outside; another possibility is to leave them in a dry quiet place at a temperature ranging from 0-4°C. If the temperature rises, the bees become excited and a cluster will disintegrate. For wintering, each colony should have 18-30 kg. of honey or sugar transferred by the bees into the honeycomb; it is also recommended that a frame with stored pollen be left. This is used as protein food for growing larvae at the end of the winter. The bees winter well on flower honey and better still on sugar, but they may perish during the winter if they use honey from the sweet secretions of plant-sucking insects; in the summer this honey is quite harmless. The increased content of dextrins in this honey, or in the flower honey with its admixture, results in the colon of the wintering bees being overfilled. This leads to the need to drink and causes diarrhea and the death of the colony.

Keeping bees is facilitated by making sure that the normal colony always maintains the necessary conditions in the beehive, such as high temperature and sufficient humidity for the brood. One should supply colonies with enough good honeycomb, space for them to build, and food; they should be protected from irritants. Colonies with only household bees should, in addition, be supplies with water. Details of apiculture are not presented here; they can be found in more specialized treatises.

GAMETES

Structure of Eggs

The honeybee egg is shaped like a slightly bent cylinder rounded on its poles, narrowing a little at the posterior part near the future ventral side of the embryo. The egg moves in the ovariole and oviduct, posterior pole forward. The laid egg is attached to the cell bottom at the posterior pole. The convex side of the cylinder corresponds to the future ventral side of the embryo. The size of the egg varies for different subspecies of the honeybee and differs with each season, too. During spring and autumn the size of the egg in European bees is 1.6 x 0.33 mm (for some queens, the length may attain 1.8 mm), while in summer, during intensive oviposition the size is 1.4 x 0.32 mm. The weight of the deposited egg average 0.13 mg, dropping by the end of its development to 0.1 mg.

The egg of the honeybee is surrounded by two membranes. The chorion, an elastic but tough external membrane, gives the egg its main protection. The chorion is produced by follicular cells, the boundaries of which are imprinted on its surface. The inner vitelline membrane is not so thick. The anterior pole contains the micropyle or "the micropylar region," a definition suggested by Snodgrass.

The nucleus is located under the egg surface on the convex side at a distance that is somewhat less than the width of one egg from the anterior pole, in a small conelike thickening of the periplasm; both maturation divisions, which are not accompanied by the extrusion of polar bodies, proceed while the nucleus is in this position. During oviposition the mature unfertilized and fertilized eggs are at the stage of the first maturation division.

Unfertilized eggs can be obtained in the way described by Tryasko. The ovipositing queen is removed from the colony for a few hours in a glass jar containing a drop of honey on the side, or placed into a microversion of the beehive which has mesh instead of a bottom, and contains 40-60 bees. The queen deposits several dozen, sometimes up to 100, eggs. They are not placed into cells and, according to available observations, are always unfertilized.

Structure of Spermatozoa and Method of Collection

Microscopic examination of the sperm, diluted with a small quantity of water, reveals the movement of spermatozoa.

Each sperm consists of a long head, neck and tail (total 240 μm).

The sperm is best obtained from mature drones just returned from a flight. A drone of known origin can be grown in a beehive equipped with a diaphragm through which only worker bees can pass; this prevents drones from flying around. Before extracting the sperm, the drones should have emptied their colons so as to prevent sperm contamination; furthermore, they should have filled their abdominal air sacs with air, which helps to turn the endophallus inside out. To achieve this they have to fly around against a window or in a special setup; alternatively, holding the drone by the thorax and head, one may sweep it in the air, allowing his wings to work.

To extract sperm, the drone should be carefully rotated with two fingers and either his thorax should be slightly pressed or his head should be torn off. As a result, his genitals turn inside out, depending on the state of drone readiness; the process occurs either at once or in two steps, the horns being initially everted, followed by the endophallus which is everted upon subsequent pressing. The surface of the endophallus may contain a creamy drop of sperm and a white droplet of mucus. The sperm is taken by using a thin pipette or capillary. In the latter case, it stays viable longer without drying up. The sperm can be kept for a long time under artificial conditions.

Obtaining drones in early summer presents no problem. By the end of summer and during autumn normal colonies do not keep drones and live only in queenless colonies. During this period the drones may be reared from laying worker bees, as described by Schasskolsky and Alber. Using this technique, drones can be obtained until October in the central regions of Russia, the last batches may be reared from a sealed brood in an incubator.

Fertilization, Artificial Insemination and Mosaicism

Egg activation in the honeybee is not associated with

fertilization and all eggs, whether fertilized or unfertilized, develop immediately after they are laid.

Fertilization

The egg is inseminated immediately prior to oviposition (0.1-0.5 min. prior to laying). As it passes through the oviduct, the egg comes into contact with a very small portion of a secretion which is produced by the accessory gland and contains spermatozoa; this secretion is delivered by a special apparatus consisting of ducts and specially controlled valves. The whole process of laying the egg, including the preceding inspection of the cell, which requires the anterior part of the body to be placed into it, takes the queen less than a minute to perform. There are reasons for believing that egg insemination takes place no earlier than 0.5-0.1 min. prior to laying.

When the queen is young and only beginning to lay eggs, the number of spermatozoa entering the egg is greater than in the case of old queens. According to Woyke 60% of the eggs receive only one spermatozoon, while the remaining 40% receive more. Petrunkevich has found up to 7 sperms in fertilized eggs. Nachtsheim believed that 3-7 sperms per egg is a normal number, and even 10 spermatozoa per egg occurs quite frequently. The first meiotic division ends immediately after the egg is laid. This is followed by the second maturation division, and, simultaneously, by the division of the first polar nucleus. The female pronucleus is located deeper, and moves further inside the egg, than the other nuclei. The remaining three polar nuclei stay near the surface, and two of them fuse. In the fertilized egg the female pronucleus meets the nucleus of one of the sperm cells which have entered the egg. The female pronucleus moves further into the unfertilized egg, almost up to the future dorsal side of the embryo. Cleavage divisions begin in the unfertilized eggs approximately 45 min after they occur in the fertilized egg.

Using sperm to wet the anterior pole of the freshly laid unfertilized egg enabled Barrat to achieve egg fertilization outside the queen's body. Queens could be reared from such eggs.

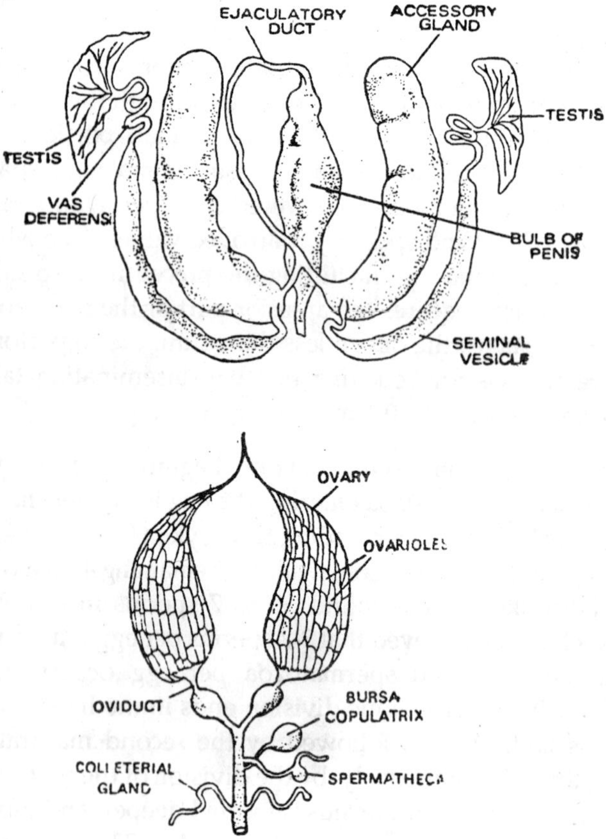

Fig. 9.4. Reproductive organs.

Artificial Insemination

The artificial insemination of queens has been developed in detail in recent years. It is usually performed by extracting sperm from several drones in accordance with the natural polyandry of the honeybee this polyandry involves mating each young queen with several drones within one or a few days.

Insemination with the sperm from 8-12 drones yields a

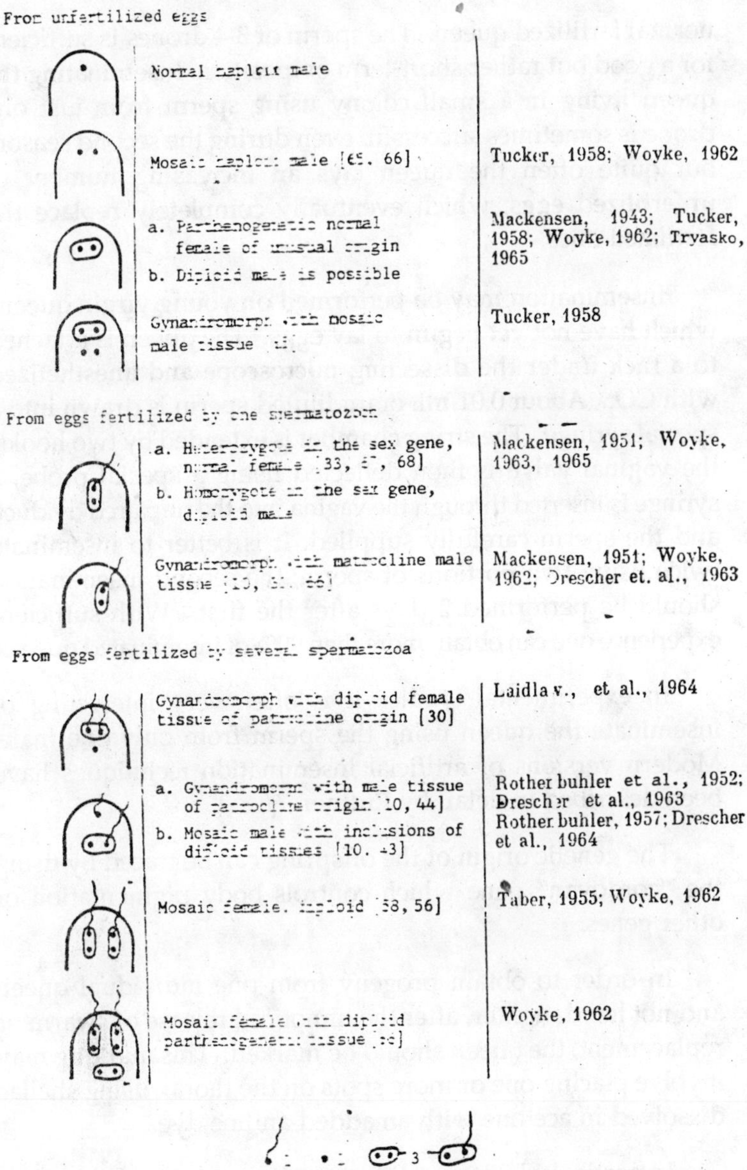

Fig. 9.5. Origins of normal and abnormal bees. 1) Spermatozoon; 2) female pronucleus; 3) zygote.

normal fertilized queen. The sperm of 3-4 drones is sufficient for a good but rather short-term oviposition. Inseminating the queen living in a small colony using sperm from just one drone is sometimes successful even during the second season, but quite often the queen lays an increasing number of unfertilized eggs, which eventually completely replace the fertilized ones.

Insemination may be performed on young virgin queens which have not yet begun to lay eggs. The queen is attached to a rack under the dissecting microscope and anesthetized with CO_2. About 0.01 ml. of undiluted sperm is drawn into a special syringe. The string chamber is extended by two hooks, the vaginal valve is then deflected using a special probe, a syringe is inserted through the vagina into the unpaired oviduct, and the sperm carefully supplied. It is better to inseminate twice using two portions of sperm; the second insemination should be performed 2 days after the first. With sufficient experience one can obtain more than 90% of inseminated queens.

In experimental studies it is often very interesting of inseminate the queen using the sperm from only one male. Modern versions of artificial insemination techniques have been described in detail by Ruttner.

The genetic origin of the offspring can be traced by using the "cordovan" gene which controls body pigmentation or other genes.

In order to obtain progeny from one individual queen and not her daughter, after the unnoticed release of swarm or replacement, the queen should be marked. This marking may involve placing one or more spots on the thorax using shellac dissolved in acetone with an added aniline dye.

In particularly demanding experimental situations, when only the descendants of one particular queen are desired, one should get rid of individuals derived from the progeny of laying workers. This is particularly important because a small percentage of unfertilized eggs develops in the females. To achieve this, one has to redistribute the frames with brood in order to establish a colony consisting of bees which differ in colour from the daughters of a particular queen; for example, a colony consisting of bees belonging to another race may be formed.

The experimental rearing of adult bees is performed in an incubator or beehive, containing a honeycomb with the sealed brood placed in an insulated box made of fine mesh. The drones may be reared in a box formed by a diaphragm. The queens may be reared in the queen cells.

Mosaics and Gynandromorphs

These have been repeatedly described in the honeybee. They are produced when more than one active sperm is found in the egg, or when meiotic divisions are impaired, or when the nuclei of polar bodies show abnormal behaviour (facilitated by a temporary drop in temperature).

Obtaining and Incubating Developing Eggs

Eggs at a known stage of development can best be obtained in the laboratory by using a special observation beehive; the queen can be observed laying eggs, the cells used noted and time of egg deposition recorded. Subsequently, the cells are taken out, and the eggs are withdrawn and placed into an incubator.

Another technique, suggested by Petrunkevich as long ago as 1903, can be used. A good honeycomb devoid of eggs is placed for one or more hours into the center of the normal (preferably strong) colony; the queen uses the comb to deposit eggs. The comb may be placed into a chamber formed by a diaphragm which does not restrict the worker bees but limits the queen's movements. Egg-laying is more successful if the adjacent frames (combs) contain abundant unsealed brood and stored pollen as well as numerous nurse bees, while the cells in the comb used by the queen are cleaned (polished) by the bees themselves. Such a frame can sometimes be taken from another colony whose queen has not yet started oviposition.

The comb with the eggs taken out of the beehive should be protected from direct sunlight, wind, and prolonged chilling. If the eggs are not used immediately, they should be prevented from drying up by wrapping the comb in a wet towel. It is more convenient to take eggs out of dark cells where they are attached to cocoon remnants rather than wax. Using a sharpened dental scalpel, part of the cell base containing the egg can be excised and picked up using small watchmaker forceps. Care should be taken not to touch the egg. The excised piece is transferred to a Petri dish covered with a layer of wax. The cell base with the attached egg is glued to the wax and a piece of wet cotton wool is placed nearby to maintain high humidity. Eggs may also be incubated in paraffin oil. For this purpose, the region of the cell base containing the egg is held in watchmaker forceps while the egg is carefully pushed by the rounded tip of a glass microneedle into an individual cup containing the warmed oil. It is better to use relatively light paraffin oil, since heavier types cause developmental abnormalities. When the eggs are submerged in oil, they are prevented from drying up. Incubation is conducted at 35.5°C. Embryos placed into oil become translucent. When the lens of a microscope is placed into oil, one can see the cells and nuclei on the surface of the embryo.

Normal Embryonic Development

The life-cycle of the honeybee consists of an embryonic period as well as larval, pupal, and adult stages.

Embryonic development has been studied in great detail using fixed preparations by Nelson and Schnetter. Schnetter distinguished four periods subdivided into 21 stages, and described both the outer and inner structure of the embryo. DuPraw traced sequential changes in individual embryos at 35.5°C using time-lapse cinematography and distinguished 10 stages of development; each was divided into early, middle, and late substages. Below the main structural characteristics of embryos, as well as their changes during each stages, are given according to DuPraw.

Stage I : Duration about 4.5 h. The stage begins with oviposition and ends when the cleavage nuclei and the surrounding cytoplasm (the energids) begin to migrate rapidly from the anterior part of the egg toward the posterior pole. Two maturation divisions, karyogamy and the first three synchronous cleavage divisions, take place during this period in the anterior part of the egg (cleavage center). There is a 30-35 min. interval between divisions. In the middle of the stage the protoplasm becomes slightly compressed in the anterior direction, leaving a clear space full of liquid under the membrane near the anterior pole.

Stage 2: Duration about 3.6 h. This state begins during cleavage division IV a a sudden rapid migration of energids from the anterior position toward the posterior pole of the egg. Synchronous cleavage divisions V-VII continue during this migration and the nuclei disperse uniformly throughout the egg. This process takes 110-115 min. At the beginning of this stage a clear space appears at the posterior pole owing to a certain compression of yolk. At the 128-nuclei stage, nuclear migration toward the posterior pole stops and the nuclei begin to move radically from the protoplasm toward the surface, while the synchronous cleavage divisions continue (VIII-X). At the 1024-nuclei stage the energids enter the periplasm at the surface of the embryo. The radial migration of nuclei takes 100-115 min. At first the energids reach the surface in the presumptive thorax region, which is the differentiation center. During this stage the embryo has a characteristic spotted appearance.

Stage 3: Duration about 2.5 h. This stage represents the period of "superficial cleavage" or blastoderm formation. It begins with the surface around each cleavage nucleus being indented; protuberances, particularly distinct in the anterior

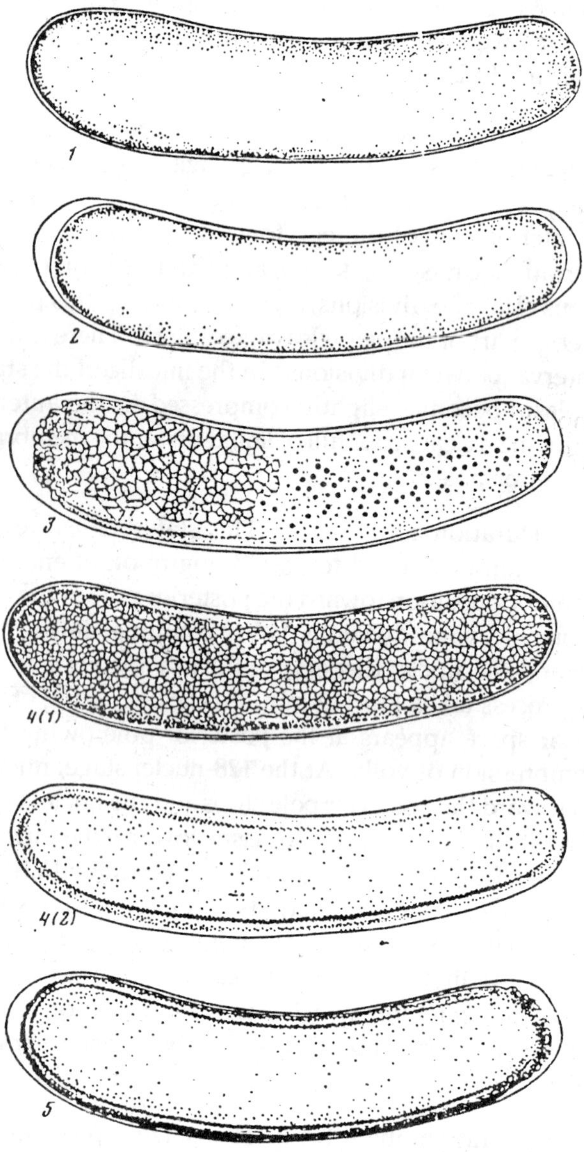
1
2
3
4(1)
4(2)
5

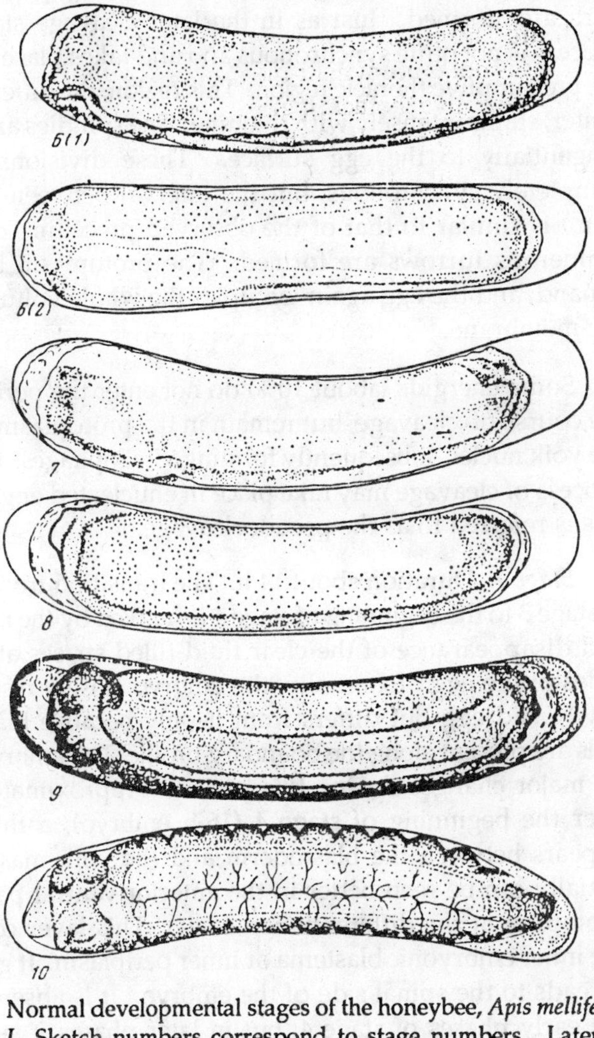

Fig. 9.6. Normal developmental stages of the honeybee, *Apis mellifera L.* Sketch numbers correspond to stage numbers. Lateral views are given of all stages; 6 (2) is viewed from above. Anterior pole is on the left; posterior pole, attached to cell, is on the right. The convex side of the embryo is ventral. 1) Beginning of stage 1; 2) middle of stage 2; 3) early phase of stage 3; 4 (1) early phase of stage 4; 4 (2)) late phase of stage 4; 5) middle of stage 5; 6 (1) early phase of stage 6; 6 (2) middle of stage 6; 7, 8, 9) middle of stages 7, 8, 9, respectively; 10) late phase of stage 10.

part, are formed. Just as in the later cleavage stages, this process is not quite synchronous and and takes place earlier in the presumptive thoracic region. This region, the differentiation center, stains intensely with thionine. The spindles are located tangentially to the egg surface. These divisions are not completely synchronous, but the interval between them (35 min.) is similar to that of the observed preceding divisions. Numerous furrows are formed. The protoplasm begins to expand, and the egg again occupies the whole volume inside the membrane.

Some energids (about 10%) do not enter the periplasm in the course of cleavage, but remain in the protoplasm forming the yolk nuclei, subsequently forming vitellophages. A similar process of cleavage may take place in enucleated eggs as well; this is referred to as the pseudocleavage.

Stage 4: Duration about 14 h. The transition from the end of stage 3 to the beginning of stage 4 is marked by the resorption and disappearance of the clear fluid-filled spaces at the two poles of the egg. Externally, the embryo closely resembles a newly laid egg but differs from it only because blastoderm cells are present along the boundary profile. The embryo shows no major changes during this period. Approximately 5.5 h after the beginning of stage 4 (16-h embryo), a thick layer appears between the blastoderm and the yolk mass, which initially can be seen better in the anterior ventral part of the protoplast at the differentiation center. This layer comprises the inner embryonic blastema or inner periplasm. It gradually spreads to the spinal side of the embryo. It is absent during the early phases of stage 4, but in later phases it surrounds both poles of the embryo.

By the end of stage 4, the protoplast again contracts anteriorly. This is accompanied by the appearance of a clear region near the anterior pole. The blastoderm cells begin to rearrange into a layer of columnar cells. They soon undergo a series of asynchronous mitoses, the spindles of which are located obliquely to the surface; this yields two cells layers. The only region of the blastoderm which does not change is the narrow longitudinal streak along the center line of the spinal side. It does not form any part of the embryo, but may contribute to the extra-embryonic membrane (amnion-serosa); in this region only one layer of spheroid or cuboid cells can be found. The phase of the two-layer embryo lasts about 4 h. During this period the cells are located in a very narrow space between the external chorion and the inner basal membrane. The total number of cells reaches 19,000 in the diploid worker bee and 34,000 in the haploid drone. This difference is equivalent to 2-3 additional mitotic divisions per cell. However, one cannot be sure that all cells participate similarity in proliferation.

Beginning at about 16 h, two major changes develop more or less simultaneously in the blastoderm structure; first, the inner periplasm makes its appearance beneath the basement membrane in the presumptive thorax region; second, in the same region, the blastoderm cells begin to rearrange themselves as a single layer of almost identical columnar cells with uniformly apical nuclei. Both of these changes progress slowly until the two layers converge, resulting in the disappearance of the basement membrane followed by the fusion between the inner periplasm and the cytoplasm of the blastoderm cells. At this stage, the blastoderm cells, which previously were discrete and typically uninucleate, are once more in continuity with each other through the inner blastoderm layer. The exception is the narrow longitudinal mid-dorsal strip. Even in late embryo the inner periplasm tapers to an end on either side of this strip. The other parts of the blastoderm at any given moment can be found in widely different transition stages until the inner periplasm formation is completed. By that time the plasma layer has a rather grainy appearance and a regular profile: it is thickened in the region of the differentiation center and tapers in thickness toward the anterior and posterior pole of the embryo and from the ventral midline toward the dorsolateral areas. This tapering may be regarded as a manifestation of anterior-posterior, medial-lateral, and ventral-dorsal gradients in the embryo. The mapping of rudiments has been performed by Schnetter and Reinhardt. Little is known about the formation of the inner periplasm and the interrelationships between its appearance and its transition to the columnar blastoderm. At the end of stage 4, the honeybee blastoderm is similar in structure to the *Drosophila* periplasm, which has incomplete columnar cells arising from the inner periplasm during superficial cleavage and are separated from each other by flask-shaped furrows. After the appearance of the inner periplasm until the latter half of stage 6, no further mitoses can be observed in the embryo. At the end of stage 4 or the beginning of stage 5 the drones can be distinguished from worker bee by the size of their cells; in the worker bee they are about twice as large as those of the drone.

Stage 5: Duration about 8.3 h. When the embryo is approximately 24-h old, a new yolk contraction begins. This is accompanied by the slow reappearance of clear fluid-filled spaces at the anterior and posterior poles. By the beginning of stage 5, the periplasm covers the whole egg with a layer of cells. No significant outer changes takes place at this stage, but many important morphogenetic movements are being prepared. They take place at the next stage (stage 6). This period signals the re-establishment of basal cytoplasmic membranes which complete the separation of blastoderm cells from the ventral yolk mass; the inner periplasm is distributed more or less uniformly between the surface cells. In the middle of stage 5, traces of periplasm are still conspicuous in the distinctly granular basal cytoplasm of the neighbouring blastoderm cells. By the end of this stage, the granular plasm can no longer be distinguished and is apparently replaced in many or all of the cells by a single large vacuole located basally to the nucleus. This vacuole persists during stages 6 and 7 and survives longer in the ectoderm than it does in the mesoderm. DuPraw assumes that the inner periplasm is not absorbed by the blastodermal cells, but incorporated as such.

Stage 6: Duration about 6.2 h. No significant changes take place at the beginning of stage 6. Longitudinal sections show an ellipse of several hundred virtually identical columnar cells enclosing the central yolk mass. The nuclei in these cells are located apically. The most differentiated cells of the embryo, which possess translucent vacuolized cytoplasm, form a cap over the anterior pole.

The map of rudiments in the blastoderm at the beginning of stage 6 was described by Schnetter. It consists of six longitudinal strips made up as follows. A mid-ventral strip occupying about one sixth of the egg circumference; its anterior and posterior tips form the endoderm, while the middle part forms the mesoderm. Two dorsolateral ectodermal strips, each with a width of about one-third of the egg circumference, move during stage 6 in the ventral direction and cover the middle part of the ventral mesoderm; these two strips form the spinal parts of the extra-embryonic membrane (amnion-serosa) which are stretched into squamous epithelium as the ectoderm migrates ventrally. A narrow mid-dorsal strip, according to Nelson forms a syncytium on the surface of the yolk mass.

The beginning of stage 6 is marked by a sudden ventrally directed migration of blastoderm cells at the anterior pole. The anterior midgut rudiment appears as a conspicuous thickening in the anteroventral blastoderm. The similar but less-conspicuous posterior midgut rudiment soon appears near the posterior embryo pole. After the appearance of the anterior midgut rudiment the cap of vacuolized cells lying over the anterior pole separates from the yolk, which is covered by the basal membrane and forms a hoodlike cell layer between the chorion and the protoplast; this is the anterior part of the amnion-serosa which is continuous with the two dorsolateral membrane rudiments. More or less simultaneously with the separation of the anterior serosa the two ventrolateral sheaths of ectoderm begin to migrate ventrally over the mesodermal plate. Early in stage 6 these ectodermal folds can be seen as short longitudinal borders lying ventrolaterally just behind the anterior midgut rudiment, thereafter the folds gradually become longer in a posterior direction while they move slowly toward the ventral midline.

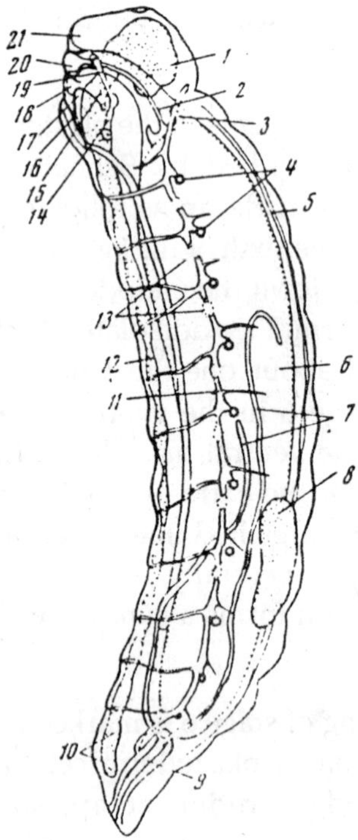

Fig. 9.7. Honeybee larva immediately after hatching. 1) Suprapharyngeal ganglion; 2) foregut; 3) mushroomlike body; 4) stigmata; 5) heart; 6) midgut; 7) Malphigian tubules; 8) ovary; 9) hindgut; 10) fused ganglia 12 and 13; 11) trachea; 12) ventral nervous chain; 13) endocytes); 14) tectorium 2; 15) subpharyngeal ganglion; 16) weaving gland, 17) maxilla; 18) labium; 19) tectoniumi; 20) mandible; 21) labium.

From the ventral side the embryo, at this stage, exhibits a conspicuous hourglass configuration which is clearly visible under oblique lighting. By the middle of stage 6 the ectodermal folds from each side meet ventrally in the anterior third of the

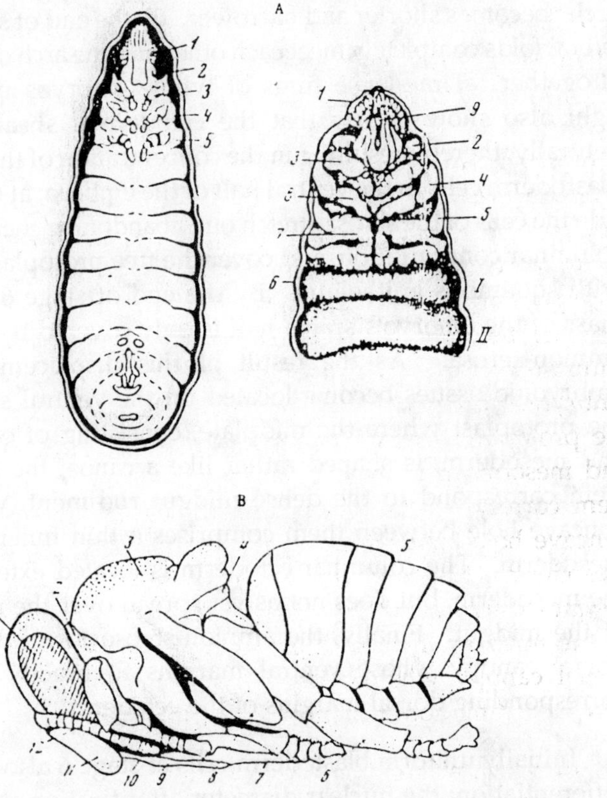

Fig. 9.8. Prepupa and pupa of the honeybee. A Prepupa, ventral view. 1) Antenna; 2) compound eye; 3) foreleg; 4) forewing; 5) hind wing; 6) larval integument; 7) hind leg; 8) mid-leg; 9) oral appendages; I, II) abdominal segments. B. Pupa, lateral view. 1) Compound eye: 2-4) thorax — anterior, middle, and posterior regions, respectively; 5) abdomen; 6) hind leg; 7) mid-leg; 8) wing; 9) proboscis; 10) antenna; 11) mandibles; 12) antennae.

embryo (differentiation center), where a full two-layer condition is first established, with the ectoderm overlying the mesoderm. Now the embryo acquires the shape of a "gothic arch." The fusion of the lateral folds spreads from the thoracic region in a posterior direction. The protoplasm contraction, which began at the end of stage 5, gradually progresses and the "gothic arch" becomes shorter and narrower. By the end of stage 6 the lateral folds completely meet each other and the arch disappears altogether. Time-lapse films of whole embryos in oblique light also show clearly that the ectodermal sheaths move ventrally, thereby resulting in the concentration of the original blastoderm cells in the ventral half of the embryo; at the spinal side the cells of the serosa stretch out, abandoning their original columnar configuration, and cover the free protoplast surface with squamous epithelium. By the end of stage 6 the yolk mass of the embryo's spinal half is only covered by this thin amnion-serosa. As the result of this displacement, true embryonic tissues become located on the ventral surface of the protoplast where the midplate, consisting of endoderm and mesoderm, is shaped rather like a canoe, the bow and stem correspond to the dense midgut rudiment, while the concave hole between them comprises a thin inner layer of mesoderm. The columnar ectoderm is located externally to the mesoderm, but does not as yet spread over the rudiment of the midgut. Finally, the amnion-serosa forms a kind of dorsal canopy whose ventral margins fit exactly into the corresponding dorsal margins of the ectoderm.

Initially uniform blastoderm cells at stage 6 also undergo differentiation; the nuclear diameter after fixation reaches 10 μm in the columnar ectoderm, 7.5 m in the anterior and posterior endodermal aggregates, but only about 6 μm in the mesoderm. DuPraw described the changes in the ectodermal and mesodermal cells during this period, changes which may relate to inductive interactions between them.

Stage 7: Duration about 7.5 h. During this period the ventral amnion-serosa is gradually detached from the embryonic tissues. In profile, embryos of this stage show a conspicuous "dent" with a transverse groove behind it. The groove is formed at the posterior margin of the amnion-serosa. The amnion-serosa completely closes the whole anterior part of the embryo up to the transverse groove on the ventral side, it is located only dorsally behind this groove, where it is still attached to the lateral margin of the ectoderm. As the amnion-serosa continues to grow, the groove slowly move toward the posterior pole. The early, middle, and late phases of stage 7 can be distinguished by their positions. By the end of stage 7, the amnion-serosa covers both poles of the embryo and is located in the fluid-filled spaces between the protoplast and the chorion.

The protoplast continues to shorted throughout stage 7. Correspondingly, the canoe-shaped embryo curves dorsally around the anterior and posterior poles to such an extent that both midgut rudiments come to lie on the dorsal surface of the yolk mass.

Stage 8: This is a short transition period lasting about 2.4 h. It begins with the amnion-serosa covering the embryo's posterior pole. The protoplast contracts to a minimum during this stage. An ectodermal thickening corresponding to an unpaired rudiment of the labium forms near the anterior pole. The midgut forms from the anterior and posterior midgut rudiments. Neither the head nor the paired structures have appeared yet.

Stage 9: Duration about 19.5 h. Stage 9 includes periods in which the larva has already developed but is not yet capable of active muscular contractions. The stage begins from the appearance of a well-defined head and 13 body segments, namely 3 thoracic and 10 abdominal. Organogenesis takes place during this period. The labrum juts our like a unicorn's horn. Three pairs of head appendages, the mandibles and two maxillae, become prominent as knobs of tissue along the ventral sides of the head. Two large ganglia are laid down as well as the ventral nervous chain. The two ganglia are connected with the circumesophageal ring. A proctodeal invagination occurs near the posterior tip of the abdomen; it does not open into the midgut until the end of the larval feeding period which precedes pupation. The tracheal system is laid down.

From the middle of this stage, the embryo starts to elongate, stretching the amnion-serosa by means of the head and posterior tip of the abdomen. The posterior end of the body becomes more pointed.

Stage 10: Duration about 3 h. This stage corresponds to hatching. It begins with active muscular movements in the head region resembling faint "nodding"; gradually this movement becomes more vigorous and extends to more anterior segments until the whole larva shows a kind of rhythmic pulsing. As soon as the larva has enough freedom of movement, it begins a 180° rotation around its long axis, bringing its ventral side to the concave surface of the chorion.

After rotation has been accomplished, the peristalsis-like movements become more vigorous and the larva expands somewhat in volume. The amnion-serosa becomes visibly stretch each time the larva moves its tail. Finally, the serosa breaks up into fragments. Soon after this fragmentation the tracheal tubes rapidly fill up with air. This makes the tracheal system opaque and the tubes become clearly visible from the outside. The larva continues its active movements; several minutes later the chorion suddenly breaks up in one or several places, and the larva creeps out from the chorion through the gradually enlarging hole.

It has been shown that in the course of hatching the chorion becomes partially dissolved by the hatching hormone which is secreted no earlier than 10 min prior to normal larva hatching.

The normal table of honeybee embryonic development has been compiled using the data of DuPraw, who studied the embryonic development in paraffin oil at a constant temperature by time-lapse cinematography.

A relative dimensionless unit of development to (which has been suggested by Dettlaff and Dettlaff according to DuPraw is 30-35 min at 35.5°C. It has not been used in Table 9.2. Since a constant temperature is maintained when the honeybees are rearing their young, το may be used to compare honeybee development with the embryonic development of other insect species.

Larval Development

The larvae of worker bees and drones after hatching are fed with the nurse bees' "milk", secreted by their salivary glands. From the third day onward they are fed with a mixture of honey and pollen. The queens larvae are fed only on the "milk," but its composition is somewhat different.

A hatched larva weighs 0.08-0.1 mg. The larval weight of worker bees belonging to two races (in milligrams) is given on the next page as a function of time of development (Smaragdova, cited from).

TABLE 9.2

Age of Larva (days)	1	2	3	4	5	6
Grey mountain Caucasian race	1.3	3.8	12.4	46.1	78.7	133.9
Central Russian race	1.6	6.7	14.5	48.0	91.0	160.5

The developing larvae of the worker bee pass through four molts, each of which lasts about 30 min. These molts take place after hatching at 12-18, 36, 60, and 71-89 h (Bertold, cited in). The larvae may be cultured outside the beehive.

The structure of the larva immediately after hatching. The midgut and Malpighian tubules are separated from the posterior gut and only become connected with it prior to pupation, thereby making defecation possible. The development of larvae and pupae has been described by Myser.

The Prepupa and Pupa of the Worker Bee

After the cell is sealed by worker bees, the larva weaves a primitive cocoon. If, for some reason, the cell remains unsealed, the larva usually falls out. After weaving the cocoon the larva becomes immobile. This is the prepupal stage. Segregation of the definitive head, thorax, and abdomen proceeds under the larval integument. Eyes, antennae, and oral apparatus are formed on the head; rudiments of both pairs of wings and of three pairs of legs are formed on the thorax. The pupal organs appear, while the larval inner organs undergo histolysis. Instead of the 4 large Malpighian tubules, up to 100 thin ones are formed. The 7 ganglia of the ventral nervous chain of the pupa and the adult insect are derived by the partial fusion of 11 larval ganglia.

The fifth molt takes place at the end of the prepupal stage, the larval integument being discarded. The pupa looks like the adult honeybee but is devoid of pigmentation. The antennae, legs, and wings are pressed to the body and are in a non-stretched state. The pupa gradually darkens, beginning with the eyes.

After the sixth molt, the adult bee perforates the cell cap with its mandibulae and emerges from the honeycomb. Its first actions are to clean the cells and feed the older larvae with honey and stored pollen. The young bees fed themselves, and their salivary glands develop, thereby making it possible for them to feed the younger larvae.

10

Development in Starfish (Asterina pectinifera)

Asterina the common sea star is free living marine animal and occurs commonly in shallow zones of the sea and can easily be collected and transported. It also adapts well to artificial conditions, and its feeding is non-specialized. The embryos and larvae, as well as the adults, can be main ained in a wide range of temperatures. The body structure is fairly simple, and this allows the isolation of gonad fragments and even whole gonads from live starfish by surgery: wounds are healed and the animals can be repeatedly used in experiments. The ovaries and/or fragments of them can be kept in sea water for many hours, and the oocytes retain their ability for maturation and fertilization. Among the invertebrates, the starfish are often used when the oocytes are needed for experimentation. The large and transparent oocytes of *A. pectinifera* are quite suitable for such studies. These and some other advantages provide the basis for the ever-increasing use of *A. pectinifera* in studies of oocyte maturation, fertilization, and embryonic and larval development.

This species was used to study the mechanisms underlying oocyte maturation and the role of 1-methyladenine in this process and, more recently, the role of gonad-stimulating substance (GSS). Polar body formation changes in the rigidity of the egg surface during cleavage the oocyte's competence to the effect of 1-methyladenine the inhibiting effects of cysteine and its derivatives on 1-methyladenine production by the follicle cells, and the fine structure of oocytes at different stages of oogenesis and after 1-methyladenine-induced maturation of the oocytes are among the notable investigations in this area. In addition, studies have been made on the effects of steroids upon gonads control of sperm motility early development after insemination of the oocytes at different stages of maturation with different concentrations of sperm, and initiation of cleavage by injecting Triton X-100-treated sperm into the eggs. Other investigations include those of the structure and localization of DNA polymerases in the oocytes, artificial parthenogenesis, and studies of early morphogenetic processes in tetraploid embryos. Many experimental studies deal with the spatial organization and cell interactions during embryonic and early larval development.

This list of references, which is far from being complete, is in itself a recommendation for using *A. pectinifera* as an experimental species for development biology studies.

Taxonomic Status, Distribution and Breeding in Nature

Asterina (Patiria) pectinifera belongs to the class Asteroidea (order Spinulosa, family Asterinidae). In Soviet scientific literature this species is usually referred to as *Patiria pectinifera* (see, for example), whereas in scientific literature of other countries (mostly Japan), it is referred to as *Asterina pectinifera* (see, for example). At present the genus *Patiria* is included into the genus *Asterina* and the species in question should be referred to as *Asterina pectinifera*.

This species occurs in the Yellow Sea and the Sea of Japan, along the far east coast of the USSR from the Posieta Bay to the Tatar Strait, near the southern coast of Sakhalin Island, and in Aniva Bay of the Sea of Okhotsk. It is the most common starfish in Peter the Great Bay of the Sea of Japan.

A. pectinifera occurs predominantly on stony ground, but can also occur on various substrata down to 40 m.

Studies in Peter the Great Bay have shown that two generations of the oocytes are produced in the ovaries during the reproductive cycle: one in the autumn and another in the spring. Spawning takes place from the second half of July to the beginning of September at surface water temperatures of 18-22°C. According to Novikova the oocytes of the autumn generation are shed during the second spawning in November. At the end of autumn to the beginning of winter the gonads are reorganized and, from midwinter until August, they are characterized by the accumulation and differentiation of gametes.

In Sagami Bay (Pacific coast of Honshu Island), *A. pectinifera* spawns from April to May and in Mutsu Bay (northern coast of Honshu Island), from August to September.

A. pectinifera females produce, on the average, 500,000 eggs per year.

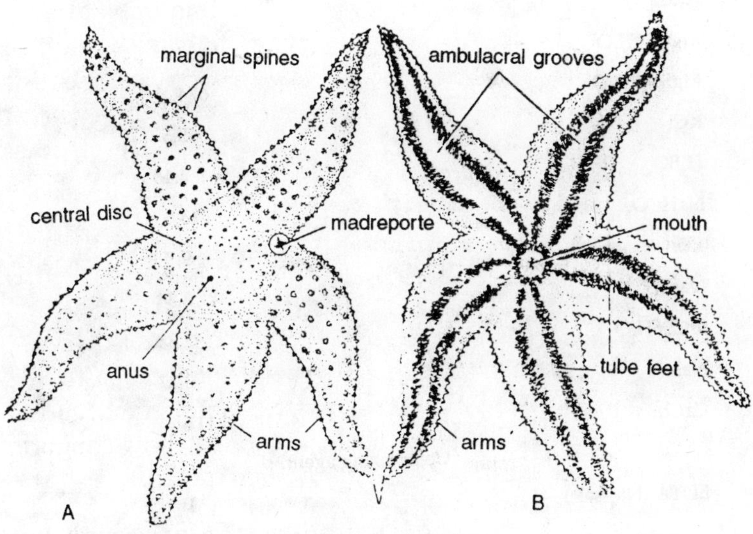

Fig. 10.1. Asterias. External features. A—Aboral view. B—Oral view.

Transport and Keeping Adults in Captivity

Before transportation, the collected animals are kept for a few days in basins with running sea water. The starfish should not be fed during this period. They are best transported in 50-liter plastic bags containing one-third of sea water and two-thirds of oxygen. For safety, the plastic bags are placed into paper bags or wooden boxes. The starfish are best transported over long distances at low temperatures (5-10°C) in foam plastic boxes or portable refrigerator bags with a layer of ice on the bottom. Under these conditions the starfish can be kept for 2-3 days. Small numbers of starfish can be transported in portable refrigerator bags on a layer of ice and under a cover of moist algae without water.

TABLE 10.1

Composition of Artificial Sea Water of 35.00% (all values in g/100 liters)

Basic constituents	
NaCl	2640
$MgCl_2 6H_2O$	491
$MgSO_4 7H_2O$	674
KCl	65.5
H_3BO_3	2.48
$NaHCO_3$	13.6
KBr	9.7
Na_2CO_3	3.34
$SrCl_2 6H_2O$	2.38
NaF	0.27
$CaCl_2$	111
Complexing agent	
EDTA (Na_4-salt)	1
Trace elements	
KIO_3	0.009
$MnSO_4 5H_2O$	0.397

$NaH_2PO_4 2H_2O$	0.397
LiCl	0.099
$Na_2SiO_3 9H_2O$	0.099
$Na_2MoO_4 .2H_2O$	0.099
$Al_2(SO_4)_3 .18H_2O$	0.172
$CoSO_4 .7H_2O$	0.0089
$CuSO_4 .7_2O$	0.0009
$ZnSO_4 .7H_2O$	0.0096
Rb_2SO_4	0.0149
$BaCl_2$	0.009

In the laboratory the starfish are kept in closed-cycle aquaria with artificial sea water. There are many recipes for making artificial sea water. *A. pectinifera* is an undemanding species and can live for a few months in sea water made of a limited number of components (only basic constituents). But in order to keep starfish for a long time (e.g., a few years) and obtain gonad maturation, a more sophisticated water is needed. We prepare artificial sea water according to the "Instant Ocean" recipe with a somewhat altered ratio of basic constituents. This recipe proved to be universally suitable for many species of marine invertebrates and fish. The water in the aquaria should be changed every month (by 20% in volume). It should be made 2-3 weeks before change and kept in darkness with slight aeration; the pH should be in the range of 7.8-8.3.

A. pectinifera can withstand a temperature rise to 26°C and fall to 8°C. At temperatures of 18-24°C (that is, in an aquarium without water cooling and heating) the starfish normally live several years. But at high temperatures spontaneous spawning may occur, this is especially true of starfish which live in captivity for more than a year and never spawn. Therefore, the starfish are better kept at moderate temperatures (up to 16°C) in aquaria with water cooling.

The starfish can feed on algae *(Ulva, Porphyra)*, various microphytes, fish and molluscs, polychaetes, and remains of various marine organisms. If they are kept on captivity for a long time and their food is not diverse enough, avitaminosis may be observed: the dark-blue colour converts to a pale-blue color. In order to avoid this, we give the starfish the following food mixture: squid meat, 500 g; sweet pepper powder, 2 g; calcium gluconate (glycerophosphate), 1.5 g. This mixture is ground and kept at — 10°C.

Since direct sunlight is detrimental for most marine invertebrates only moderate illumination should be used when keeping the starfish in captivity.

Sex Identification, Obtaining of Gametes and Artificial Fertilization

A. pectinifera is a dioecious species but has no external sexual dimorphism. The sex of starfish can be identified by several methods; one of them is the gonad biopsy. A hollow needle is introduced into the arm base closer to the central region, thus obtaining enough gametes to identify the sex. The advantage of this method is that the starfish is practically uninjured. However, it is rather difficult to isolate gametes from immature gonads and it is therefore more useful to cut the starfish open. It is placed on a Petri dish with its aboral side up and a cut is made across the interradius (1.5-2 cm). A piece of gonad is then cut off. The mature ovaries are orange, and the testes cream coloured. Under normal conditions the cut is usually healed within 5-6 months. Unlike in some other starfish species no autotomy of injured arms takes place.

A. pectinifera has short arms; therefore, in order to obtain a large number of oocytes, a circular cut should be made and whole gonads should be isolated. An oocyte suspension can be obtained as follows the ovaries are rapidly washed with an ice-cold Ca^{2+}-free sea water and they are delicately eased apart for 1-2 min. with fine forceps. The suspension is then filtered through a double layer of cheesecloth and centrifuged at low speed for 0.5-1 min. As a result, the follicle cells remain in the supernatant and are removed. The suspension is washed

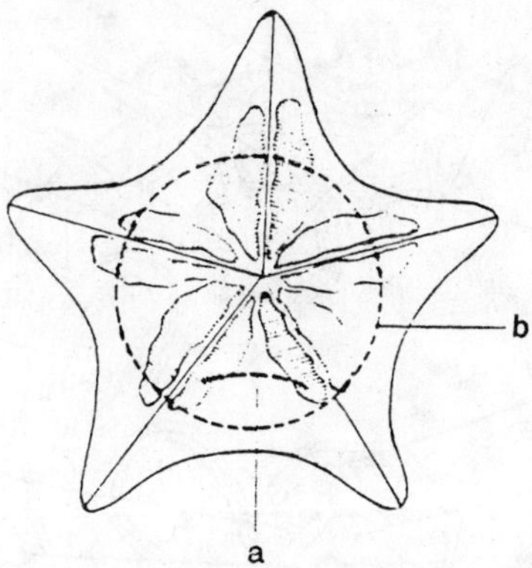

Fig. 10.2. Sketch of a starfish section with isolating a piece of gonad (a) or the whole gonad (b). Aboral body side.

until it becomes free of follicle cells. The use of Ca^{2+}-free water prevents the secretion of 1-methyladenine by the follicle cells and the adhesion of the follicle cells to the oocyte's surface; the low temperature increases the necessary contact time for 1-methyladenine to induce maturation. Various recipes for making Ca^{2+}-free water are provided by Meijer et al.

Isolation of gonads, by the method described above, will inevitably cause the death of the starfish; therefore, this method is used mostly at the marine stations. However, a part of the ovary, or even the whole gonad, can be isolated through a small cut between the arms. The operated animal is then placed back into the aquarium.

Mature gametes can also be obtained by the hormonal stimulation of spawning. The starfish are injected in each arm with 0.5-1 ml of 1.10^{-4}M solution of 1-methyladenine in the filtered sea water and placed in separate vessels. The males start spawning first (10-20 min. after the injection), and they are immediately placed into a refrigerator in dry Petri dishes. The spawning of females starts 30-60 min. after the injection. The intensity of spawning varies and depends, above all, on the state of ovaries.

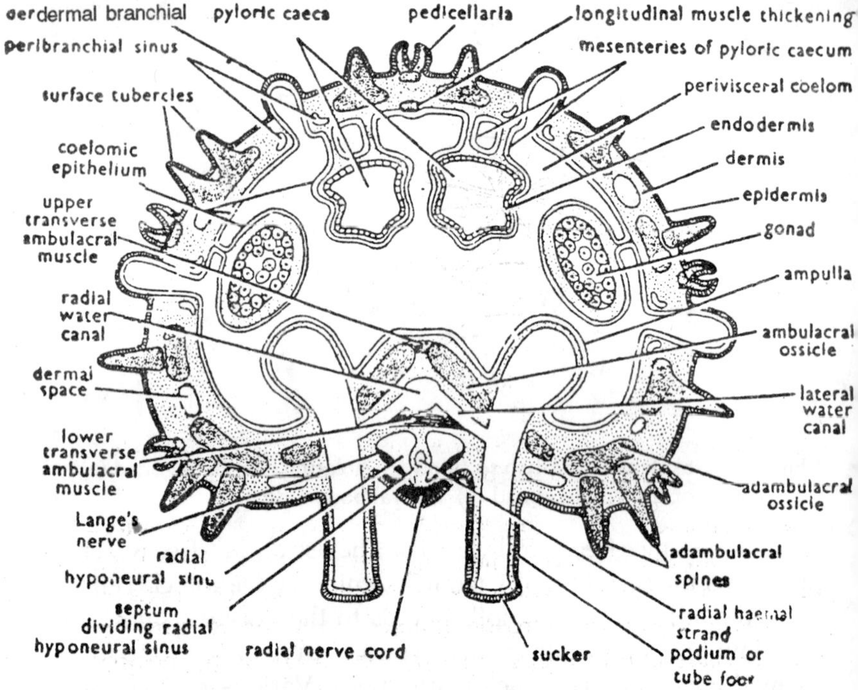

Fig. 10.3. *Asterias*. Diagrammatic T.S. of an arm showing the gonads.

The fertilized eggs can be obtained by placing the injected males and females into the same vessel. For artificial fertilization, a piece of testis is cut off, minced on a dry glass, and kept in a refrigerator. When needed, the "dry" sperm is diluted with sea water and the activity of spermatozoa is monitored under the microscope. The method of artificial fertilization proposed by Busnikov and Podmarev for sea urchins can be used: the diluted sperm is mixed with the oocytes (0.1-0.3 ml of sperm per 50-100 ml of dense suspension of oocytes), the volumes is brought to 1 liter within 1-2 min., and, after the oocytes have settled, the water is changed. Fertilization can be monitored by the rate of membrane elevation.

Although the starfish oocytes are fertilizable at any stage of maturation after germinal vesicle breakdown, the period from the germinal vesicle breakdown until the extrusion of the first polar body is considered to be optimal for artificial fertilization. When the spawning is induced, the oocytes shed by the females are at this stage of oocyte maturation. Therefore, oocytes are best fertilized 30-45 min. after the beginning of spawning (with respect to its intensity). The first polar body is extruded 60-70 min., and the second part polar body 90-105 min after 1-methyladenine treatment of the oocytes at 20°C. A depression is formed on the egg surface at the site of the polar body extrusion some 10 min before its appearance.

It is advisable to gradually increase the temperature, by 0.5-1.0°C a day, in the aquarium for 1-2 weeks before spawning to 18-20°C.

We induce spawning of *A. pectinifera* by hormonal stimulation during the autumn-winter period of its gonad cycle, when no spawning is observed in nature. The embryos and larvae developed quite normally.

In addition to 1-methyladenine, dithiothreitol methylglyoxal-2-(guanylhydrazone), arachidonic acid, and some substances have also been used to induce oocyte maturation.

Egg Structure, Removal of Egg Envelopes and Isolation of Germinal Vesicles

The mature eggs of *A. pectinifera* are transparent, of brownish colour, about 170 μm in diameter, their shape is typically spherical, but it can be irregular when the oocytes are densely packed in the ovaries. The eggs are heavier than the sea water. As in most echinoderms, the egg is surrounded by a thin hyaline membrane, adjacent vitelline membrane, and outer jelly coat; the eggs are liberated from the follicular layer during spawning.

The membranes can be removed by the methods used for many starfish species. The jelly coat is dissolved by washing the eggs in acidified sea water (pH 4.5-5.5). The vitelline membrane can be removed by short-term treatment with pronase (0.05-0.1% w/v) or trypsin (0.05% w/v) with subsequent

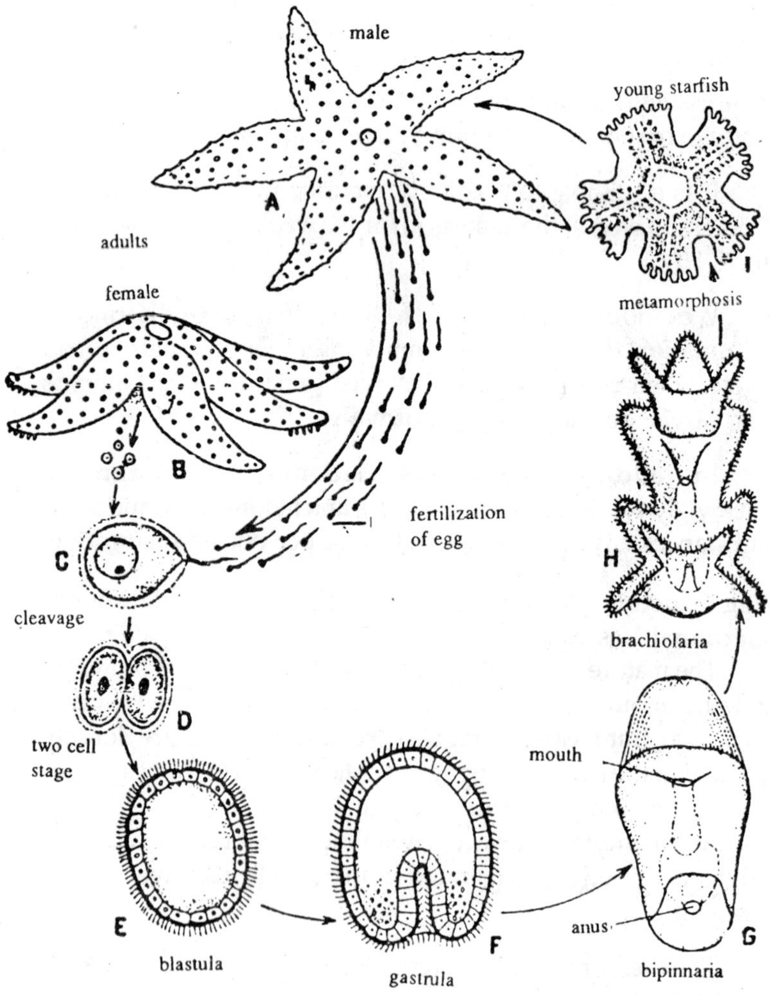

Fig. 10.4. Sea star. Life history and development.

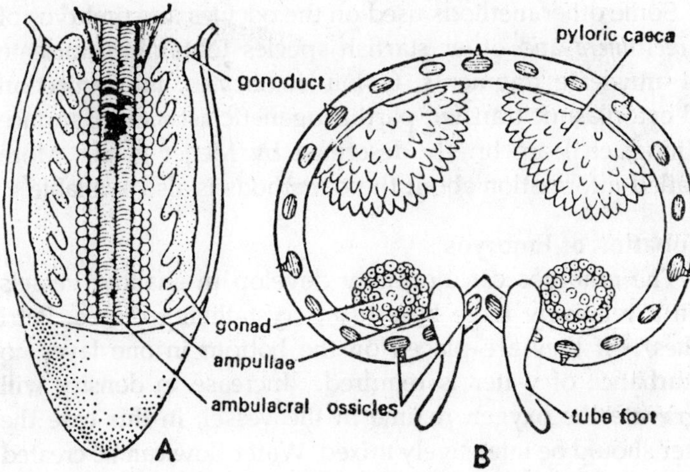

Fig. 10.5. *Asterias.* Gonads. A—An arm dissected to show gonads. B—T.S. of an arm showing gonads.

washing in 1 M urea. Chiba and Hoshi removed the vitelline membrane by 5-min. treatment with trypsin (7 mg/70 ml of Ca^{2+}- and Mg^{2+}-free sea water). The trypsin effect was arrested with disoprophylfluorophosphate (1 ml of 1mM solution).

A simple and effective method has been developed for mass isolation of germinal vesicles from the oocytes of *A. pectinifera*. After the vitelline membranes have been removed, the oocytes are washed in Ca^{2+}- and Mg^{2+}-free sea water and suspended in 100 ml of 0.2 M sucrose in the presence of 50 mM Tris-HCl (pH 7.5). If the oocyte are not lysed within 5 min they are passed through a glass column (1.5 mm x 5.0 cm) three times. Within 5 min., 80 ml of 0.2 M sucrose is added dropwise to the homogenate and gently mixed. Aliquots (30 ml) of the homogenate are layered with a pipette on the sucrose gradient (10 ml of 1.7 M + 5 ml of 1.4 M) and centrifuged at 750 g for 8 min. The germinal vesicles settle to the bottom. The isolated germinal vesicles are gently suspended in 1.7 M sucrose, and they are ready for further treatment.

Some other methods used on the oocytes and embryos of *A. pectinifera* and other starfish species (obtaining nucleate and anucleate fragments, fusion of oocytes, microinjection and cytoplasmic transfer, parthenogenetic activation, surface staining, etc.) are briefly described by Meijer et al. More detailed information about these methods is also available.

Incubation of Embryos

The embryos can normally develop in shallow vessels with a relatively large base (e.g., crystallizing dishes, Petri dishes). If they are placed on the bottom in one layer no disturbance of water is required. Increase in density will aggravate the oxygen regime in the vessel; in this case the water should be intensively mixed. Water flow can be created by stirrers made of a plastic pipette and a cover of the Petri dish. The cover is glued to the pipette at an angle of 30° and the pipette is connected via a rubber tube to the spindle of an electric motor set at about 60 rpm. The embryos are best incubated by mixing in high vessels, for example in 1-liter beakers. The water in the vessels should be changed 2 or 3 times a day. The light regime appears not to affect markedly the embryonic development in *A. pectinifera* but the developing embryos are best kept shadowed. The blastulae are characterized by positive phototaxis, and this can be used for their catching and transferring to another vessel. When incubating the embryos for experimental studies, an artificial sea water composed of 5-6 basic constitutents can be used. But in order to obtain the late larval stages, settling, and metamorphosis, we recommended the recipe given in Table 10.1.

Rearing of Larvae

The technique of rearing the planktonic larvae of *A. pectinifera* is much more complicated than that of incubating the embryos. This is related to the ever-increasing requirements of the larvae kept under artificial conditions, including their active feeding on phytoplankton.

TABLE 10.2

Duration of Development at Different Temperatures, Expressed in Absolute (hours) and Relative (τ_n/τ_0) Time Units

τo*	τo *min*	*To rotation stage*				*to hutching stage*				*To gastrulation stage*			
		τ1 *hours*	τ1/τo	τ1* *hours*	τ1/τo	τ2* *hours*	τ2/τo	τ2* *hours*	τ2*/τo	τ3 *hours*	τ3/τo	τ3* *hours*	τ3*/τo
11.8	83	30.0	21.7			32.8	23.7						
14.8	57	22.8	24.0			21.5	24.7			25.5	20.8		
15.7	52	20.8	24.0			21.3	24.6			23.3	26.9		
16*	49			19.5	23.9			20.5	25.1			23.5	29.8
19.3	38	15.3	24.2			15.8	25.0			16.7	26.3		
20*	33			13.8	25.1			14.5	26.4			16.5	30.0
20.4	35	13.3	22.9			14.0	24.0			14.3	24.6		
23	28	12.3	26.4			12.3	26.4			13.3	28.6		

One-liter beakers are suitable for rearing the larvae. They are compact; the high water column ensures the vertical distribution of the larvae. The development and behaviour of the larvae can be easily observed, and they are easily caught. The permissible density of larvae is determined first of all by the stage of their development. At the early larval stages (to middle bipinnaria) the larvae can be reared at high concentrations (up to 10,000/liter) using a plastic stirrer. Later, after the larvae have started active feeding, a shaker is best used for water mixing since it does not interfere with the movement of the larvae in the vessels. Strathmann suggests that the concentration at the later stages should be reduced to 100/liter. We have reared *A. pectinifera* larvae to the brachiolaria stage at a concentration of 100-150 larvae per liter. The late brachiolariae should be isolated from the less-advanced larvae and reared separately.

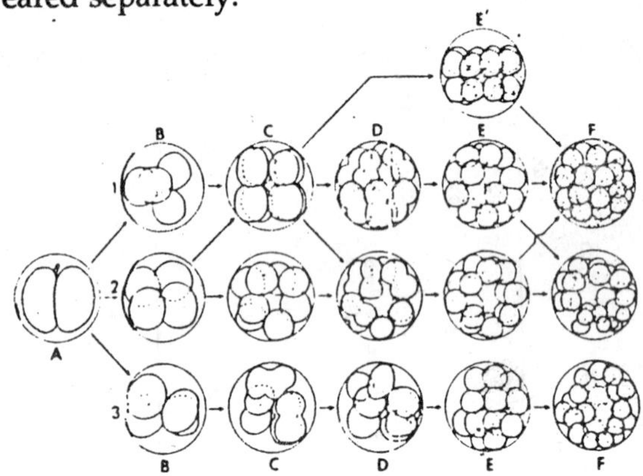

Fig. 10.6. Various cleavage patterns in *Asterina pectinifera.* Side view. A) 2-Cell stage; B) 4-cell stage; 1C, 2C) 8-cell stage; 3C) beginning of 8-cell stage; 1D, 2D) beginning of 16-cell stage; 3D) 8-cell stage; 1E-3E, E) 16-cell stage; 1F-3F) 32-cell stage.

There are different recommendations as to the frequency of water change when rearing the starfish larvae (see, for example). We obtained the best results while changing the water in the containers by one-third every 2 days and, at the

TABLE 10.3

Normal Development of Asterina pectinifera

StageStage No.			Time after insemination day hour	min	τ_n/τ_0	Size µm	Observations
1	2	3	4	5	6	7	8
1a	Immature egg (oocyte)	25±2				170	Germinal vesicle with a nucleous is seen; the egg is surrounded by the licular envelope
1b	Mature egg	25±2					Germinal vesicle and follicular envelope disappear within to 40 min after 1 methyladenine treatment
1c	Extrusion of the Ist	25±2					Takes place at the animal pole within 80 min after 1 methyladenine treatment polar body
2	Formation of fertilization membrane	20±1 23 25±2		2-3 3 2	0.1		The envelipe is clevated at the site of sperm entry. The oval egg assumes a regular spherical shape
3	Fusion of pronuclei	25±2		40			Zygote nucleus enlarges
4	2 blastomeres	20±1 23 25±2	1 1 1	40 30 25	3.2		Ist cleavage is meridional, equal. The furrow is laid down in the animal pole region
5	4 blastomeres	23 25±2	2 1	00 40	4.3		2nd division, Blastococle appears

1	2	3	4	5	6	7	8
6	8 blastomeres	25±2 23	2 2	99 35	 5.5		3rd cleavage is more often equal. Unequal blastomeres can form
7	16 blastomeres	23 25±2	3 2	10 20	6.8		4th cleavage is equal. Blastomeres of spherical shape. Blastocoele increases
8	Early blastula	25±2	4	30			Blastomeres surrounding the blastocoele are located in one layer. Blastoderm starts to invaginate into the clastocoele
9	Midblastula	25±2	6-8	30			As the number of cells increases, the invaginating folds thicken and reach the blastocoele center
10	Late blastula	25±2	9	30			The blastoderm folds return to the initial position; they are superimposed and fuse—in these places the blastoderm becomes multilayered. The egg surface is smoothed out. A large blastocoele is again seen in the center
11	Hatching	20-22 25±2 23	11 10 11 12	30 50 20 20	 26.4		The embryo starts to rotate inside the fertilization envelope, the direction of rotation varies. In the vegetal pole region the blastula wall thickens. Within 10-40 min after the beginning of rotation, the envelope is attenuated and breaks down; the envelope is not shed at once but remains as a cap on the em-

1	2	3	4	5	6	7	8
							bryo for sometime. Folds can be seen in some embryos at the moment of iverse, posterior-transverse, and longitudianl bands.
16	Early bipinnaria	25±2 23	3		350		The anterior body part is elongated. The preoral ciliary band is closed into a ring, which limits the frontal field of the preoral lobe. It is separated from the rest of the band. The body shape becomes more complicated, lateral protrustions appear. The larva starts to feed on microalgae. Cocloms grow in the anterior—posterior direction
17	Mid-bipinnaria	23	7		650		Cocloms fuse in the preoral lobe before pharynx
18	Late bipinnaria	23	12-14		1000		Rudiments of transient attachment organs of brachiolaria, brachioles, and attachment disk appear in the preoral region. The medioventral protrusion transforms into the medial brachiole. The ciliary band configuration becomes more complicated, lateral protrusions of the larval body are elongated
19	Brachiolaria	23	16			1000-	Three brachioles form (two lateral and one)

1	2	3	4	5	6	7	8
						1100	medial) which carry attachment papillar Five to six days before settlement the attachment disk becomes distinctly seen between the lateral brachioles (24-25 days of development). The developing coelom gets into the medioventral protrusion and lateral brachioles. In the coelom region, which corresponds to the left hydrocode, five lobe shaped coelomic protrusions appear on the 20th day: rudiments of the ambulacral system's radial channels
20	Juvenile starfish	23	30			410-485	The attachment stalk is first clearly seen, and then gradually (in 3-4 days) disappears. The primary tube fect touch the substrate. The developing starfish have five primary tube fact. Lobe-shaped radial channels of ambulacral system get into the arms

later larva stages, every 4-5 days. The *A. pectinifera* blastulae and larvae are characterized by positive phototaxis; therefore, when the water is changed, the glass containers are covered with black paper or dense tissue to shield the upper two-thirds and the source of light is left to one side. Within 1.5-2 h the larvae sink to the illuminated part of the vessel, and the upper part of water can be removed using a rubber tube.

Water mixing is desirable for all larval stages since it helps to disperse the larvae and reduces the possibility of bacterial infections. There are different methods for creating a water flow in the vessels (see, for example). Water mixing by air bubbles leads to the death of sea urchin and starfish larvae, and thus cannot be used.

The feeding of the echinoderm and, in particular, starfish larvae, has been studied by many authors. At the early stages the larvae can survive without feeding, but their settling and metamorphosis are possible only if suitable food is available. For the larvae of some species of the family Asterinidae, including *A. pectinifera,* microalgae *Nitzchia* sp., *Dunaliella tertiolecta, Isochrysis galbana,* and *Phaeodactylum tricornutum* are suitable. Strathmann suggests algal species as food for the larvae of many starfish species: *Phaeodactylum tricornutum, Dunaliella tertiolecta, Isochrysis galbana, Cyclotella nana, Thalassiostra fluviatilis, Crisosphaera carterae, Amphidinium carteri, Diryium brightwelli.* We have fed larvae on microalgae *Platymonas viridis, Exuviella sp., Glenodinium sp., Pavlova lutheri,* and *Bodo marina;* the late brachiolariae were given diatomeans a few days before settling. The food was given during water change. According to Strathmann, starfish larvae in a suspension of algae with a concentration of 5000 cells/ml cease to feed within a few minutes: the ciliary band becomes saturated with algal cells; therefore, an excess of microalgae in the vessels for rearing the larvae should be avoided. For feeding, we prepare a dense mixture of the five above-mentioned microalgal species and introduce it into water on a pipette tip.

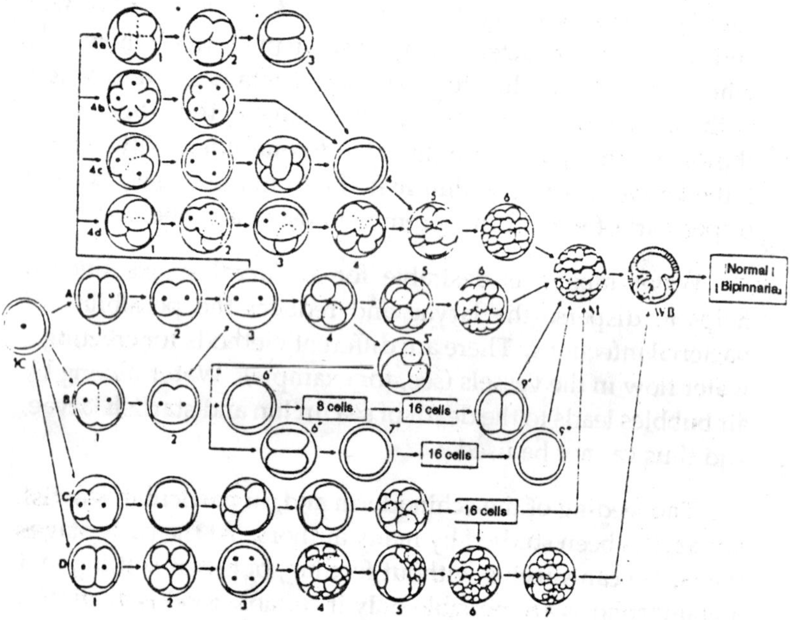

Fig. 10.7. Development of *Asterina pectinifera* in the case of full blastomere fusion. A-D) Various developmental patterns; 1C) 1-Cell stage. 4 in A-4a pattern; A3, A6', A7", A9", C2, D3 — apparently 1 cell. 3 in A-4a pattern; A9', A6", C4—apparently 2 cells. M, morula; WB, wrinkled blastula.

In addition to planktonic algae, the larvae require bacteria for successful rearing: in their absence the growth of larvae is delayed. That is why it is better to take water for change from an acting closed-cycle aquarium, since fresh water is practically sterile (it is better to add this to the aquarium). The role of bacteria and dissolved organic matter in the feeding of starfish larvae has as yet been insufficiently studied.

When settling, some starfish species are characterized by substrate selectivity: for example, *Stichaster australis* brachiolaris settle only on algae of the genus *Mesophyllum*. The larvae of other starfish species, such as *Coscinasterias calamaria* and *Acanthaster placi*, settle on practically all substrates covered with a primary bacterial film, substrate selectivity also appears to be absent in the *A. pectinifera* larvae. A few days before settling, the larvae are placed into clean vessels, with mussel or oyster shells covered with a bacterial-algal film (for this film to be formed, the shells are placed for a few days into an illuminated aquarium). This suffices to stimulate the settling of brachilariae on the shells.

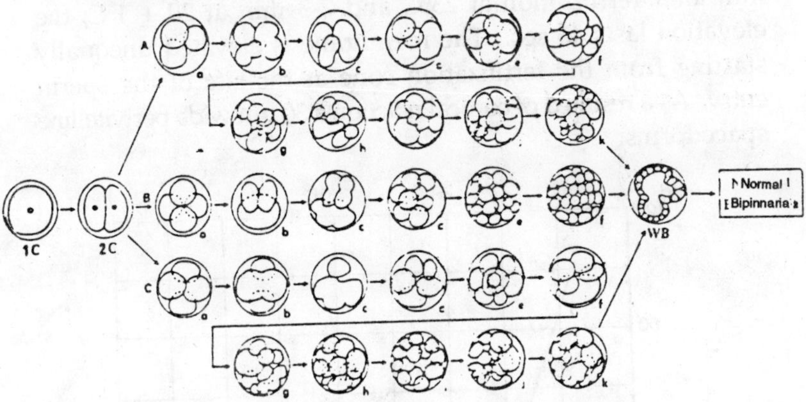

Fig. 10.8. Development of *Asterina pectinifera* in the case of partial blastomere fusion. A-C) Various developmental patterns. 1C) 1-cell stage; 2C) 2-Cell stage. WB, wrinkled blastula.

Continuous illumination inhibits the metamorphosis of the starfish larvae therefore, the larvae are better kept in darkness during this night.

It is advisable to rear the larvae at a temperature of no less than 16°C, since the development is markedly delayed at lower temperatures. We obtained settling and metamorphosis at a temperature of 23-24°C.

Development of Embryos and Larvae

The normal development of *A. pectinifera* is poorly documented. There are some incomplete descriptions of embryonic development and Teshirogi and Ishida compiled the normal tables for *A. pectinifera* development until the early bipinnaria stage. The embryonic and larval development of

A. pectinifera has been briefly described by Kasyanov et al. using materials collected from Peter the Great Bay. We have also studied the development of embryos and larvae, but under artificial conditions the gonad development in juvenile starfish has been outlined.

In *A. pectinifera* the fertilization membrane is elevated 3 min after fertilization at 23°C and 2-3 min at 20 ± 1°C; the elevation lasts 45 sec. The membrane is elevated unequally starting from the fertilization cone at the site of the sperm entry. As a resulted of cortical reaction a 30 µm wide perivitelline space forms.

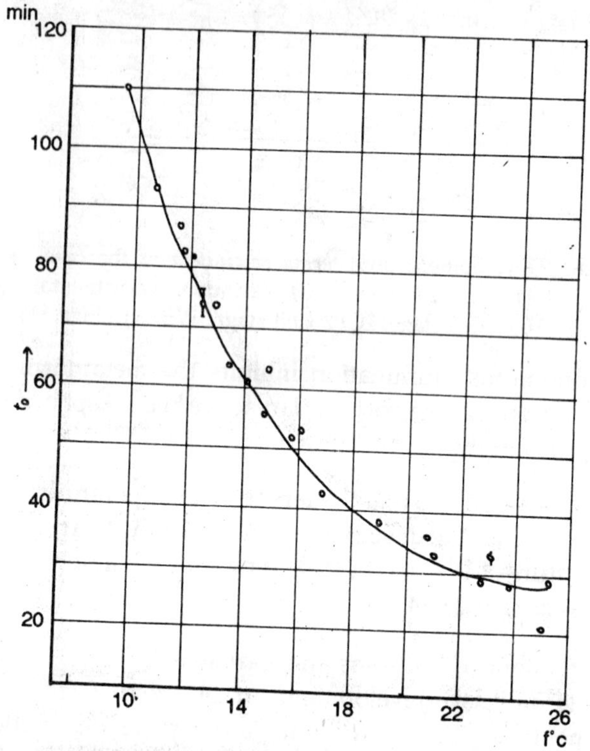

Fig. 10.9. Duration of one mitotic cycle during synchronous cleavate divisions in *asterina pectinifera* as a function of temperature.

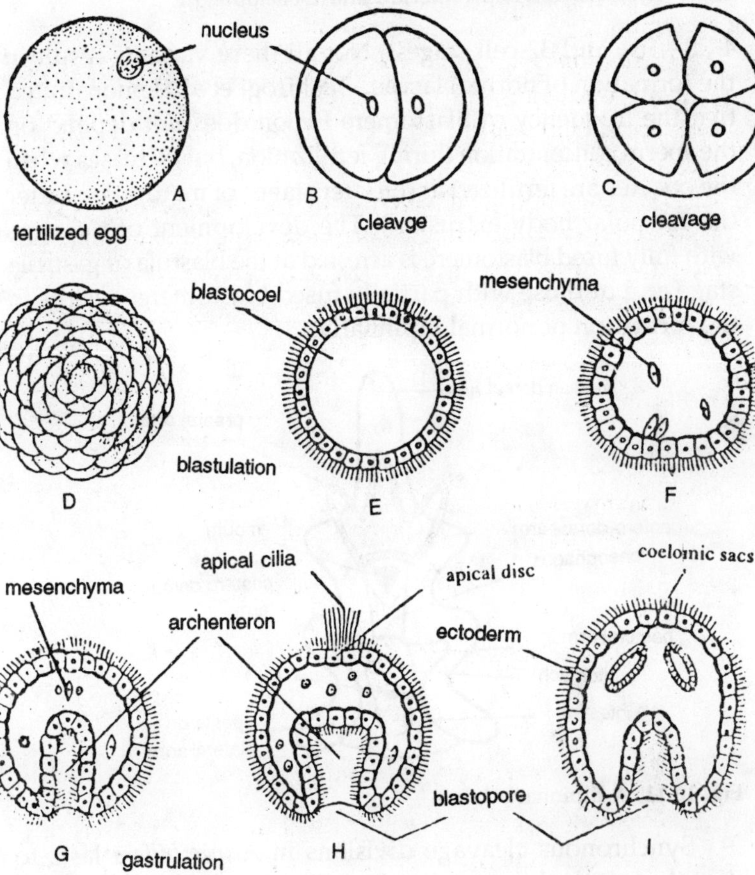

Fig. 10.10. *Asterias—Stages* in the development. (A) Unfertilized egg. (B) Two cell stages, (C) Four cell stage, (D) Biastula (E) Biastula in section, (F) Early gastrula, (G) Formation of mesenchym, (H) Older grastula, (I) Formation of coelomic sacs.

The cleavage is holoblastic radial to a weak extent bilateral, and somewhat asynchronous. At the same time, according to Teshirogi and Ishida marked deviations from this scheme can be observed. For example, during cleavage the furrows may disappear and the blastomeres may completely partially fuse. In addition, there are three possible patterns of cleavage at the

4-, 8-, 16-, and 32-cell stages. Not all these variants result in the formation of normal larvae. Teshirogi et al. demonstrated that the frequency of blastomere fusion does not depend on the sperm concentration during fertilization, but increases when the oocytes are fertilized at the later stages of maturation (after the 1st polar body extrusion). The development of the eggs with fully fused blastomere is arrested at the blastula or gastrula stage and of those with partially fused blastomeres results in the formation of normal bipinnariae.

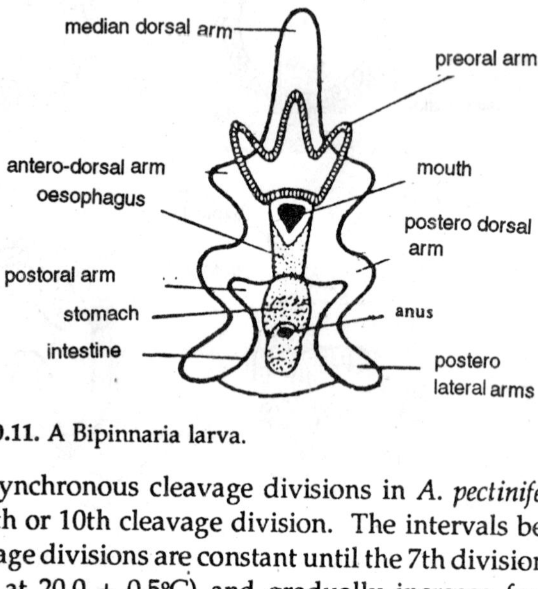

Fig. 10.11. A Bipinnaria larva.

Synchronous cleavage divisions in *A. pectinifera* lasts to the 8th or 10th cleavage division. The intervals between the cleavage divisions are constant until the 7th division (about 33 min at 20.0 ± 0.5°C) and gradually increase from the 8th division (more than 1 h between the 10th and 11th divisions. According to Dan-Sohkawa and Satoh the intervals between the cleavage divisions until the 5th division are roughtly 35 min (at 20 ± 1°C). There were no published data on the cell cycle duration during cleavage at any other temperatures.

We measured the values of τ_0, the duration of one cell cycle during synchronous cleavage divisions, at different temperatures and plotted the τ_0 against temperature curve. The duration of the 2nd cell cycle is the easiest to measure.

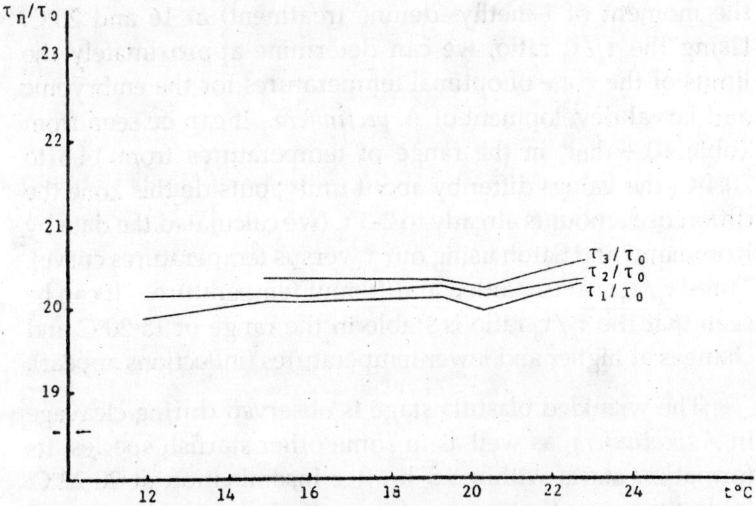

Fig. 10.12. Duration of certain developmental periods in relative time units (τ_n/τ_0) at different temperatures. τ_1, rotation; τ_2, hatching; τ_3, gastrulation.

Using the plotted curve, we can determine the value to t_0 at any temperature in the zone of optimal temperatures. According to our data, the range of temperatures at which the cleavage of *A. pectinifera* from Peter the Great Bay is possible, is limited by 10-11 and 24°C. At a temperature above 25°C, the mass death of embryos is observed; at 25°C the amount of abnormal embryos rapidly increases to 80-90% at the 8- and 16-cell stages. At low temperatures the development proceeded in one of the experiments at 9°C, although at a very slow rate and with a high percentage of defects. No cleavage was observed at 7.6-7.8°C. The embryos of *A. pectinifera* from Mutsu Bay (Honshu Island) could develop at higher temperatures; the cleavage was arrested at temperatures above 28 and below 12°C.

We determined the time from fertilization to the stages of rotation (τ_1), hatching (τ_2) at six temperatures, and gastrulation (τ_3) at five temperatures (Table 10.2). In addition, Table 10.2 includes the data obtained by Kominami and Satoh for the stages of rotation, hatching, and gastrulation (τ_1* τ_2* τ_3*) (from the moment of 1-methyladenine treatment) at 16 and 20°C. Using the τ_n/τ_0 ratio, we can determine approximately the limits of the zone of optimal temperatures for the embryonic and larval development of *A. pectinifera.* It can be seen from Table 10.2 that, in the range of temperatures from 14.8 to 20.4°C, the values differ by about unity; outside this zone the difference amounts already to 2-3 τ_0 (we calculated the data by Kominami and Satoh using our τ_0 versus temperatures curve). Three τ_n/τ_0 curves plotted at different temperatures. It can be seen that the τ_n/τ_0 ratio is stable in the range of 15-20°C and changes at higher and lower temperatures (inflections appear).

The wrinkled blastula stage is observed during cleavage in *A. pectinifera,* as well as in some other starfish species. Its formation starts within 5.5 h after insemination at 20-22°C: folds appear on the surface of the spherical coeloblastula and multiple blastoderm invagination takes place. The invaginating regions reach the blastocoel center. In a few hours the adjacent folds fuse and the blastula's surface is smoothed out; in some places it becomes multilayered. There are three different types of blastula wrinkling. 1) Invaginating folds increase in size during a few hours and fill almost the whole blastocoel. The blastula's surface is then smoothed out. 2) The folds invaginate to a small extent and the blastula develops into the gastrula without the typical wrinkled stage. 3) The wrinkled blastula develops from the wrinkled morula. The normal development of *A. pectinifera.* In compiling the normal table we gave ordinal numbers to successive stages (1-20). Stages 1-16 are described mainly according to Teshirogi and Ishida. Our changes concern only the stage numeration (the number of stages is diminished since some stages are divided into designated substages, for example, as 1a, 1b, 8a, 8b, etc.), and additions to the description of some stages. The timing of stages after fertilization at different temperatures is given (where possible) according to Teshirogi and Ishida (25 ± 2°C), Dan-Sohkawa and Satoh (20 ± 1°C), Komatsu (20-22°C), and our data (23°C). Stages 17-20 are described on the basis of our observations.

11

The Development of Echinoids

There are about 900 species of sea urchin. All echinoids are dioceans, and the majority display no sexual dimorphism. In *Echinas* there are five gonads suspended along the interombulaeral on the inner side of the test.

The sea urchin egg has been studied more extensively than any other invertebrate system. The reason for the popularity of sea urchin material lies in the ease with which the adult animals are induced to shed their gametes. Animals will shed eggs or sperm in response to the injection of a small volume of isotonic KCl into the coelomic cavity. Fertilization is external and can be easily controlled. Adult sea urchins may be maintained in the laboratory in sea water aquaria and may be injected repeatedly, provided they are well fed between spawnings. In nature, spawning is seasonal and dependent mainly on tides and temperature. Development is indirect i.e., through *Echinoplutes larva.*

GAMETO-GENESIS

Egg Production

The echinoids (sea urchins and sand dollars produce new oocytes from oogonia throughout life. The oogonia are produced in the germinal epithelium which surrounds the ovary beneath the muscle layer. Young oocytes in early meiotic prophase deptotene and zygotene) are found among the larger oocytes which have acquired a layer of follicle cells and may be attached to the basement membrane lying just beneath the germinal epithelium. Following zygotene (synapsis), the homologous chromosomes begin to move apart as diplotene chromosomes. During the early portion of the diplotene stage, the chromosomes become highly extended and stain as very faint, thin strands; a process called decondensation. The chromosomes remain in this state during the previtellogenesis (pre-yolk forming) period. This portion of diplotene is one of maximal RNA synthesis and is characterized by a large nucleolus and possibly by "lampbrush" chromosomes, though the existence of this type of chromosome structure has not been demonstrated conclusively in sea urchin oocytes. Most of the ribosomal RNA of the mature egg is formed during this period, along with some messenger RNA. Prior to vitellogenesis, the diplotene chromosomes condense and remain in this state until the first meiotic division. During vitellogenesis small oocytes become enlarged and filled with yolk platelets. The polar body divisions take place following vitellogenesis. The nucleolus disappears during meiosis and does not reappear in the pronucleus. The mature eggs are spawned from the ovary as ootids, leaving the polar bodies in the ovary. The eggs acquire a vitelline membrane and an external gelatinous coat while in the ovary. The jelly coat is possibly secreted by the follicle cells surrounding the large oocytes, though it is also possible that the oocytes themselves may synthesize and extrude this material.

The cytoplasm of the oocyte also contains DNA, in association with large numbers of mitochondria. The role of this DNA in embryogenesis, if any, is not known.

Ultrastructure of the cytoplasm: oogonium, oocyte, and mature ootid : Oogonia posses, like young and older oocytes, a prominent nucleolus. The Golgi apparatus is well-developed and positioned near the nucleus, a position it also occupies in young oocytes. The cytoplasm contains many monoribosomes and a few mitochondria. A rough endoplasmic reticulum, studded with ribosomes, appears in the young oocytes. The absence of the endoplasmic reticulum serves to distinguish oogonia from young oocytes.

In young oocytes, the Golgi complex increases in size, and mitochondria apparently multiply. Structures terms annulate lamellae, derived possibly from the nuclear envelope, appear. The precursors of the cortical granules appear as clear vesicles randomly scattered throughout the cytoplasm. The gelatinous coat material is not visible in electron microscope preparations in either oocytes or mature eggs. The small cortical granules later acquire an electron dense component, grow in size, and take up a peripheral position in the outer (cortical) cytoplasm.

In older oocytes, the plasma membrane elaborates a number of protrusions, the microvilli. an external membrane, termed the primary envelope or vitelline membrane, is found external to the plasma membrane. It has a matted appearance and appears in variable amounts in different species, being most prominent in starfish oocytes.

Electromicrographs of ootids reveal more details of cytoplasmic organization of the mature eggs. In the unfertilized egg, the surface layer is folded into widely spaced microvilli. Immediately below the surface layer (the vitelline membrane) and the plasma membrane, the layer of cortical granules is encountered. These granules are bounded by a bilaminar unit membrane (200-300Ä thick) resembling the plasma membrane and contain a matrix of electron dense material as well as less dense spherical units, which are arranged at the periphery of each granule. The entire cortical granule is 0.5-1.0 micron in diameter. Scattered through the cytoplasm are dense spheroidal vesicles measuring around 1.5 microns at the longest axis. These are yolk platelets containing stored protein. Other "empty" (electron clear) vesicles about the same size as yolk platelets are visible and are presumed to represent glycogen containing granules. Sprinkled uniformly throughout the cytoplasm are small black dots of electron dense material about 100 Ä in diameter. These are single ribosomes (monosomes). Mitochondria are scattered through the section (diameter about 0.5 micron) and are frequently found grouped around less dense yolk platelets. A typical endoplasmic reticulum is about from the mature ootid.

Morphology and Ultrastructure of Sperm

A diagram of sea urchin sperm is shown. The sperm is divided into a head, midpiece, and tail, measuring about 3.0, 0.7 and 45.0 microns, respectively, in length. The tip of the head contains the acrosome and acrosomal filament, the nucleus makes up most of the main body of the head. The t ail (in cross section) is seen to be composed of nine peripheral and two central fibrils. The midpiece contains tightly packed mitochondria.

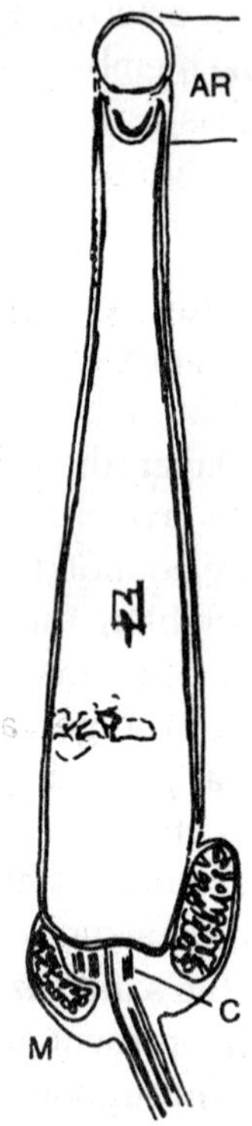

Fig. 11.1. Diagrammatic view of sperm of the sand dollar. Echinarachnius parma. AR, acrosomal region: N, nucleus; M, mitochondria, fused of midpiece; C. centriole and tail filaments extending back into the axial filament.

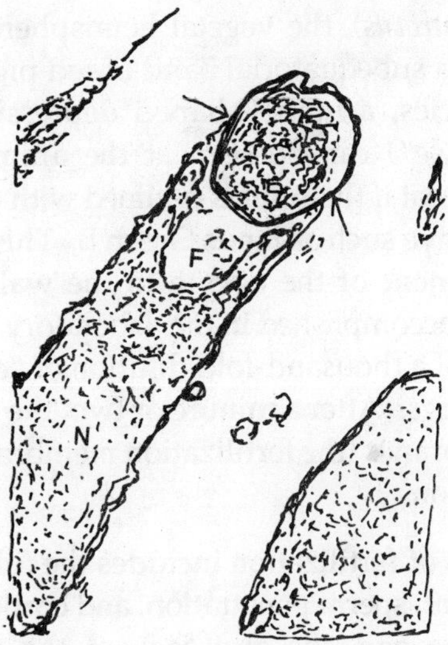

Fig. 11.2. Electronmicrograph of Echinarachnius sperm, (about 30,000x). G, acrosomal granule; F, subacrosomal fossa; N, nucleus.

Early Fertilization Development

Spawning: After injection with isotonic KCl (or at normal spawning times), the ripe gametes are shed into the sea water through five gonopores. Sexes in most species are indistinguishable before spawning. The pigments of the egg are of two sorts, a reddish pigment called echinochrome, which is substituted naphthoquinone, and a yellowish carotenoid pigment.

The unfertilized egg appears to be homogeneous: In the light microscope, the egg appears to be evenly filled with yolk (isolecithal). A small light spot may be observed in some lightly pigmented eggs and corresponds to the position of the female pronucleus. The egg is approximately 100 microns in diameter and is surrounded by a halo of gelatinous coat material, which is virtually invisible because it has about the same refractive index as sea water. There is no outward indication of polarity in most species since the markers of the animal pole (the polar bodies) were left behind in the ovary. In some species (*Paracentrotus*), the vegetal hemisphere is marked by the presence of a subequatorial band of red pigment. In fresh eggs of all species, a funnel-shaped depression in the jelly layer ("micropyle") can be seen at the animal pole in the unswollen jelly coat if the latter is outlined with carbon particles (India ink) or a dye such as Janus Green B. This represents the point of attachment of the oocyte to the wall of the ovary. Insemination is accomplished in the laboratory by the addition of a few drops of a thousand-fold dilution of "dry" sperm to a suspension of eggs. After a minute or two, the eggs will begin to elevate membranes (the fertilization membranes) indicating sperm penetration.

The process of fertilization includes the following events: sperm attachment, sperm penetration, and cortical and internal responses of the egg, the approach of the pronuclei, and pronuclear fusion.

Fertilization dramatic ultrastructural changes: In the vicinity of the egg, the acrosome granules of a number of sperm discharge producing an acrosomal secretion (probably an enzyme), which aids in dispersing the jelly layer, and an acrosomal filament. The acrosomal filament contacts and fuses with the plasma membrane of the egg, effecting a functional connection between the two gametes through which the sperm nucleus will pass. The egg responds to sperm attachment and penetration by the production of a protuberance, called the fertilization cone, at the point of sperm attachment. Starting at this point, the cortical granules "explode" to the outside of the egg beneath the surface layer (vitelline membrane) and proceed to fuse with this layer which elevates to form the fertilization membrane. The space between the fertilization membrane and the new egg surface is called the perivitelline space. The remainder of the cortical granule material (less the electron dense spheres) becomes distributed over the egg surface as the hyaline layer, or intercellular cement. The membranes of the exploded cortical granules fuse with the plasma membrane of the egg creating in part a new surface membrane. Now microvilli project into the perivitelline space from the areas between the spaces formerly occupied by cortical granules.

Fertilization activates egg metabolism: Sperm penetration sets off a chain of events which drastically alters the physiology of the eggs. These changes reflect the change from an essentially senescent cell to a cell poised for replication and differentiation. There is an abrupt increase in the rate of oxygen consumption (three -to five-fold). The precise mechanism involved is not clearly known, but increased permeability to oxygen can be ruled out since early experiments have shown that oxygen tension may be changed widely without affecting the respiratory rate of fertilized eggs. There is a change in membrane potential at the time of sperm penetration. The immediate effect of contact between egg and sperm is an action potential, a depolarization and repolarization that includes a temporary reversal of membrane polarity due to an influx of Nations. This overall sequence converts the impermeable membrane of the unfertilized egg which has a potential of about - 10 millivolts to a more typical K^+ permeable membrane with a potential of about 60 millivolts, inside negative. The potential of the fertilized egg can be reversibly abolished through addition of potassium ions to the surrounding sea water. Intracellular free calcium increases. Small changes in membrane resistance (decrease) and membrane capacitance (increase have also been noticed. The latter change might be due to an increase in surface area upon fertilization, in turn provided by the opening of the cortical granules and the development of the new microvilli. Increases in permeability to water and to sparingly ionized substances may also be a consequence of increased surface area.

Fertilized eggs show a marked increase in uptake of orthophosphate from sea water following a lag period of about 15 minutes, during which time an active phosphate transport system is developed. Permeability also increases to nucleotides (such as uridine) which are phosphorylated in transport by an energy-requiring process which is deficient in unfertilized eggs. Amino acids, in contrast to other precursor molecules, penetrate both unfertilized and fertilized eggs readily. The rate of uptake nonetheless increases on fertilization, and more accumulated amino acid is incorporated into protein in fertilized eggs than in unfertilized eggs.

The increased protein synthesis observed in fertilized eggs represents activation of the translation process and precedes increased RNA synthesis. The RNA involved in this translation to protein is stored maternal messenger RNA produced during oogenesis. Proof of the lack of participation of nuclear RNA, which is newly synthesized after fertilization, can be found in experiments where enucleate halves of eggs were activated artificially. In the absence of the nucleus, the same increased protein synthesis was noted following activation.

Finally, the fertilized eggs prepares to replicate the DNA of the pronuclei in anticipation of the first cleavage division. There is a preformed DNA polymerase, which is localized in the unfertilized egg cytoplasm. As DNA is replicated during cleavage the polymerase becomes progressively more localized in the nuclei. By the gastrula stage, new polymerase must presumably be made in order to continue DNA synthesis, since the maternal polymerase is exhausted by this time.

The consequences of polyspermy are deleterious: In sea urchins, normally only two sperm is admitted to the egg cytoplasm, the others are excluded. The occasional entrance of more than one sperm is called dispermy (two) or polyspermy (three to many). The presence of one extra sperm nucleus and centriole leads to abnormal development. The developmental abnormalities can be traced to irregularities in chromosome distribution at the cleavage. Assuming a chromosome number of 36 as the normal diploid value, the egg nucleus contributes 18 single strands, the two sperm nuclei 18 each for a total of 54 strands. During DNA synthesis, following fertilization, each strand is replicated in preparation for the first cleavage division. This yields 108 strands. Since there are two sperm centrioles and since both usually divide, there are four division centers, and division will be into four cells at the first cleavage. Arithmetically, one cannot divide 108 by four and come out with the normal diploid number of 36. In actuality, the resulting four blastomeres are heteroploid because of random assortment of dyads on the spindle fibers. If one sperm centriole fails to divide, the resulting arrangement of spindle fibers causes the production of three cells at the first cleavage. It is arithmetically possible to divide 108 by three and come out with 36 per cell. But the assortment on the division apparatus is still random and the chances are very low that the three cells will all have 36 of the right kind of chromosomes (at least one member of each of 18 homologous pairs). Polyspermy leads invariably to abnormal development in a high proportion of cases. In fact, the necessity for a balanced set of chromosomes for normal development was deduced at the turn of the century from these observations on dispermic sea urchin eggs.

Nature of the block to polyspermy: Some species allow the penetration of more than one sperm to the egg (physiological polyspermy). In these instances, only one of the sperm participates in the formation of the fusion nucleus with the egg pronucleus. An example of physiological polyspermy is found in the urodele amphibia. In sea urchins and others, including placement mammals, extra sperm (supernumary sperm) are prevented from entering the egg. (It has been found in sea urchins that the breakdown of the cortical granules following sperm penetration release a protease that attacks the remaining species-specific surface combining sites for sperm on the egg surface). The extra sperm adhering to the egg surface are thus disengaged during the cortical reaction, before they have had time to penetrate. Polyspermy might be expected when eggs are subjected to high sperm concentrations, causing a number of simultaneous or near simultaneous collisions between egg and sperm. Near simultaneous collision could result in the entrance of more than one sperm, since the release of protease from the cortical reaction could not progress rapidly enough to disengage the extra sperm. In reality, polyspermy has been shown to increase as predicted with increasing frequency of sperm-egg collisions.

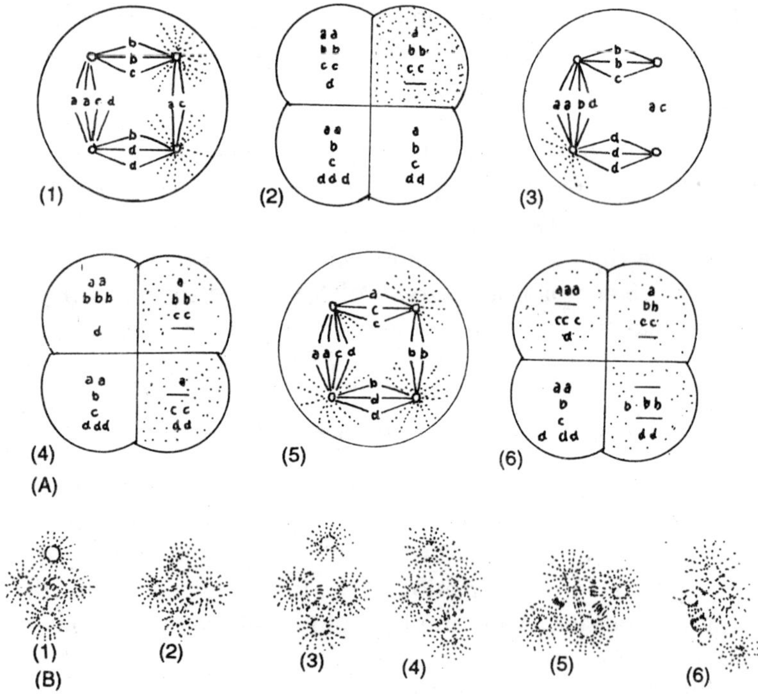

Fig. 11.3. (A) Diagram of results of Boveri's observations on dispermic sea urchin eggs. Unshaded blastomeres develop normally. Shaded blastomeres do not contain at least one member each of the homologous pairs and will develop abnormally.

The conclusion of the fertilization process: The final event in fertilization is the fusion of the male and female pronuclei to form the zygote nucleus, a process called syngamy. The cytological details, with time-table, are shown at low magnification in sectioned material. Replication of DNA in the pronuclei begins about 45 minutes after fertilization in *Strongylocentrotus* and lasts for approximately 15 minutes at 12°C. The sperm centriole initially produces a monastral figure, which fades during pronuclear fusion and is replaced by an amphiaster at metaphase, as the sperm centriole divides. The mitotic apparatus (astral rays and spindle fibers) is provided by the sperm centriole. The egg centriole is not demonstrable with known techniques.

Cleavage, or Segmentation

Importance of microfilaments and microtubules: The actual furrowing of the cytoplasm (cytokinesis) appears to be the function of the microfilaments. These organelles are located just beneath the plasma membrane, in the egg cortex, and form a contractile ring that completely girds the egg. The filaments effect division of the cytoplasm by contraction. The contractile process does not involve thickening of the microfilaments, and it has been proposed that contraction is brought about by means of the structural elements of the filaments sliding over each other, a occurs in the contractile molecules of striated muscle. The furrowing of the cytoplasm is inhibited by agents which are known to dissolve the microfilaments (cytochalasin B).

Fig. 11.4. Photograph of living sea urching egg containing an injected oil drop which has displaced the spindle fibers (mitotic apparatus). Constricting microfilaments are furrowing the egg, having been previously positioned by the astral rays of the mitotic apparatus before its experimental displacement.

The other elements in cell division are the microtubules. These fibers are composed of subunits found preformed in the cytoplasm in the form of the protein, tubulin. They are assembled at each division into the mitotic apparatus (spindle fibers and astral rays) under the influence of the centrioles. The mitotic apparatus is responsible for chromosome movements and is in some way responsible for initiating division by the microfilaments. Inhibition of assembly of the subunits of the mitotic apparatus can be brought about by low temperature. Such inhibition also inhibits cytokinesis. If the mitotic apparatus is displaced at metaphase, prior to any constriction of the cytoplasm, the cell will divide anyway and at the same place where division would have occurred in an undisturbed cell. It is evident that the astral rays of the mitotic apparatus are of primary importance in positioning the microfilament ring.

Replication of DNA proceeds without detectable interphase: The chromosomes form small nuclear vesicles (micronuclei) upon arrival at the poles (telophase), and replication of DNA in preparation for the next division begins immediately, there being no demonstrable interphase during the early cleavage divisions. In the unfertilized egg, the amount of cytoplasmic DNA contained in the mitochondria and to a lesser extent in the yolk is approximately 20 times the amount found in the pronucleus. By the 32-cell stage, the amount of DNA per embryo has increased through chromosomal replications, and the proportion, of the total DNA contributed by cytoplasmic components is much less, the mitochondria are divided between daughter blastomeres without replication.

Blastula

Changing geometry and cell migration: hatching. Arbitrarily, the early blastula stage can be considered to contain 64-cells. The blastocoel is evident and the embryo is still encased in the fertilization membrane. During the blastula stage, cell division continues with some loss of synchrony. The micromeres migrate into the blastocoel just inside the vegetal pole and arrange themselves in a circular pattern around the future blastopore.

The blastocoel continues to enlarge. Just before hatching (300-400 cells), the cells of the blastula become ciliated. A tuft of longer cilia (the apical tuft) is present at the animal pole. The embryo now rotates inside the membrane and slowly hatches from the membrane as the latter is dissolved by a hatching enzyme secreted by the embryo into the perivitelline space.

LATER DIFFERENTIATION

Gastrula and Pluteus

Germ layers are established; more changes in form: The vegetal pole becomes flattened, forming the gastral plate, which shortly thereafter indents to form the beginning of the archenteron. The micromeres form a circle of cells around the archenteron and begin the elaboration of calcite spicules which will form the larval skeleton. Since the micromeres now occupy a position between the archenteron and the blastula wall (epidermis), they are called mesenchyme cells. Since this mesenchyme is the first evidence of a middle cell layer, it is called primary mesenchyme. The term mesenchyme is used, rather than mesoderm, since the cells are loosely arranged and tend, initially, to be independent of each other. During the elaboration of the skeleton, these cells are connected by "cables" of microtubules, which form the matrix for deposition of calcite.

The archenteron continues to deepen by invagination and by contractions of slender pseudopodia (filopodia) which originate from secondary mesenchyme cells at the archenteron tip and adhere at intercellular boundaries of the blastula wall. As the archenteron deepens, it bends toward the future oral side. The archenteron makes contact with the ectoderm of the blastula wall, meeting a depression in the ectoderm, the stomodeum. The fused ectoderm and archenteron endoderm form a bilaminar layer, the oral plate, which will soon rupture to form the mouth opening. The blastopore becomes the anus. The larva now has assumed a prism shape, flattened on the oral side and on the blastopore side; it is bilaterally symmetrical. From the anterior third of the archenteron, two lateral outpocketings appear, the enterocoelic pouches. These pouches form the coelomic mesoderm. The spaces in the pouches are the coelomic cavities. The micromeres produce a bilateral skeleton on the right and left of the archenteron, composed of enlarged triradiate spicules.

Only a portion of the pluteus gives rise to the adult: The pluteus shape (typical larval form of echinoids) develops as a pair of arms grows out anterior to the mouth region and a second longer pair appears anterior to the blastopore (anus). The pluteus is free swimming and feeds on plankton until the time of metamorphosis (a matter of weeks or months, depending upon the species). At metamorphosis the larva becomes geotropic and seeks the bottom. The adult animal is formed from a small patch of cells on the left side of the embryo, which is called the echinus rudiment and includes epidermis and underlying mesoderm. A bilateral larva thus develops into a radially symmetrical adult. In the experimental studies, which will be considered next, normal development is followed only to the early pluteus stages. Prior to the opening of the mouth, feeding is impossible and the embryo represents a closed system uncomplicated by the addition of exogenous food materials or the elimination of waste products.

Early development is a closed system: No detectable nitrogen excretion takes place during the period of development to pluteus. Many precursor molecules are able to penetrate the sea urch in egg. This enables the investigator to label RNA, DNA, and proteins for experimental purposes. Since there is no nitrogen excretion, one is dealing with a system involving turnover and repartition of materials within the egg. A large portion of the nitrogen must be finally localized in DNA, which is a "sink" in the sense that nitrogen once put into DNA never comes out. The sea urchin egg carries a reserve of yolk material that is sufficient for development through an extensive period of morphogenesis and differentiation. The medium in which the embryos are cultured is simply sea water, which can be filtered from the ocean or made artificially from reagent-grade chemicals in the laboratory.

Experiments on Egg Polarity

The experimental analysis of development is designed to show cause and effect at the cellular level. Most experiments require careful controls, such that only one variable is studied at a time. Two groups of eggs or embryos are set up, and one procedure is imposed on one set and not on the other. Differences between the experimental group and the control is said to be due to the experimental condition. Unfortunately, in experimental embryology, the processes being studied are most complex, and experimental results often merely describe what will happen to developing eggs when an experimental situation is imposed from without; the basic "cause" may remain a mystery. Out of such results grow concepts, which can further be tested. There are very few "laws" of development that can be verified by simple experimental procedures. Still, our understanding is broadened by this approach far beyond the simple description of normal developmental events.

The necessity for fate maps: A fate map of the sea urchin egg can be produced by noting the final position in the pluteus of the various regions of the uncleaved egg. The extreme vegetal pole (micromeres) migrates inward and forms the skeleton. The cells of the gastral plate and a short distance beyond invaginate to form the archenteron and the enterocoelic pouches. The remaining cells form ciliated epidermis and the apical tuft found at the extreme animal pole. The fate of the animal and vegetal hemispheres is thus quite different. Knowing these fates, one is in a position to discover whether or not the fate of a region can be altered experimentally (see section immediately below). Obviously, if a fate map is not available, the normal fate is unknown and interpretation of experimental manipulations is impossible.

The unfertilized egg is really polarized: One of the first questions one can ask about embryonic development is whether or not the animal and vegetal poles exist as functionally distinct portions of the egg. If the animal and vegetal regions are predetermined (polarized) to form specific and distinct structures during development, they would be expected to develop into different portions of the larva when isolated from each other and allowed to develop separately. The development of isolated animal and vegetal halves has been followed using the eggs of *Paracentrotus*, the species which contains a red pigment band restricted to the equatorial region of the vegetal hemisphere. This band forms a natural marker for microsurgical procedures. The unfertilized egg is cut with a glass microneedle, either parallel to the polar axis or at right angles to it. The hemispheres derived from eggs cut parallel to the polar axis will form complete but half-sized plutei after fertilization. An isolated animal hemisphere forms only ectoderm and an exaggerated apical tuft; an isolated vegetal hemisphere forms only endoderm and mesoderm, an exogastrula. Micromeres are formed only on the vegetal hemisphere and appear at the normal time following fertilization.

The formation of complete embryos from halves cut parallel to the polar axis (meridional halves) is an example of regulative development. These embryos also serve as a control, proving that the microsurgery itself does not cause abnormal development. The formation of only mutually deficient larvae from eggs cut at right angles to the polar axis (equatorial halves) is an example of mosaic development. These deficient larvae also answer the original question regarding the functional reality of the animal and vegetal poles: The sea urchin egg is indeed polarized prior to fertilization, and the outward signs of polarity (the pigment band, for example) may well be secondary manifestations of the underlying differences between the two polar hemispheres.

The question of the nucleus: When an egg is cut in two, as in the above experiments, the egg nucleus obviously must end up in one part of the egg or the other — one egg half is nucleated and the other is non-nucleated. Following fertilization,

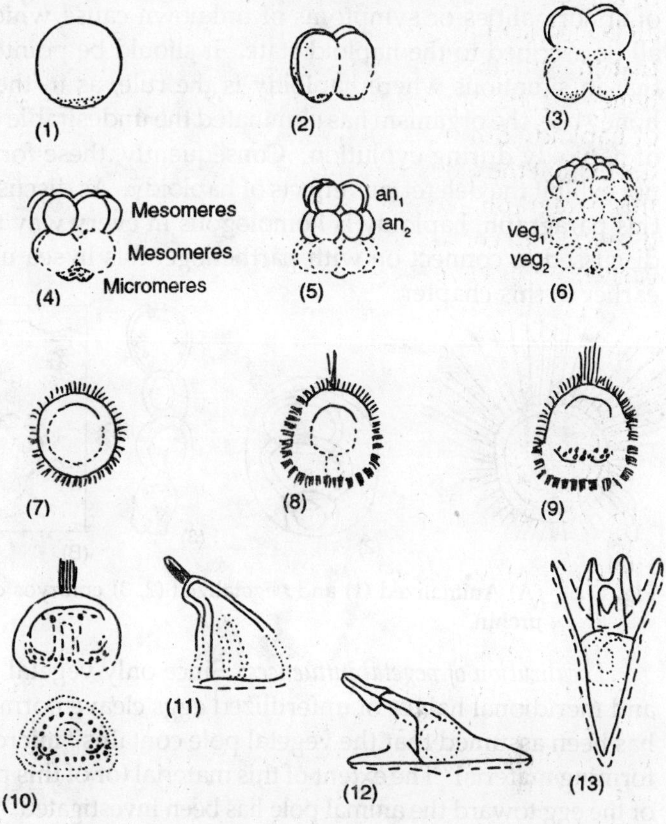

Fig. 11.5. Cell lineage or fate map of the sea urchin. Paracentrotus. The only cells to invaginate are the Veg_2 layer and the micromeres. The former forms the archenteron and secondary mesenchyme; the latter, the skeleton.

the nucleate half will be diploid and the non-nucleate half will be haploid. The ploidy question can be resolved most easily by observing only those halves which are diploid assuming (reasonably) that the egg nucleus will be in the vegetal hemisphere half the time and half the time in the animal hemisphere. Haploid meridional halves do regulate and develop, but more slowly and with more frequent abnormality. This condition has been found in many species and is referred to as the "haploid syndrome," a phrase meaning a collection of abnormalities or symptoms of unknown cause which can all be ascribed to the haploid state. It should be pointed out that in situations where haploidy is the rule, as in the male honey bee, the organism has eliminated the undesirable effects of haploidy during evolution. Consequently, these forms do not exhibit the deleterious effects of haploidy. As discussed in this paragraph, haploidy is homologous in every way to that discussed in connect on with parthenogenesis in sea urchins earlier in this chapter.

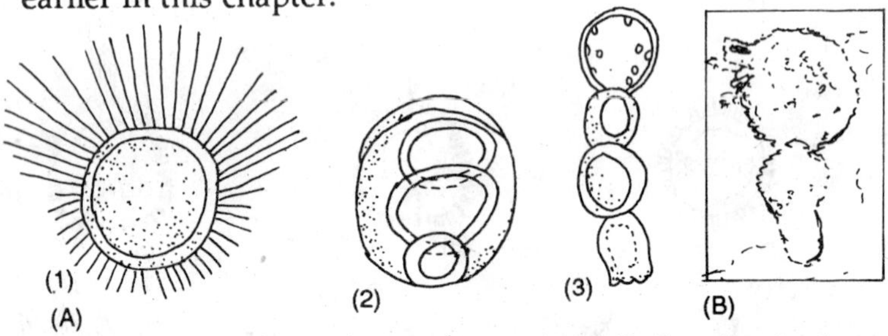

Fig. 11.6. (A) Animalized (1) and vegetalized (2, 3) embryos of a sea urchin.

Localization of vegetal influences: Since only vegetal halves and meridional halves of unfertilized eggs cleave normally, it has been assumed that the vegetal pole contains "micromere-forming material "The extent of this material (or of this portion of the egg toward the animal pole has been investigated. Again, using the pigment band for orientation, fragments of increasing size were removed from the vegetal pole region. Four micromeres is the normal number. As more and more vegetal cytoplasm is removed, fewer and fewer micromeres are formed after fertilization of the operated egg, until some eggs form only one micromere or none in the extreme experimental condition. It was therefore found that the material capable of forming micromeres extends well into the pigmented vegetal region, though it diminishes considerably beyond the extreme vegetal region where micromeres normally form.

Micromere are derived from the egg cortex: In pigmented eggs it can be seen that the micromeres contain little pigment as they are formed. This indicates that the micromeres contain primarily cortical material of the cytoplasm and very little of the inner, or endoplasmic, material since pigment granules are localized in the endoplasm.

Micromere removal and addition indicates role in development: Micromeres form intercellular bridges with neighbouring macromeres, creating a syncytium or common cytoplasm. They also form similar bridges between each other. It is possible that materials are passed to the macromeres from the micromeres through the intercellular bridges. If the micromeres are removed before the establishment of such bridges, the embryo will fail to form an archenteron and will be without the skeletal spicules usually provided by the micromeres. If the micromeres are removed late in the 16-cell stage following the connection with the macromeres, the macromeres are capable of forming an archenteron and the skeletal elements as well. If an additional set of micromeres is placed toward the mesomere side early in the 16-cell stage a blastula will develop which later has a secondary invagination and a secondary archenteron induced by the micromeres. Intercellular connections are made here as well, between the implanted micromeres and the mesomeres.

Additional evidence for the cortical location of the polarity of the egg and micromere activity: Under the influence of relatively mild centrifugal force, the contents of the egg cytoplasm become layered or stratified according to relative densities and particle sizes. It is necessary to support the egg by means of a dense (but isomotic) sucrose solution to prevent their being thrown to the bottom of the centrifuge tube. The procedure is diagrammed, along with a diagrammatic view of a stratified egg. The relative centrifugal forces required are in the range of 5,000 to 15,000 times the force of gravity. As lower forces, longer times are required to achieve stratification. It is possible to locate the "micropyle" in the gelatinous coat of the eggs following centrifugation. After centrifugation it is observed that the stratification is produced by centrifugal force acting at random with respect to the polar axis of the egg, as indicated by the location of the micropyle with respect to the axis of stratification. (Since pigment granules are moved to the centrifugal pole by the experiment, pigmentation patterns, such as that of *Paracentrotus* eggs, are of no use in locating the polar axis).

Cleavage of stratified eggs is most frequently perpendicular to the stratification pattern. This, of course, means that the first cleavage plane is *at random* with respect of the polar axis. The second cleavage is usually parallel to the stratification and equal giving two cells relatively free of stratified material and two containing most of the yolk and pigment. The third cleavage is at right angles to the first two. The fourth cleavage results in the production of mesomeres and macromeres, as well as micromeres — these *always* appearing opposite the micropyle at the vegetal pole. It can be concluded that the position of the micromeres is unaffected by the stratification or by the position of the first three cleavages. In other words, the original polarity of the egg is unaltered by random shifting of the sedimentable components of the cytoplasm (yolk, mitochondria, pigment granules) with respect to the polar axis, and therefore, these cytoplasmic components have nothing to do with egg polarity. The cortical granules are not shifted by centrifugation. It is therefore reasonable to postulate that the micromere-forming properties of the vegetal pole are in some way dependent on the cortex, which is not visibly affected by centrifugation. Recall that during the micromere division (fourth cleavage), the granular elements of the cytoplasm of the vegetal pole are excluded from the micromere end of the cleavage spindle, leaving only cortical material and/or clear protoplasm as the cytoplasm of these small cells.

The development of stratified eggs beyond the 16-cell stage is completely normal, giving rise to plutei having abnormal pigment distribution which is gradually obscured by the manufacture of new pigment by the larvae. It is most probable that plutei derived from centrifuged eggs would, if properly fed, metamorphose into normal sea urchins.

If centrifugation is continued after stratification, the unfertilized egg will be pulled into "light" and "heavy" halves. Further centrifugation yields quarters from the halves. All of these fragments can be fertilized and will cleave (those without the female pronucleus will be haploid). Development of such fragments is varied, since the eggs have been separated at every possible plane with respect to the original polar axis. Considering only the light and heavy halves, one would predict that normal development of both the light and heavy halves of a single egg would take place only if the plane of separation happened to be parallel to the polar axis. Accordingly, normal embryos, as well as varying degrees of animal and vegetal development, are found among both light and heavy halves. The development of light and heavy halves parallels the development of egg fragments produced by microsurgery, a technique used in earlier experiments. The centrifuge experiments provide the additional information that none of the materials sedimenting at relatively mild centrifugal forces is responsible for the anima-vegetal polarity of the egg.

Lithium imitates action of micromeres: If fertilized eggs are grown in solutions of sea water containing about 3 parts 0.55 M lithium chloride to 97 parts sea water, the embryos will exogastrulate. The archenteron is enlarged at the expense of the ectoderm and pushes outward instead of into the blastocoel. The cleavage pattern and numbers of micromeres is the same as found in control embryos. Though the mode of action of lithium remains unclear, the conclusion remains that lithium in some way augments vegetal factors. In other experiments, animal halves of *Paracentrotus* eggs exposed to similar concentrations of lithium as employed for fertilized eggs have been found to develop in a more normal fashion. It is therefore concluded also that lithium can induce vegetal activity in the animal hemisphere. The effect is not limited to lithium as the lithium effect can be mimicked by other unrelated compounds. For example, a compound known as cadaverine (diaminopentane) will also induce exogastrulation in whole eggs, an effect virtually indistinguishable from that caused by lithium.

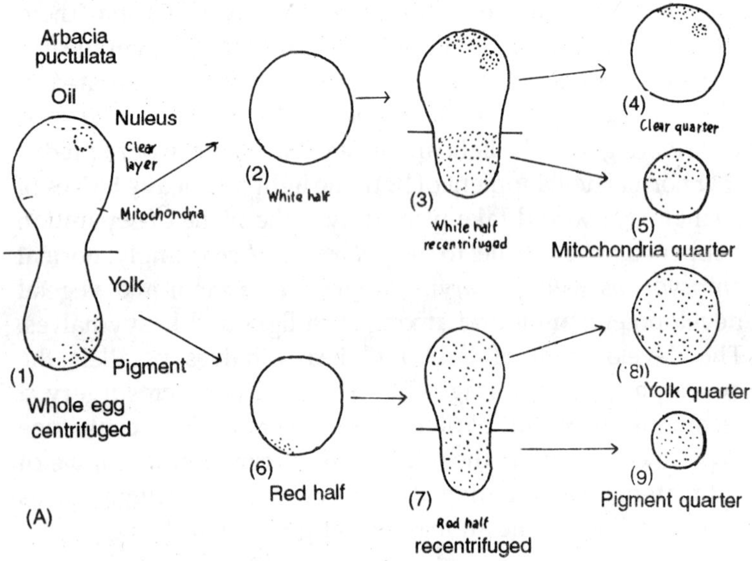

Fig. 11.7. (A) Diagrammatic view of stratification and separation of unfertilized *Arbacia* egg into halves and finally into quarters. The clear quarter contains (presumably) no mitochondria (400x). (B) Development of stratified sea urchin eggs. Note that the micromeres appear opposite the micropyle funnel regardless of the orientation of the polar axis with the axis of applied centrifugal force.

Self-differentiation of isolated micromeres: Isolated micromeres from dissociated 16-cell stages can be cultured apart from the mesomeres and macromeres. The micromere division pattern *in vitro* parallels that found in intact embryos. They form aggregates of cells within which spicule rudiments develop. The spicules develop into skeletal elements which are characteristic of the species. Micromeres are thus both inductors and capable of self-differentiation.

Gradient theories of polarity: The evidence presented so far would favour a theory of polarity involving only the micromeres of the vegetal pole. To summarize, it could be theorized that the vegetal cortex of the unfertilized egg contains some factor determining embryonic organization. This cortical cytoplasm imparts special properties to the micromeres, which are self-differentiating (skeleton) and also are responsible for the formation (by induction) of the archenteron during normal development.

Evidence from additional experiments indicates that two gradients might be involved. One gradient is an "animal" gradient and is responsible for determination of animal hemisphere structure; the other is a "vegetal" gradient and responsible for determination of vegetal structures. This concept is more complex than the one involving only the micromeres. The evidence to follow is strong enough to warrant the adoption of a double-gradient hypothesis by some investigators.

Evidence from recombinations of blastomeres: It is possible to separate the three type of cells from one another, as layers, at the 16-cell stage. These cell layers (micromeres, macromeres, mesomeres) can be recombined maintaining their original orientation with respect to the polar axis. The number of each cell type used can also be varied by teasing the cells of the different layers apart. A combination of eight mesomeres and no micromeres or macromeres results in a strongly animalized permanent blastula. Eight mesomeres plus two micromeres gives a more normal embryo, but still animalized. The normal number of eight mesomeres plus four micromeres produces a nearly normal pluteus of reduced size. Decreasing the number of mesomeres from eight to four and adding four micromeres produces a vegetalized larva with an overdeveloped archenteron in proportion to the size of the embryo. Interestingly, the four macromeres by themselves frequently develop into normal embryos. These experiments have been interpreted as evidence for two opposing gradients, which can be titrated against each other almost as one titrates an acid with a base. Normal development would be dependent on a balance between the extremes of animal and vegetal influences — a situation which would be expected in the middle of the two influences, in the region of the macromeres. These experiments led to the proposal of a double-gradient theory of development for sea urchin eggs. In concept, the theory differs considerably from the simpler theory of micromere action alone. The double-gradient theory implies two active, mutually anatagonistic processes which must be balanced for normal development. The other requires a single influence only from the vegetal micromeres, animal effects being due to the absence of vegetal effects. Though the simpler theory is attractive because of its simplicity, the development of the macromeres alone, along with the other recombination results, present ample reason for consideration of the double-gradient hypothesis.

Independent support for the double gradient from dye reduction patterns: Some nontoxic dyes can act as hydrogen acceptors in biological oxidations. Under anaerobic conditions, the dyes act to accept hydrogen from the various electron transport systems in the absence of oxygen as hydrogen acceptor. In becoming reduced, these dyes frequently change colour (e.g., from green to colourless). Such a dye is Janus Green B. Living sea urchin embryos are permeable to this dye and will reduce it to a colourless form in a given time in the absence of oxygen. It was noted that the pattern of due reduction in the blastula begins at the vegetal pole and spreads towards the equator. At a somewhat later time a second center of reduction appears at the animal pole and spreads toward the equator. Both reduction patterns suggest coincidence with animal-vegetal polarity. Additional observations support this view. Lithium ions, which suppress animal development, suppress the dye reduction of the animal pole. Isolated animal halves show only the gradient associated with the apical cilia. Isolated vegetal halves show only the vegetal reduction. And finally, the implantation of an extra set of micromeres in the equatorial region of the blastula produces an extra gradient in association with the induced secondary archenteron. In all instance, it appears that the dye reduction pattern parallels the addition or substraction of animal or vegetalizing influences. It is tempting to assume a causal relationship between dye reduction and the double gradients. The connection, however convincing, is obscured when one realizes that the organelles which would be expected to be active in electron transport (the mitochondria) have been previously shown not to be involved in polarity. The interpretation of the dye reduction studies in relation to polarity remains difficult. The safest procedure this kind of circumstance is to keep the observed facts in mind concerning recombination experiments, micromere influences, and dye reductions. Perhaps with more information from future experiments a theory can be proposed which will account for all data.

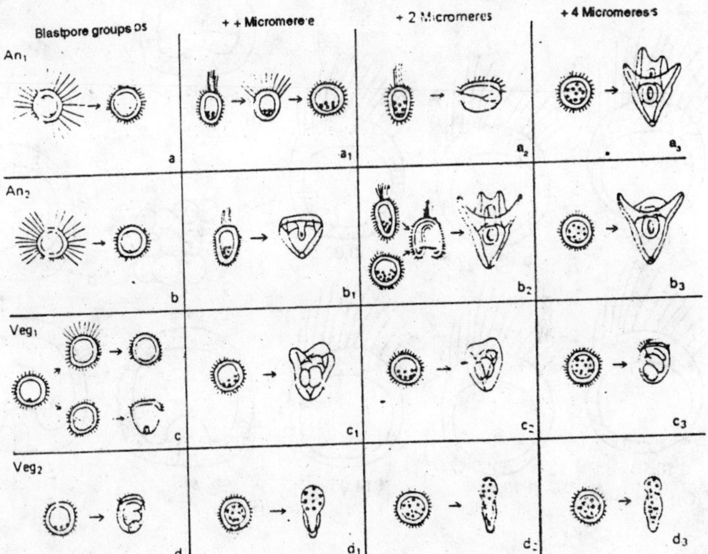

Fig. 11.8. Development of An_1 An_2 Veg_1 Veg_2 layers, from the fate map in Figure 11.5, in isolation and with increasing numbers of micromeres. The extremes of animal-vegetal polarity are balanced using cell layers and appropriate numbers of micromeres.

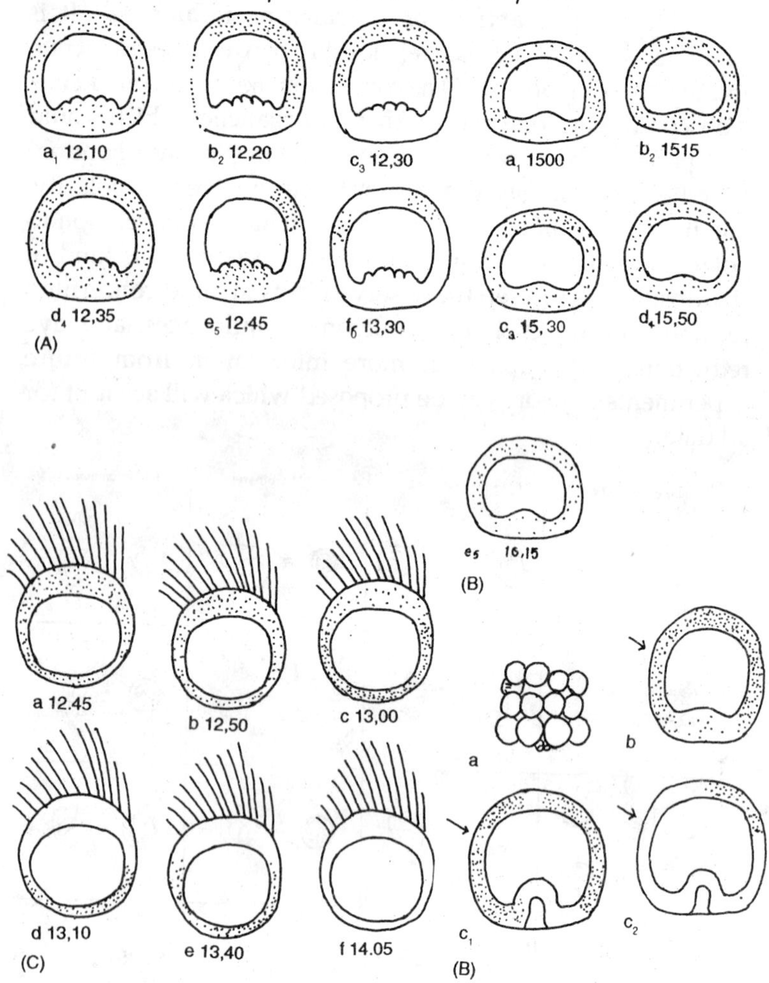

Fig. 11.9. (A, B, C, D) Pattern of reduction of the vital dye, Janus Green B, by normal and by animalized and vegetalized larvae. Stippling indicates dye, clearing indicates reduction of the dye to a colourless form. In normal embryos, the reduction begins in the vegetal region, followed by a second reduction in the animal region (A). In vegetal halves, only the vegetal gradient is seen (B). In animal halves, only the animal reduction gradient is seen (C). Implantation of extra micromeres (in D) creates a new vegetal gradient in association with the implant, as well as the expected animal and vegetal reduction gradients.

The importance of cleavage pattern and the equivalence of nuclei during development: Fertilized eggs of sea urchins can be compressed in a small drop of sea water between the slide and cover slip. Under such conditions, cleavage always proceeds parallel to the direction of mechanical pressure (the spindle orients perpendicular to the direction of pressure). If the eggs are placed under pressure at the four-cell stage, one can obtain flat plates of 8, 16, or 32 cells before the pressure is released thus altering the normal animal and vegetal tiers at the eight-cell stage. After release from pressure the next division will be parallel to the compressing plates and will produce two tiers of cells. These cells will obviously contain nuclei that would not have been in their present location had the cleavage proceeded along a normal pattern. The nuclei have been relocated with respect to the polarity of the egg in a completely random fashion. The development of such compressed embryos is totally normal. Micromeres will appear during pressure, provided the vegetal pole has not been flattened against one of the compressing pieces of glass. It, therefore, follows that the usual sequence of cleavage is not essential for the formation of the normal embryo and furthermore, that the nuclei of the cleavage stages are equivalent.

Cell Interactions

The role of specific cell adhesion in morphogenesis: Cell adhesion was first studied in the early 1900s using sponges. It was discovered that adult sponges forced through bolting silk could be dissociated into single cells and small clumps. Over the period of one day, these dissociated cells were able to reorganize and reform a number of small sponges, with all tissues in their proper places. Even more remarkable, it was found that two species of sponge could be dissociated together and could reaggregate, not only with tissue specificity, but in species-specific fashion as well. This points to an important role of the cell surface in differentiation and morphogenesis. The surfaces of cells are able to recognize patterns in each other and, on the basis of these patterns, to sort out specifically according to tissue or species. In sponges the sorting is both species and tissue specific. In vertebrate embryo (Chapter 4) the sorting is not strongly species specific, but is tissue specific to a high degree.

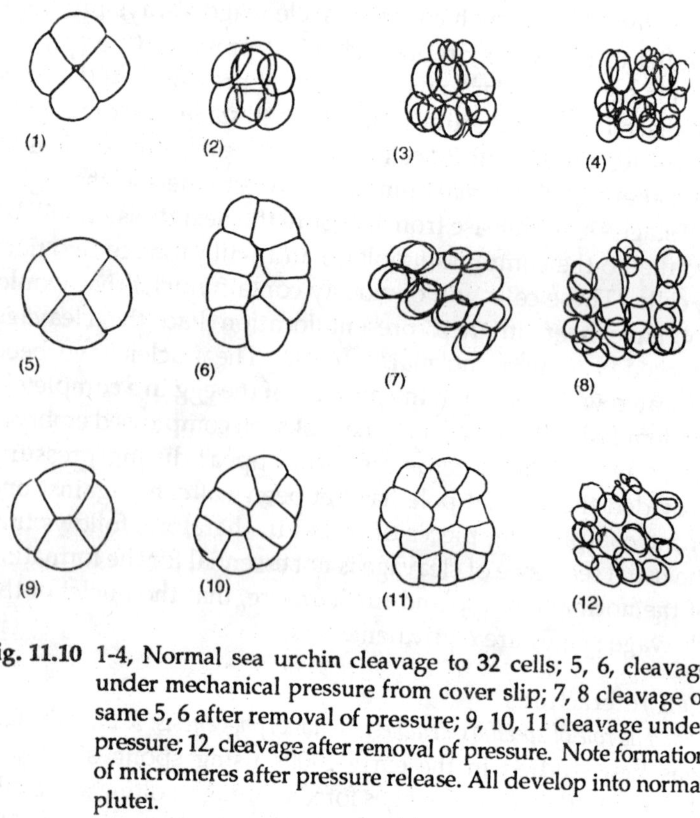

Fig. 11.10 1-4, Normal sea urchin cleavage to 32 cells; 5, 6, cleavage under mechanical pressure from cover slip; 7, 8 cleavage of same 5, 6 after removal of pressure; 9, 10, 11 cleavage under pressure; 12, cleavage after removal of pressure. Note formation of micromeres after pressure release. All develop into normal plutei.

Sea urchins have also been studied to see whether or not embryonic cells have the ability to recognize differences in cell surfaces according to tissue and species. Tissue-specific cell sorting could obviously play an important role in morphogenesis. For example, micromeres might be thought of as migrating into the blastocoel because of a loss of affinity for the cells of the vegetal pole of the late blastula. Specific cell adhesion or affinity can also be viewed as a molecular event in differentiation. Recognition among cells is presumably based on molecular components in the surface coat of cells, which, in turn, must be synthesized at sometime during development and located on the cell surface when required for cell adhesions. In some cases, when cells lose their affinity for other cells, the molecules may need to be withdrawn from the surface. In any event, differential or selective adhesion is a molecular event in the differentiation of the embryo, and is of considerable importance in understanding the overall process of formation and maintenance of the integrity of specific embryonic tissues.

The sea urchin system: The most recent work with sea urchins is also the most complete, and the most remarkable. Cleavage stages (16-cell stages) of one or another species of sea urchin were used. The fertilization membrane was removed and the cells dissociated in Ca^{++} free sea water which dissolved the intercellular cement of the hyaline layer. The three sizes of blastomeres were separated by gravity, yielding quite pure suspensions of micromeres, mesomeres and macromeres. These cells were recombined in normal proportion of 4 : 8 : 4, respectively, in normal sea water containing calcium. Other proportions, as well as pure suspensions of one or another cell type, were also used in some studies. The procedure, following return of the mixture of cells to sea water, was to follow closely the subsequent events using time-lapse cinematography. This technique involves taking one frame of a motion picture at spaced intervals (about 30 minutes) and running the final film at the usual 16 frames per second. This results in an acceleration of the natural process of aggregation and sorting, which can be better visualized. It also results in a record, on each frame, of the happenings during filming. When species specificity was studied, two species of sea urchin were employed. One, *Arbacia*, has heavily pigmented eggs; the other, *Lytechinus*, has nearly transparent eggs. The position of each species in cell mixtures was readily evident. The two species also have differences that can be used to identify the source of cells in the electron microscope. These differences involve the appearance of yolk platelets after fixation. The platelets of *Lytechinus* cells are invariably surrounded by a clear "halo," which is not produced in Arbacia cells.

Single species reaggregation and reassortment: The time for reaggregation in mixtures of micromeres, mesomeres, and macromeres of *Arbacia* (4 : 8 : 4) is about 150 minutes. Appreciable clumping occurs at the beginning, since the cells become sticky upon return to sea water. The clumps settle and flatten against the bottom of the glass dish. The areas grow larger as time passes. During reaggregation, the cells divide at a rate comparable to cells in intact embryos. The cells are capable of movement, as are larger aggregates of 50-100 cells. Fusion of aggregates takes place as well as additions to aggregates by single cells.

The aggregates are two-dimensionally flat, and cell-cell contacts are made actively by the formation of clear pseudopods. After three to four hours, the flat aggregates contact and form three-dimensional compact masses of cells.

The emergence of a complete pluteus: There are three routes taken by the compact aggregates in the formation of a normal pluteus. In the first instance, the aggregates round up and lose their adhesion to the glass dish. They become ciliated and swim off as spherical "blastulae." The blastulae gastrulate in the normal way and, frequently, develop into normal plutei. Pure suspensions of micromeres, mesomeres, or macromeres can also be followed to the aggregate phase. Micomeres form clumps and, under these conditions, eventually redisperse into single cells. No skeletons are formed. Under other culture conditions, it is possible to observe species-specific formation of skeletons in isolated micromeres. Pure mesomeres form blastulae which never gastrulate, in keeping with their normal polarity in the intact embryo. Macromeres form larvae varying from vegetalized to nearly normal, as found in recombination experiments. The dissociation treatment has not altered the original animal-vegetal properties of these cells.

When the concentration of cells is higher than above, long chains of strands are spaced over the bottom and interlace at various levels. By 12 hours, the chains of aggregated cells have developed swellings, such that the appearance now is like a string of beads. The beads become blastulalike, develop cilia, break from the chain, and swim away. Sometimes the "blastulae" remain within the chain. Either the free-swimming blastula or the blastula in the chain may develop into a normal pluteus.

In the third mode of development, a ciliated clump may return to the substratum, settle down, reattach, and spread over the surface. In time, the cells spread to form essentially a monolayer of cells looking very much like a typical plate of tissue culture cells. The spreading takes about one day for completion. About 12 hours after formation of the monolayer, typical "A" frames of plutei are visible in the monolayer. During this period, cells pile up around the developing skeleton and a pluteus begins to take shape. The pluteuslike larva is undercut from the substratum, eventually detaches, and swims off leaving a bare spot in the cell monolayer. Use of radioactive thymidine label suggests that the micromere derivatives line up and form typical cables of microtubules between themselves and deposit the calcite skeleton in a manner similar to normal spicule deposition. This third mode of pluteus formation proved to be particularly useful in studies involving cell sorting and in studies on regulation of size of the resulting larva. The initial sorting out of micromeres provides a size framework around which the other cell derivatives aggregate. These results strongly suggest that the three cell types of the 16-cell embryo can give rise to cells capable of tissue-specific reaggregation in accordance with their presumed fates.

The 16-cell stage also sorts according to species: Mixtures of *Lytechinus* and *Arbacia* cells will sort from one another, even though they stick together temporarily in the original mixture. The aggregates formed are ultimately of one species or the other. When aggregates of the two species are occupying the same place, the *Arbacia* cells move to the upper side and the *Lytechinus* move toward the substratum. The movements of the cells of the two species are thus non-random. In addition, movements of cells with respect to other cells of the same species are also non-random; *Arbacia* cells actively seek out *Arbacia cells; Lytechinus* seek Lytechinus.

Electron microscope studies, the role of microvilli: *Arbacia* cells form prominent, branched microvilli on surfaces about to be juxtaposed; the microvilli become confluent at several points. Material of the hyaline layer is also found on the cell surfaces and on the surface of the microvilli. A similar situation obtains in the *Lytechinus,* save that the microvilli are somewhat shorter. The microvilli are presumably the means by which the cells of like species adhere to one another. In mixtures of cells from the two species, similar interactions between cells of the same species are noted. When cells of different species approach, no microvilli protrude toward the future area of contact and no hyaline material is produced.

The appearance of the septate desmosome: The septate desmosome is a structure binding cell membranes together in intimate fashion. In normal development, they appear in the late blastula stage. In cell aggregates of a single species, these form between the ouer cells of the spherical clump two or three hours after reaggregation has begun. Cells to two different species do not remain in contact long enough for these structures to be formed. The transitory contact between species may be a function of the intercellular cement, which acts in intraspecies combinations as well, but is augmented by the formation of microvilli in the latter instance.

The nature of the hyaline intercellular cement: The hyaline layer was seen to arise originally from the cortical granules at fertilization. It was shown to be soluble in calcium-free sea water in early experiments involving the isolation of the first two blastomeres. If the hyaline layer is removed in calcium-free sea water, it is rapidly replaced upon return to sea water containing calcium. The material has been called hyaline and has been isolated from all stages, from fertilization of gastrula. The material is a protein, containing a high proportion of acidic amino acids. It has a minimum molecular weight of about 100,000. Hyaline forms a two-layered structure at gastrulation, sometimes referred to as the hyaline membrane. Amino acids are incorporated into the hyaline layer during development. The incorporation into the hyaline layer is inhibited by puromycin, a translation inhibitor, but not by actinomycin. Strangely, the amount of hyaline material released at fertilization agrees quantitatively with the total amount found on the cells of the gastrula, an indication that this material is not increased during development.

In attempting to assess the role of hyaline in specific reaggregations of cells, it becomes difficult to visualize anything more than a non-specific role for this substance, perhaps merely to keep cells in proximity long enough for microvilli to form. It is reasonable to state that the hyaline layer may be responsible for the initial, non-specific stickiness between blastomeres, regardless of species.

The microvilli, on the other hand, seem to offer many possibilities in the area of specific cell adhesion. These structures appear whenever reaggregation between cells of the same species is taking place and only along the future intercellular junctions. The appearance of specific molecules in the microvilli might account for the abilities of cells to associate and dissociate in the remarkably specific ways just described.

Throughout the above discussion, it has been assumed that specific cell-cell adhesion is probably not related in any direct way to the stickiness of cells in the presence of an inert substrate, such as glass or plastic dishes.

Cell polarity maintained throughout dissociation and reaggregation: In an interesting study independent of the above study of reaggregation, it was found that embryos which had been animalized (by treatment with zinc sulfate) or vegetalized (using lithium) retained their animal or vegetal properties through the reaggregation procedure. Animalized or vegetalized embryos, dissociated and allowed to reaggregate, gave rise to embryos showing the effects of whichever treatment had been applied.

Gene Activity During Oogenesis and Early Development

Early experiments with hybrids implicate the embryonic genome: The embryo begins life with a large endowment of cytoplasm plus a nucleus from the mother, and only a nucleus from the father. Early development shows many strikingly maternal influences, particularly the programme of cleavages. One of the ways to discover if and when the zygote genome becomes important has been through the use of hybrid embryos. Hybrids are created at fertilization by utilizing the sperm of one species to fertilize the eggs of another species. The eggs can be complete or they can be non-nucleate egg fragments. Fertilization hybrids may be viewed as one of the earliest methods of nuclear transplantation, wherein a foreign nucleus is introduced into a particular body of cytoplasm to study nucleocytoplasmic interactions.

Early hybrid development is universally maternal in nature: Cleavage pattern and cleavage rate is maternal in sea urchins, frogs, and other organisms where such studies have proved possible. Even in hybrids of non-nucleate eggs, where the foreign sperm provides the only nucleus, cleavage follows the schedule of normal embryos of the maternal species. The first paternal influences, judging from morphological data, appear at the mesenchyme blastula in the sea urchin and at neurulation in frogs. Some hybrid sea urchin blastulae show a paternal influence in the placement of the primary mesenchyme cells prior to skeleton formation. Other paternal traits, mainly the form of the skeleton, appear during and after gastrulation.

Paternal proteins in hybrid sea urchins: Molecular markers are somewhat less equivocal indicators of the activity of the paternal genome than are morphological traits, since morphology may be environmentally influenced. Enzymes and antigens, showing the influence of paternal DNA, make their appearance at times when it is known from other sources that the embryonic genome is active in transcription of new information. Hybrid studies thus provide evidence of the emergence of paternal genetic traits utilizing a foreign cytoplasm for expression of paternal DNA through protein synthesis. These hybrid studies also serve as indicators of the activation of the normal, non-hybrid embryonic genome, since it can be assumed that both maternal and paternal elements are activated at the same time, which is presumably also the time the embryonic genome would have been activated in homospermic embryos.

A note should be added as a caution. Sometimes foreign sperm activate eggs, but sperm chromosomes may subsequently be eliminated selectively from the developing zygote. Such hybrids are called false hybrids. Even in true hybrids later development is sometimes heteroploid and abnormal due to loss of some of the paternal chromosomes or to fragmentation of paternal chromosomes. These difficulties seem to arise from the inability of paternal chromosomes in some crosses to keep pace with the DNA replication signals supplied by the cytoplasm of the egg. As a consequence, hybrid systems showing slower development, or lower enzyme activity, are difficult to interpret since a negative finding, or a loss of activity, can always be due to loss of portions of the genome or to vaguer problems which are sometimes lumped under the heading of "genetic incompatibility."

Studies of hybrids have proved to be consistent with more recent studies on the molecular biology of development, and the data from each kind of study are generally mutually supportive.

Transcription during oogenesis: During the early diplotene stage the oocytes will actively incorporate precursor molecules involved in RNA synthesis. Most commonly, the precursor supplied is uridine labeled with radioactive hydrogen (^{3}HU). Labeled uridine is injected into the coelomic cavity and finds its way into all body tissues, including the ovary. In the ovary, the uridine label is found initially in the RNA of the nucleus and nucleolus of all oocytes, but not in mature ootids. Some of the label will later be found in the cytoplasm. The location of the radioactive material is revealed by the technique of autoradiography. Chemical analysis confirms that the bulk of the radioactivity is present in the sub-units of the ribosomes, which are passed into the cytoplasm and assembled into monosomes. A small amount is present as messenger RNA, which represents the maternal messenger RNA found in the cytoplasm of the unfertilized egg. The lack of a dense population of silver grains over the (single) nucleolus is evident. The grains reflect primarily RNA synthesis associated with the diplotene chromosomes.

Transcription - the unfertilized egg and development of blastula. The unfertilized egg is singularly inactive in transcription and protein synthesis. Turnover rates for RNA and protein, though definite, are very low. Fertilization brings about changes : First, an increase in protein synthesis; second, the beginning of RNA synthesis at a low level. Thus, oocyte activity in RNA and protein synthesis is relatively high. The mature egg is virtually inactive in both categories. Fertilization brings about the activation of both transcription and translation. The protein synthesis which is manifest at fertilization is not dependent on RNA synthesis by the zygote nuclei. This has been shown in two direct ways :

1. Non-nucleate fragments show increased protein synthesis in the absence of any nucleus, following artificial activation.
2. Actinomycin does not inhibit the early burst of protein synthesis.

The template for translation of early protins is clearly

derived for the oocyte. This mRNA has been stored in cryptic (masked) from up to the time of fertilization, at which point it was "unmasked." This template is then used for early protein synthesis. Some of the RNA synthesized by the embryonic genome is also used for protein synthesis, since newly synthesized RNA is found in active polysomes of cleaving eggs. The mRNA transcripts from unique-sequence DNA found in the oocyte represent 5-6 per cent of the single-copy DNA sequences to be found in the genome. The complexity corresponds to about 3.0 x 10^7 nucleotides. This level is comparable to values reported for *Urechis* (3.0 - 5.0 x 10^7 nucelotides), and for *Xenopus* (2.7 - 4.0 x 10^7 nucleotides). The coding ability of this complexity, if all is used as mRNA, is sufficient for 1.0 x 3.0 x 10^4 different proteins, depending on the average length of the protein molecule.

Transcriptional changes after the early blastula : At the mesenchyme blastula-gastrula stage, a large increase in RNA synthesis begins. Nucleoli appear and all classes of RNA are transcribed. With respect to rRNA, new ribosomes appear with the appearance of the nucleoli. These new ribosomes are to replace those of the original unfertilized egg and are, of course, essential for continued translation of messages. This burst of RNA synthesis is essential for further development, since embryos treated in actinomycin, which develop to early blastula, do not undergo the increase in RNA synthesis and do not develop further. Since actinomycin effects are reversible, it is a simple matter to show that these treated blastulae are not dead. After removal of the actinomycin, the treated embryos will develop through gastrula to normal larvae and show an RNA synthesis pattern similar to controls. Clearly, the burst of new RNA synthesis is essential for new development.

The nature of postblastula transcripts - selective gene activation: A number of possibilities exist regarding the kind of mRNA produced during the stages beginning with the mesenchyme blastula. Either more of the same RNA is made, or new sequence are transcribed, or a mixture of both kinds of RNA is produced.

A recent study has approached the question of the operation of different sets of genes during early and later development. Early development, to gastrula, is characterized by transcription of mainly reiterated-sequence DNA, possibly largely histone message. Later development is characterized by the near exclusive transcription of unique sequences. The question of homology between the later embryonic stages and earlier stages, as well as with adult tissues, was approached using cDNA made complementary to mRNA from gastrula polysome. This labeled single-copy cDNA was reached, in molecular hybridization experiments, with mRNA from blastula, gastrula, pluteus, adult tube feet and intestine, ovary, and mature oocytes. A control cDNA was prepared that was entirely devoid of sequences found in gastrula mRNA. The results indicate that each stage of differentiation is characterized by the transcription of a distinct set of structural (single-copy) genes, each stage being generally characterized by several thousand different sequences. A small set of sequences (mRNA) was shared by several tissues and may represent a set of "household" functions which are ubiquitous among cells and essential to maintenance of normal cell metabolic machinery.

Parallel observations on protein synthesis: As would be expected from mRNA studies, a comparison of proteins made before and after gastrulation shows a clear difference between the kinds of protein synthesized, as evidenced by the electrophoretic mobility of proteins in gels. Comparing actinomycin-treated embryos and controls (up to blastula), one sees little difference between these two, but it is noted that some changes in the population of proteins with them are to be found in both. Two important deductions can be made from this last observation. First, pregastrula proteins must be largely translated from pre-existing maternal template, since their translation is insensitive to inhibitors of transcription of the embryonic genome. Second, since changes in the protein profile are observed in the period prior to gastrulation, the appearance of new molecules during this time must be governed by control mechanisms operating at the level of translation and not at the level of the nuclear genome of the embryo. Translational level control has been described in a number of embryonic systems and seems to be a general feature of early embryonic development, in plants as well as animals.

Molecular information agrees with that obtained from hybrids: Both of the above sources point to the mesenchyme blastula and gastrula stages as the time for the embryonic genome to asset itself. At the molecular level, this is the time of appearance of new transcripts new kinds of protein. In hybrid embryos, this is the time when paternal genes make their presence known, either by the appearance of a paternal marker trait or protein, or sometimes, by evidence of genetic incompatibility and developmental failure. In the case of development failure, it is presumed that the paternal genome would first have to become active in order to later become "incompatible."

The organization of the sea urchin genome: The organization of the animal genome as regards the interspersion of unique- and reiterated-DNA sequences has been studied both in sea urchins and in *Xenopus*, an amphibian. It has been concluded that repetitive and non-repetitive sequences are intimately interspersed in the majority of DNA of the sea urchin. Labeled sea urchin DNA was denatured and sheared to various lengths and reannealed with an excess of 450 nucleotide long fragments to Cot 20. The percent of double-strandedness was measured. Repetitive sequences measured in this way are found on about 40 per cent of the 450 nucleotide long fragments. As the DNA fragment length is increased, more and more of the fragments contain repetitive sequences. It seems that about 50 per cent of the genome consists of the short period pattern of 300-400 nucleotide length repetitive sequences interspersed with about 1000 nucleotide length unique sequences. It is clear that the repetitive DNA is responsible for histone message. Both repetitive and unique sequences are transcribed as heterogeneous nuclear RNA. Heterogeneous nuclear RNA is of high molecular weight and the average size is larger than that of cytoplasmic messenger RNA. There is, however, an overlap in size and it is certain that some heterogeneous nuclear RNA eventually finds its way to the cytoplasm as messenger RNA. The bulk of the heterogeneous nuclear RNA is turned over quite rapidly within the nucleus, never leaving the nuclear envelope. It is felt that rather than being simply a read-through of redundant information, the heterogeneous nuclear RNA sequences have a regulatory function within the genome.

Some proteins are performed: Besides the DNA polymerase mentioned earlier, a number of proteins pre-exist in prefabricated form in the unfertilized egg cytoplasm. It is possible that the enzyme responsible for the digestion of the fertilization membrane at hatching may be preformed within the egg either as maternal template RNA or as stored hatching enzyme, already translated. Also, prominent among preformed proteins is the subunit of the microtubules, tubulin. This protein goes to make up the repeated assembly of the mitotic apparatus and eventually, ciliary proteins which appear on the blastula surface. Translation inhibitors such as puromycin do not inhibit the formation of astral rays and spindle fibers. Some tubulin is synthesized using maternal message during early development. The translation of these messages is under regulation, but the nature of translation regulation is not known.

The cytoarchitecture of the cells, sometimes called the cytoskeleton, depends in large measure on the presence of organized sets of microtubules. One can see micromeres giving rise to primary mesenchyme cells, which in turn form long syncytial cell processes containing cables made of microtubules. These cables extend between the cells responsible for deposition of the calcite skeleton of the pluteus larva. The shape of the skeleton will ultimately depend upon the distribution of these cytoskeletal elements.

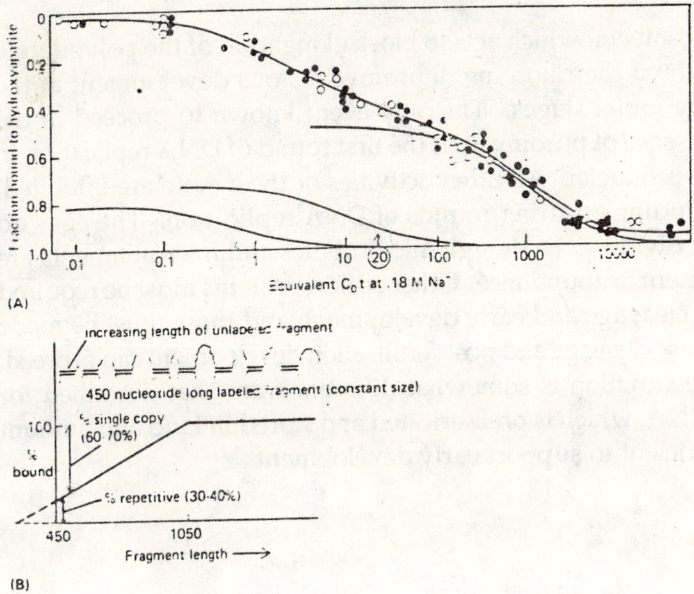

Fig. 11.11. (A) Reassociation curve of 450 long nucleotide fragments of sea urchin DNA. Two classes are evident, and it has been calculated that reannealing of reiterated sequence is complete by Cot 20. Interspection measurements were made by incubation to Cot 20. (B) Theoretical effect of fragment length on the fraction of DNA fragments containing reiterated sequence elements, where interspersion of unique or single copy with reiterated DNA is present.

Inhibitors of translation stop development immediately: Application of transcription inhibitors (actinomycin D) eventually stops development after a number of cell divisions and indicates the presence of preformed transcripts in the egg cytoplasm. Some preformed proteins have just been described. These proteins must also have been synthesized during oogenesis. The question then arises as to whether or not these preformed proteins are sufficient in kind and amount to support early phases of development. The answer can be found by inhibiting translational activity by use of such inhibitors as puromycin, which acts to block elongation of the polypeptide chain at the ribosome. Puromycin stops development at the very initial stages. The only event known to proceed in the presence of puromycin is the first round of DNA replication in the pronuclei. All other activities of the zygote are inhibited, including ensuring rounds of DNA replication. The eggs do not divide, even though microtubules and microfilaments are present in abundance. Other critical proteins must be required for cleavage and early development, and these must be made before cleavage and post-fertilization development can proceed. The situation is somewhat different from that described for mRNA, which is prefabricated and stored in kind and amount sufficient to support early development.